Modern 1
and Manu
Techniques

The text provides the reader with an in-depth understanding of the need for next-generation materials and manufacturing, especially in terms of designing process, manufacturing, upscaling, and finally selection for industrial applications. It further discusses path-planning strategies for robot-based additive manufacturing.

- Discusses synthesis, modelling, and analysis of green composites and functionally graded materials.
- Explains hybrid manufacturing processes to address the challenges faced by the manufacturing industries.
- Covers additive manufacturing of advanced materials for smart products.
- Presents applications of lasers for sensing, characterization, and material processing.
- Illustrates principles and applications of additive manufacturing and cold spray-based additive manufacturing.

The book focuses on sustainability in material and manufacturing processes. It covers important topics such as material recycling, optimal utilization of resources, green materials, improving surface inhomogeneity, stable material properties, and utilization of renewable energy sources. The text highlights the applications of deep learning for diagnosis and analysis in materials and manufacturing technologies. It is primarily written for senior undergraduate students, graduate students, and academic researchers in the fields of manufacturing engineering, industrial and production engineering, materials science and engineering, and mechanical engineering.

Modern Materials and Manufacturing Techniques

Edited by Ravi Kant

CRC Press
Taylor & Francis Group
Boca Raton London New York

CRC Press is an imprint of the
Taylor & Francis Group, an **informa** business

Front cover image: Andrei Armiagov/Shutterstock

First edition published 2024
by CRC Press
2385 NW Executive Center Drive, Suite 320, Boca Raton FL 33431

and by CRC Press
4 Park Square, Milton Park, Abingdon, Oxon, OX14 4RN

CRC Press is an imprint of Taylor & Francis Group, LLC

© 2024 selection and editorial matter, Ravi Kant individual chapters,
the contributors

ISBN: 978-1-032-56633-7 (hbk)
ISBN: 978-1-032-70303-9 (pbk)
ISBN: 978-1-032-70304-6 (ebk)

DOI: 10.1201/9781032703046

Typeset in Times New Roman
by Apex CoVantage, LLC

Dedicated to my Parents, Wife and Son

Contents

3 An overview of power factor correction techniques   65

JAGDEEP KAUR BRAR AND YADWINDER PAL SHARMA

Preface

The manufacturing sector is facing issues of high production cost, harm emission to the environment, need of customized production and low efficiency. These issues arise not only because of limitations of the manufacturing processes but also due to less effective design and materials. Advancement in both materials and manufacturing processes is important for the development of the manufacturing sector. The new, improved and efficient design of a product and its cost-effective sustainable manufacturing cannot be imagined without compatible and supportive materials and manufacturing processes. Development in these sectors can lead to customized and flexible manufacturing of components at competitive price and less emission to the environment. It will have positive economic, societal and environment impact. The requirement of sustainable materials and manufacturing processes are changing continuously. Therefore, a large number of researchers across the globe are working for long in these sectors. This book bridges the knowledge gap by providing an overview of recent development in materials and manufacturing processes. This book has a total of 11 chapters that cover literature review and original research in the field of materials and manufacturing processes.

Chapter 1 discusses the development of sustainable biocomposites with natural fibres including plant-based fibres and animal-based fibres, and epoxy or resins. It also discusses the naturally available epoxy and resin, which can be a sustainable alternative to the commercially available synthetic counterparts. Chapter 2 presents theoretical understanding of functionally graded materials by summarizing the current modelling and optimization methods, fabrication, post-processing techniques and characterization tools. In addition, the potential strategies to overcome various technological barriers and future research opportunities are demonstrated.

Power factor is important when considering the generation, transmission and distribution of electric current. Poor power factor causes an increase in current flow and hence more power consumption. Chapter 3 presents the causes of low power factor and the techniques to improve it. The advantages and limitations of various power factor improvement techniques and the advantages of improving the power factor is discussed in this chapter.

The next two chapters focus on the non-traditional finishing and joining processes. Chapter 4 presents nano-finishing of non-ferromagnetic material using magnetorheological (MR) finishing process. This process consists of a flexible tool of permanent magnet, electrolytic iron particles (EIPs) and abrasive particles. Various parameters during MR finishing, including workpiece rotational speeds (400 rpm, 600 rpm and 800 rpm), tool linear speed (10 cm/min, 30 cm/min and 50 cm/min), mesh sizes of SiC abrasives (600, 800 and 1000) and mesh sizes of EIPs (200, 300 and 400), are varied to see the change in surface roughness. The minimum surface roughness obtained was 97 nm from initial roughness of 361 nm with 40 minutes of finishing time. Chapter 5 explores various hybrid friction stir welding processes for joining of hard materials. The application of secondary heat sources like plasma, ultrasonic laser, induction, arc and electric current are explored to improve friction stir welding by reducing the plunging force, improving material flow and enhancing tool life.

Chapter 6 to 8 represents additive manufacturing as an important pillar of Industry 4.0. Different aspects like application of robotics and machine learning in additive manufacturing are explored. These chapters will provide readers with a practical understanding of how different pillars of Industry 4.0 operate, emphasizing their integration and synchronization with each other. Chapter 6 focuses on the use of robotic systems in the manufacturing industry, specifically in the field of additive manufacturing. It covers the major additive manufacturing processes that can be enhanced with advanced robotic systems and explores the potential of using robotic systems with advanced materials. This chapter also explores recent advancements in industrial applications and future perspectives, considering the integration of robots with additive manufacturing processes. Chapter 7 presents the usage of cold spray coating technique for additive manufacturing of components without melting of metallic powders. The history of cold spray, bonding mechanisms, the effect of process parameters, the post-processing involved in cold spray and strategies to turn cold spray additive manufacturing into a smart manufacturing system are presented in this chapter. Chapter 8 discusses the integration of machine learning algorithms with additive manufacturing systems. Various case studies demonstrate the effective utilization of deep learning in the field of additive manufacturing and quality control which can improve product quality, reduce waste and optimize the manufacturing process. It can make the additive manufacturing more competitive, effective and efficient in the age of Industry 4.0.

Modern industries are looking for cost-effective, flexible and sustainable solutions for precision manufacturing. Laser has evolved as a flexible manufacturing tool which can be a promising solution for meeting such demands. Chapter 9 focuses on the applications of laser technology in the academic, industrial, defence and medical sectors. Applications related to material processing, measurements, sensing and characterization are explored in this chapter. Chapter 10 discusses various techniques to improve sustainability in machining processes by reducing the running cost and limiting harmful emission and waste. Various kinds of machining processes including dry machining, minimum quantity lubrication, cryogenic machining, and hybrid machining processes are discussed in this chapter. This chapter also discusses laser and ultrasonic vibration assisted machining, which can improve the machining performance without utilizing cutting fluids.

The last chapter presents the machining performance and optimization of electrochemical machining of difficult-to-cut Monel 400 alloy. Studied are the effects of voltage, electrolyte concentration and inter-electrode gap on the material removal rate, tool wear rate and surface roughness; the parameters are optimized using a Box–Behnken design generated response surface methodology (RSM).

The chapters in this book highlight research gaps and directions for future research. The chapters are a good combination of review articles and original research in the domain of advanced materials and manufacturing processes. The volume will help readers to gain broader and deeper knowledge in the field of materials and manufacturing. The content of this book will be useful for academicians, researchers and practicing engineers who are looking for sustainability in materials, design and manufacturing processes.

Editor: Ravi Kant

Acknowledgements

I, the editor of this book, had a great experience while working on this book. Many people have inspired, motivated and helped me during the entire course of the work. It is my heartfelt desire to acknowledge their immense goodwill and valuable support.

I express my indebtedness and gratitude to all of my teachers and friends whose support and guidance helped me in editing this book. My students helped me in reviewing and formatting the book chapters. Their selfless efforts and services are highly appreciated.

I sincerely acknowledge all the authors for contributing their valuable work for this book. The authors submitted the chapters and responded with revisions promptly, which made it possible to publish this book on time.

I express my heartfelt gratitude to my parents, wife and son, who are always understanding and supportive. They have made many compromises in their life for my professional growth. They give me emotional strength and take care of duties that are otherwise to be performed by me. Their patience, goodwill, caring nature and encouraging words keep me driving and help me to focus on the work. I dedicate this book to all of them.

Last, but not the least, I express my sincere gratitude to the staff members of CRC press, Mr. Gauravjeet Singh Reen, Ms. Isha Ahuja and Mr. Mayank Sharma, for their dedicated support during the publication process of this book.

About the editor

Dr. Ravi Kant is Assistant Professor in the Department of Mechanical Engineering at Indian Institute of Technology Ropar, India. He received his bachelor's degree in mechanical engineering from Maharshi Dayanand University, Rohtak (Haryana, India). He completed his M. Tech. from the Department of Mechanical Engineering at Indian Institute of Technology (IIT) Guwahati, India, with specialization in computer-assisted manufacturing. He worked on the "Investigation on Formability of Adhesively Bonded Sheets" for his M. Tech. project. He also obtained his doctorate from IIT Guwahati in the field of laser forming process. His research interests include laser transmission welding; hybrid machining; laser forming; cold spray coatings; additive manufacturing, hybrid joining, and sustainable materials. He has completed many research projects and consultancy works in these research areas. He has contributed around 100 research articles in peer-reviewed journals, conferences, and edited books. He has edited several books in the field of industrial design, materials and manufacturing processes. He has also guest edited special issues in reputed journals. He has developed and taught advanced courses such as modern manufacturing processes; sustainability science and technology; analysis of casting, forming and joining processes; advanced welding technology; micromanufacturing; and manufacturing. He has conducted various international conferences, workshops, symposiums, colloquiums and faculty development programs in the field of advanced materials and manufacturing technology.

Contributors

Talwinder Singh Bedi, RIMT University, Mandi Gobindgarh

Papiya Bhowmik, Indian Institute of Technology Ropar, Rupnagar, Punjab, India

Jagdeep Kaur Brar, GZSCCET MRSPTU Bathinda, Punjab, India

Rajiv Chaudhary, Delhi Technological University, Delhi, India

Neeraj Deswal, Indian Institute of Technology Ropar, Rupnagar, India

Hema Gurung, Bara Phool, IIT Ropar, Rupnagar, India

Ravi Kant, Indian Institute of Technology Ropar, Rupnagar, Punjab, India

Avneesh Kumar, Indian Institute of Technology Ropar, Rupnagar, India

Krishan Kumar, Indian Institute of Technology Ropar, Rupnagar, Punjab, India

Rahul Nair, Indian Institute of Technology Ropar, Rupnagar, India

Atul Pandey, Indian Institute of Technology Ropar, Rupnagar, India

Shashwath G. Patil, National Institute of Technical Teachers Training and Research (NITTTR), Chandigarh, India

Ayush Pratap, Indian Institute of Technology Ropar, Rupnagar, India

Khushboo Rakha, Indian Institute of Technology Ropar, Rupnagar, Punjab, India

Ajay Singh Rana, RIMT University, Mandi Gobindgarh, India

Sumitkumar Rathor, Indian Institute of Technology Ropar, Rupnagar, India

P. Sudhakar Rao, National Institute of Technical Teachers Training and Research (NITTTR), Chandigarh, India

Shahriar Reza, Moghbazar, Ramna, Dhaka, Bangladesh

Neha Sardana, Indian Institute of Technology Ropar, Rupnagar, India

Yadwinder Pal Sharma, GZSCCET MRSPTU Bathinda, Punjab, India

R. C. Singh, Delhi Technological University, Delhi, India

Harpreet Singh, Indian Institute of Technology Ropar, Rupnagar, Punjab, India

Akhilesh Sinha, National Institute of Technical Teachers Training and Research (NITTTR), Chandigarh, India

Ekta Singla, Indian Institute of Technology Ropar, Rupnagar, India

Sri Phani Sushma, University College of Engineering Narasaraopet, JNTU Kakinada, Andhra Pradesh, India

Ankit Mani Tripathi, SRM University Delhi-NCR Sonipat, Haryana, India

Gidla Vinay, Indian Institute of Technology Ropar, Rupnagar, India

Chapter 1

Green composites for sustainable applications

Papiya Bhowmik, Ravi Kant and Harpreet Singh

1.1 INTRODUCTION

Contamination of the soil, water and air by plastic pollutants is increasing globally, raising severe concerns among environmentalists. The outbreak of SARS-CoV-2 has elevated the use of one-time plastics due to the sudden adoption of improved hygiene practices [1]. As a result, post-Covid pollution control will be a major concern for different government bodies worldwide. Scientists are continuously working towards developing a biodegradable alternative to plastic that is non-toxic, renewable and affordable [2]. Over the past decade, green composites have replaced many synthetic composites and plastics in various applications [3–5]. This is possible because of the specific qualities of the organic polymer and fibres used to develop green composites over conventional components of synthetic composites [6]. Developing natural fibre-based composites is now well practised worldwide because of their easy availability, renewability, outstanding mechanical properties, non-toxicity and excellent biodegradability [7–11]. Natural fibres have the potential to replace conventional petroleum-based synthetic fibres. They possess impressive qualities like low cost, higher flexibility, renewability, non-toxicity, low density and better mechanical properties [12–15]. The wide availability in the market has also triggered their wide application. However, the low degradation temperature of the fibres does not allow them to be used in high-temperature applications or curing processes [16, 17]. Another significant drawback is the super-hydrophobic property of the natural fibres, which opposes their natural ability to adhere to hydrophobic polymer matrices. It leads to poor adhesion between the fibre and the composite matrix [18]. To tackle this problem, researchers have taken different approaches to modifying the surface texture of the fibres and matrix material. The modified fibres and matrix have better adhesion because the changed surface texture allows better bonding [9, 19]. Other approaches to enhance the physicochemical properties of the developed composites include adding cross-linkers and plasticisers to them [20–24]. Various researchers have explored the extrusion of proteins under different thermal conditions. Generally, protein extrusion or moulding occurs at a relatively low-temperature range of 105 to 120 °C [25]. To add flexibility to the protein, the glass transition temperature is reduced

DOI: 10.1201/9781032703046-1

1

from 220 °C to below 100 °C. The plasticiser influences the protein to form a clay-like structure, allowing the protein to be thermally extruded or moulded at a temperature range of 100 to 120 °C. Various plasticisers are used for different natural polymers, considering their plasticising capacity, hydrophilic compatibility with the polymer, and the capacity to produce many hydrogen bonds. Plasticiser enhances the elasticity of the protein by enabling more fluid melt, flexibility and elongation [26, 27]. Besides plasticisers, crosslinkers are also used to modify natural polymers' microstructural rheological features. Crosslinking increases the mechanical properties of the polymer by making longer polymeric chains connected with intermolecular bonds or links [28–32].

In recent years natural fibre-reinforced polymers have provided lifelong benefits over petroleum-based fibre-reinforced composites [33]. The commendable ecological benefits of natural fibres include their suitability for manufacturing a wide range of products while being biodegradable [5]. If these fibres can be wisely combined with natural biodegradable polymers, it can significantly reduce the carbon footprint. In India's current plastic management policy, plastic carry bags are the most significant contributor to littered waste. Millions of plastic bags end up in the environment vis-à-vis soil, water bodies, watercourses and so on. Each bag takes an average of one thousand years to decompose completely [34]. In India, the Plastic Waste (Management and Handling) Rules, 2011, included plastic waste management as its priority to address the issue of scientific plastic waste management. The government has announced the Plastic Waste Management Rules, 2016, to replace the earlier Plastic Waste (Management and Handling) Rules, 2011, focusing on developing alternative biodegradable plastic with a faster degradation rate. An eco-friendly product that serves as a complete substitute for plastic in all uses has not been found to date [35]. In the absence of a suitable alternative, it is impractical and undesirable to impose a blanket ban on the use of plastic all over the country. The real challenge is to improve plastic waste management systems or develop a feasible alternative to plastic disposables [2, 36]. So, the dire need of the hour is to furnish a viable low-cost alternative to plastic with zero or minimised carbon footprint. In this chapter, we will discuss further developments in the field of green composites and their application to reduce the carbon footprint on our environment.

1.2 NATURAL POLYMERS AND FIBRES

1.2.1 Natural polymers

The prime goal of a polymer is to hold together the fibres and shield them from surrounding chemical changes, uneven mechanical loading and ambient moisture [14, 37–39]. Upon loading, it distributes the load uniformly throughout the structure of the composite, thus reducing the chances of early failure. Apart from the polymer, the strength of the composite also depends on the orientation of the fibres [40, 41].

There are two types of polymers: synthetic and natural. Synthetic polymers are derived from petroleum oil and made by scientists and engineers. Examples of synthetic polymers include nylon, polyethylene, polyester, Teflon and epoxy [42]. Natural polymers occur in nature and can be extracted. They are often water-based. Naturally occurring polymers are silk, wool, DNA, cellulose and proteins. Natural polymers are of two types: thermosets and thermoplastics. Thermoplastics have poor wettability and high viscosity, while thermosetting polymers possess low viscosity and better wettability with the fibre surface [42–44]. Adding coupling agents/compatibilisers to the thermoplastic polymers improves interfacial adhesion. Thermoplastic polymers like polyvinyl chloride, polyethylene and polypropylene become softer at higher temperatures and regain their original properties when cooled down [6, 45, 46]. Table 1.1 depicts the properties of the most common, renewable and widely used natural polymers reported by researchers.

1.2.2 Natural fibres

The main constituents of natural fibre-reinforced composites are plants, minerals and animals. Of plant fibres, mainly five types are used: seed fibre, leaf fibre, bast fibre, fruit fibre and stalk fibre. Of animal fibres, mostly animal hair and silk fibres have been used, and of mineral fibres, amosite, crocidolite, tremolite, actinolite and anthophyllite fibres are popularly being employed. The performance of green composites or natural fibre composites mainly depends on their length, shape and interfacial adhesion with the matrix. These fibres have good thermal and acoustic insulation, specific mechanical properties and the most desirable property – biodegradability. As per renewability concerns, most researchers prefer plant-based fibres for developing biocomposites. The popularly used plant fibres and their relevant mechanical properties are shown in Table 1.2. Concerning animal fibres, they are mostly found as waste material from meat factories, textile factories or fish markets. Table 1.3 depicts the popularly used animal fibres and their mechanical properties reported by different researchers.

1.3 COMPARISON WITH MMCs, CMCs, SYNTHETIC NPMCs

Metal matrix composites (MMCs) have a lot of advantages over standard alloy-based materials. They are lightweight with outstanding specific modulus and strength. They also have a lower coefficient of thermal expansion and resist wearing off. Because of all the advantages of MMCs over alloy-based materials, they are widely used in aerospace, transportation and many high-end machineries [80, 81] MMCs also have the potential to be used in almost any application. Their thermal properties, however, confine them to be used around 1000 °C only. Beyond this temperature, generally, ceramic matrix composites (CMCs) are used because of their favourable electric, thermal and magnetic properties; in general, MMCs have superiority over polymer matrix composites (PMCs) or CMCs [82, 83].

Table 1.1 Properties of different natural polymers

Name	Scientific name	Properties		Significant property	Ref.
		Tensile strength	Modulus		
Wheat gluten	Triticum aestivum	4.5 MPa	25 MPa	Wheat gluten can be used as a natural polymer for developing composites. The property of the polymer can be altered using different crosslinkers and plasticisers.	[47]
Soya protein	Glycine max	20 MPa	500 MPa	Soya protein is available in abundance and renewable in nature. It provides an excellent moisture barrier when used with gelatin.	[48, 49]
Tapioca starch	Manihot esculenta	45 MPa	2.4 GPa	It is superior to other starches due to its amylose content compared to other lower amylose-containing starches. It also has a higher molecular weight due to amylose and amylopectin.	[49, 50]
Potato starch	Solanum tuberosum	7 MPa	70 MPa	The polysaccharides of potato starch deliver unique edible-film-forming properties and chemical stability.	[52, 53]
Corn starch	Zea mays	40 MPa	2.6 GPa	From forming thin films for different applications to making composites; starch is an economic candidate as a natural polymer due to its biodegradable properties, low cost and renewability.	[52, 53]
Gelatine	NA	95 MPa	3.5 GPa	For packaging and shipping food-grade products, gelatine coatings or thin films are used because of their excellent forming capacity. It is also blended with other polymers to increase the tensile strength and elongation.	[54, 55]
Agar-agar	Aquilaria malaccensis	36.7 MPa	1.1 GPa	Agar-agar is a chemically inert and non-toxic hydrophilic polysaccharide. It forms thin films with better and more flexible mechanical characteristics.	[54, 56]
Arrowroot starch	Maranta arundinacea	16 MPa	1.2 GPa	Starch from arrowroot has been used recently, which is a less studied starch source. Arrowroot starch is cheaper and readily available everywhere.	[57, 58]

Table 1.2 Properties of different commercially used plant-based fibres to manufacture composites

Name	Scientific name	Properties		Significant property	Ref
		Tensile strength	Modulus		
Hemp	Cannabis sativa	1200 MPa	60 MPa	Industrial hemp is a source of high cellulose fibres. It possesses superior properties such as high specific strength, stiffness and low density.	[59]
Ramie	Bohesmeria nivea	615 MPa	20.5 GPa	Ramie fibres are widely known for long, silky and strong plant fibres, making outstanding textile fibres with impressive mechanical properties.	[60]
Kenaf	Hibiscus cannabinus	692 MPa	10.94 GPa	Kenaf is considered an economically viable natural fibre commercially available on a large scale. The impressive properties of kenaf fibre make it a suitable replacement for glass fibres in polymer composites as a reinforcement.	[61]
Jute	Corchorus	399 MPa	24.70 GPa	Because of its superior properties, such as high tensile strength, low thermal expansion and high strength-to-weight ratio, it is being widely used for biocomposite manufacturing.	[62]
White coir	Cocos nucifera L.	192 MPa	3.44 GPa	White coir fibres possess low strength, low modulus of elasticity and high strain to failure due to their low cellulose content and large micro fibrillar angle.	
Brown coir	Cocos nucifera L.	343 MPa	4.94 GPa	On the contrary, brown coir has an advantage over white coir. It improves the hybrid composite by improving the impregnation of small-diameter fibres.	
Bamboo	Bambusa vulgaris	206.2 MPa	20.1 GPa	Bamboo fibres are famous primarily for their easy availability, low cost, easy and fast renewability, high strength-to-weight ratio, high tensile strength, non-toxicity and biodegradability.	[63]
Bagasse	Saccharum officinarum L.	70.9 MPa		Bagasse fibres have the advantage over other natural fibres and traditional reinforcement due to their excellent thermal property, energy recovery, good specific strength and low density.	[63]

(Continued)

Table 1.2 (Continued)

Name	Scientific name	Properties		Significant property	Ref
		Tensile strength	Modulus		
Rice husk	Oryza sativa	135 MPa	1.8 GPa	The ultimate tensile strength of rice husk is compared to nylon fibres. The micro-arrangements of cellulose and hemicellulose help rice husk to attain high strength.	[64]
Napier grass fibre	Pennisetum purpureum	12.4 MPa	2.2 GPa	Napier grass fibres hold the lowest densities compared to other natural fibres. The tensile strength of the napier fibre is greater than bamboo and coir but lower than flax, hemp, jute, kenaf and banana.	[65]
Sisal	Agave sisalana	68 MPa	3.8 GPa	When used as composite reinforcements, sisal fibres showed outstanding wear resistance, indicated by lower mass loss, specific wear rate and coefficient of friction.	[66]
Abaca	Musa textilis	308.7 MPa	6.186 GPa	The high cellulose and low water content of abaca fibres have a high value of mechanical properties with impressive tensile strength.	[67]
Banana	Musa acuminata	550 MPa	3.5 GPa	With high strength and high flame resistance, banana fibres are preferred for making fire-retardant panels, strong garments and yarns.	[68, 69]
Betelnut	Areca catechu	128.79 MPa	2.56 GPa	Dried betelnut fibres exhibit good Young's modulus compared to ripe betelnut fibres. Betelnut fibres are preferred for applications with high dimensional stability because of their low moisture content and water uptake.	[70]

Table 1.3 Properties of different animal-/insect-based fibres

Name	Scientific name	Properties Tensile strength	Modulus	Significant property	Ref
Human hair	Homo sapiens (trichology)	280 MPa	4.1 GPa	Hair executes high tensile strength, but it varies depending on the strain rate and humidity. The strain rate of hair is comparable to keratinous materials and other common synthetic polymers.	[71]
Human nails	corpus unguis	120 MPa	4.5 GPa	At low humidity, nails are more brittle and flexible than at high moisture. The reason behind this is a reduction in torsional stiffness with increasing relative humidity.	[72]
Chicken feathers	Gallus gallus domesticus	203 MPa	3.59 GPa	The study showed chicken feather fibres have high resistivity and low dielectric constant, which helps them to be used as composite reinforcements to develop electric insulator panels.	[73]
Silk	Bombyx mori	623 MPa	14.96 GPa	Natural silkworm silk fibres consist of two core of fibroin filaments bonded together by a layer of sericin. Silk fibres are abundantly studied because of their several unique properties such as good processability, desirable biocompatibility and biodegradability.	[74]
Pig hair	Sus scrofa	135 MPa	6.39 GPa	Being coarser and thicker than natural silk fibres pig fibres can only be used in non-woven applications. As the tensile properties of pig fibres are comparable with other natural fibres, they can be used as low-cost environment-friendly biocomposites.	[75, 76]
Alpaca	Vicugna pacos	190 MPa	2.7 GPa	Alpaca fibre is a soft, durable, luxurious and silky natural fibre. It is also valued because it is lustrous, extremely strong and very warm.	[77]
Camel hair	Camelus bactrianus	212.15 MPa	3.87 GPa	Camel fibre adheres to tremendous breaking stress, higher breaking strain and significant rupture energy. The outstanding texture and mechanical properties make it a viable natural fibre for producing composites.	[78]
Spider silk	Araneae	1 GPa	10 GPa	Spider frame silk can be considered the best fibre in terms of stiffness and strength, making it stronger per unit weight than high-tensile steel.	[79]

On the other hand, ceramics are generally made of inorganic materials and non-metallic substances. Ceramic matrix composites usually have high heat, chemical and abrasion resistance, making them advanced materials for some particular applications. They possess high strength, impeccable hardness, resistance to wear, chemical inertness and the ability to withstand high temperatures. These properties make ceramics a good candidate for high-end applications, but they have certain disadvantages. Due to the inherent properties of high brittleness and low thermal expansion coefficient of ceramic materials, they are tough to process and manufacture. Under tensile and impact loading, they tend to undergo brittle fracture, propagating quickly [84]. Thus to increase the toughness of ceramics, various reinforcements like carbon fibres, silicon carbides, silicon nitrides, alumina, and titanium diboride have been added to them. These reinforcements improve the toughness and load-bearing capacity of the CMCs. CMCs can withstand temperatures of nearly 1500 °C under different working conditions. They offer increased toughness compared to traditional ceramic materials [85].

Due to easy mouldability and easy manufacturing processes, polymer matrix composites are more widely manufactured for various applications. Their light weight, high strength and high stiffness promote PMCs for lightweight vehicles, turbine rotors and blades, aerospace applications and more. But PMCs also come with disadvantages [86, 87]. They have high sensitivity to moisture and radiation. They work in low temperatures, that is, around 200 °C, which is very low compared to MMCs and CMCs. Regardless of these disadvantages, PMCs are widely popular because they don't require high pressure and heat for processing, saving the composite from fibre damage and thermal degradation. PMCs are generally made of polymeric resin and fibre reinforcements [3]. The polymeric resins are byproducts of petroleum, and the fibre reinforcements are mainly made of glass or carbon. The increasing use of PMCs thus increases soil pollution. However, the growing popularity of PMCs has sparked great interest among researchers to work on sustainable PMCs. The green PMCs consist of biodegradable polymeric resin and fibres collected or procured from nature. The easy availability and competitive properties and pricing make green PMCs a sustainable option in the global market [88, 89].

1.4 BIO-COATINGS FOR THE COMPOSITE

Coatings are essential to prevent moisture, water or toxic environmental contaminants from entering the composite body. They allow for minimal friction and dust build-up on the substrate, thereby enhancing the lifespan of various composites [90]. With the growing popularity of natural fibre-based polymer composites, the need to develop biodegradable natural coatings is increasing too. Natural fibres have strong non-polar characteristics, which leads to poor adhesion of fibre with the matrix materials because it induces an uneven distribution of fibers within the polymer matrix [91]. In addition, they are prone to moisture absorption when exposed to a humid atmosphere, which leads to early failure of the material during mechanical loading or thermal stress [90]. Many studies reported improving the fibre-matrix adhesion and different processes to coat the composite naturally. These coatings make the composite less susceptible to moisture or a toxic atmosphere [92]. One of the most popular methods to form a biobased coating on the fibres and the biocomposite is hygroscopic ageing. In a recent study, it has been shown that the moisture absorption of a composite reduces to a great extent when coated with biobased polyurethane (PU) and polyfurfuryl alcohol (PFA) [93]. The study shows that the flex/phenolic laminates coated with PU and PFA showed lower water absorption when compared to uncoated laminates. Among PU and PFA, PFA displayed better hydrophobic properties as a coating material. It also showed that a little moisture absorption has no significant effect on the mechanical and thermal stability of the composite with coatings on it [94]. Recent studies have shown that biocoatings can also be applied to metallic substrates [90, 95]. Biobased materials like hemicellulose are extracted from plants and are also a byproduct of the pulping industry. Hemicellulose is a water-soluble polysaccharide with low molecular weight and hydrodynamic radius [96]. However, the material is sensitive to moisture under high humidity conditions. Water absorption reduces the binding capacity and weakens the hemicellulose-based film's structural stability and respiratory barrier. Alkyl ketene dimer (AKD) is used as a crosslinker and a hydrophobic modifier to prepare a novel hemicellulose-based nanocomposite coating to improve the moisture barrier [94]. Another polysaccharide biocoating, sodium alginate, is generated from brown seaweed. It is constituted of α-D-mannuronic and α-L-glucuronic acid [97]. Sodium alginate is used as a thickening agent in the food industry because it is non-toxic and biodegradable and has a high consumption acceptance rate. If applied as a coating, it improves the shelf life of food during storage [98]. Another promising candidate in the field of biobased coating is propolis. Propolis is a plant-derived resin product collected by bees to develop and protect the hive. Studies have shown that green propolis produced by *Apis mellifera* bees from plant-extracted nectar can be used as a coating for its excellent antioxidant and antimicrobial properties [98, 99]. In a recent approach, commercial epoxidised cardanol from cashew nutshell liquid (CNSL) has proven to be an outstanding biocoating when blended with sucrose epoxy derivative, sorbitol and isosorbide. Bisphenol A diglycidyl ether (BADGE) is a popular coating because of its unique molecular structure. Finding a substitute for BADGE is very difficult, but various biobased epoxy reactants can tailor the required properties of the biobased coatings for different applications [100]. Preserving fish or fish byproducts using green coatings is of tremendous interest to the food industries. Sodium alginate bilayer coating combined with green propolis can increase the shelf life by up to 11 days [98]. In a similar approach, to enhance the shelf life of exotic fruits like green asparagus, they are coated with a dense hemicellulose-based coating containing hemicellulose, cellulose, nanofibres and montmorillonite. AKD was added as a hydrophobised coating to enhance the respiratory barrier property of the asparagus [94]. Natural coatings are also used to protect the indoor microbial colonisation of higher places that require a high level of environmental hygiene. Melamine formaldehyde as a capsule shell and lavandin-tea tree essential oils as core material were reinforced into painting as a biocide agent [101]. A novel UV-curable waterborne polyurethane acrylate, comprising polymerisation of hydroxyl telechelic natural rubber (HTNR), was developed as a greener approach. The material provides enhanced wettability with the surface, thus increasing the mechanical and thermal properties of the substrate [102]. Coatings are also used to enhance the anti-crease properties of cotton fabric: α-lipoic acid (ALA) has improved the fabric's crease resistance and hydrophilic properties without loss of strength or the use of formaldehyde. ALA attaches itself to the cotton fibre through esterification between the carboxyl group and disulfide bond of the materials [103]. Neem gum is used as an effective biodegradable glue. It has hydrophilic polysaccharides obtained from exudates of neem tree and can be used to coat nanoparticles of palladium nano particles (PbNPs) using the sonochemical method [104]. For locking in the soil's nitrogen content and reducing nitrogen loss due to irrigation, urea is granulated using duel layers of two natural oils and hydrogels, namely epoxidised soybean oil, linseed oil and crosslinked co-polymeric hydrogel of polyacrylic acid and polyacrylonitrile (PAN/PAAc) hydrogels [105]. Natural edible coatings are gaining popularity in the food and packaging industries. Agar-agar is a plant-derived polymer with promising results when used as a coating. It has worked wonderfully to develop effective coatings for different fresh fruits, vegetables, fish and meat [106]. Edible coatings, which are thin and easily soluble, are used on the surface of the food products. The coatings provide better protection to the substrates against oxygen, moisture and other toxic elements from the food surface. Composites of edible coatings containing antimicrobial compounds can also reduce bacterial proliferation. Adding vegetable oil to edible food coating also increases the food's moisture barrier, which helps the food stay juicy [104]. Though the concept of biodegradable coating is gaining interest globally, these coatings aren't practical considering the current solid waste disposal process. As we know, in landfills or dump yards where solid waste is disposed of, it does not receive a compostable environment to degrade naturally. So it is also a global concern to find an appropriate method for managing solid degradable waste. However, biocoatings are still required for food preservation and creating moisture barriers for end products [107]. In an approach to enhancing bananas' shelf life from physical and biological deterioration during transport and storage, an active nanocomposite is used. The nanocomposite consists of polyvinyl alcohol (PVOH), agar-agar and maltodextrin with silver nanoparticles. This coating protects the fruits against *Escherichia coli* and *Staphylococcus aureus* bacteria. The addition of silver nanoparticles enhances the flexibility of the composite films, allowing them to coat different surfaces of complicated shapes [108]. Agar-agar is one of the most commercially used coating materials. Being a porous and inexpensive polymer, agar-agar is used as a promising absorbent material. Being hydrophilic, it is often used as the skeleton adsorbent. Agar-agar is used to prepare edible polymeric composite sheets in various applications with these many properties [109].

Agar-agar usually produces a thick homogeneous gelled structure, and it takes a long time to dry off completely. But when mixed with xanthan gum or locust bean gum, it produces a translucent flexible film that is easy to dry and prevents water for a long time [110]. In another approach, agar-agar (1%) based coatings with 0.2% chitosan and 0.2% acetic acid showed promising properties when used as an edible coating. The study showed that when applied to raw garlic cloves, the coating delayed decolouration, fungal and bacterial growth, moisture loss and respiration rate, thus increasing shelf life [111]. The cohesive structural layer of agar and ethanol reduces the water vapour permeability of polymeric composites from 75% to 95%, depending on the layer thickness. The gelation of agar creates a dense network which increases water vapour resistance and mechanical resistance [109].

As part of natural and sustainable coatings, gelatine plays a significant role. Gelatine is a biopolymer with a multifaceted nature. It is a filmogenic polymer with a sound barrier against moisture, oxygen and other atmospheric elements. It makes transparent films with antimicrobial properties, which can be used as a highly functional biodegradable coating. The adhesive property of gelatine helps it to stick firmly to any surface [105, 107, 112–115]. Researchers have combined gelatine with other substances to make an effective coating for various applications. In one of the approaches, gelatine-based coating is used to increase the shelf life of strawberries. The coating only consisted of gelatine because starch doesn't offer good antioxidant properties. To overcome this problem, plant-derived essential oils are used to improve the coating's mechanical, antimicrobial and barrier properties. A study showed that to further enhance the barrier properties of gelatine-based biopolymer coatings, *Mentha pulegium* can be used. It further enhances the barrier and antioxidant properties of the coating, thus providing better resistance to the strawberry's surface [112, 113]. The adhesion of gelatine is different for different surfaces. So, other substrates need a different approach to be coated with gelatine-based coatings. In a different approach, gelatine and chitosan are used to develop an edible coating to delay meat discolouration in retail displays. The chitosan-gelatine edible coatings are incorporated with tarragon essential oil (TEO). Nanoparticles of chitosan gelatine were prepared with the gelation method. The nano-encapsulated TEO helps in the controlled release of TEO in the system, thus improving the meat's antioxidant and anti-microbacterial properties [116]. Another study showed that glycerol instead of TEO can be used with different chitosan and gelatine blends, reducing weight loss due to water shrinkage and lipid oxidation of the meat in display storage [116]. To reduce microbial infections on the surface of fish fillets and delay their discolouration, gelatine-based coatings are used widely [105]. Gelatine and native cassava starch (NCS), sorbitol and a different fraction of *Tetradenia riparia* extract are used to develop an advanced edible coating for strawberries. The films produced provide better protection against bacterial growth and antioxidant activity losses [112]. But with increased thickness, the solubility reduces, directly influencing the water vapour transmission between surfaces. Thus, using gelatine-based antimicrobial edible coating is gaining popularity in the food industry. To protect rice grain from atmospheric conditions, folic acids or folates are used to fortify the nutrients of the rice by coating it. Adding folic acid also increases the amount of folic acid intake among the consumers, which is suggested by government bodies. Using any acid for rice fortification is not suitable for creating a protective barrier to minimize the loss of folic acid from within. To tackle this problem, different materials were sprayed or press-coated against the fortified rice, including locust bean gum, agar, xanthan gum–coated premix, rice gruel, and rice gruel and salic acid combined. At the end of the process, calcium phosphate and starch powder were evenly sprayed as a final layer while rolling the rice. The final result showed that locust bean gum, xanthan gum, and agar coating showed successful masking of rice, which further reduced the loss of folic acid [110]. Different biopolymer-based coatings act differently with added oil, fats and acids, improving shelf life, quality, physical properties and chemical barrier, delaying discolouration of the substrate it has been applied on. These edible green coatings help food-grade products last longer, thus keeping the system sustainable.

1.5 BIODEGRADABILITY AND ENVIRONMENTAL ASPECT

Biodegradability tests can be performed using ASTM D6003–96. In popular practice, the natural soil burial test is considered one of the most effective methods to check the rate of biodegradability [117–120]. Apart from that, the simulated solid waste aerobic compost test is also used for the same [117]. In the soil burial test, the specimens are buried 6 to 7 cm deep inside the soil surface. The soil is kept moist by watering it in a regulated manner. The compost soil is kept in a controlled environment in the compost test. The temperature is generally controlled at 30 °C with 50%–60% water content. The pH level of the compost soil is maintained at 7. As per the design of the experiment, the samples are removed from the soil and washed with distilled water until clean. The samples are then air-dried in a hot air furnace at a temperature around 65 to75 °C for 24 hours. The weight loss of the sample during the soil burial test and composite tests signifies the rate of biodegradability of the sample [120].

Polymers are divided into two categories, biodegradable and nonbiodegradable. Though nonbiodegradable polymers have better properties, they face challenges during disposal. Synthetic, nonbiodegradable polymers are a matter of concern globally as they do not degrade fast. So, the popularity of degradable and recyclable polymers is increasing. Poly(lactic acid) (PLA) is one of the most commercialised degradable polymers. PLA is a starch-derived natural polymer that reserves the exclusive properties of commercially used plastics [121]. Chitosan is another feasible option to be used as a biodegradable polymer. It is derived from shrimp shells and is completely non-toxic. Being biodegradable and having all the necessary qualities, shrimp shells are used mostly for biomedical studies [121, 122]. Starch is another naturally found biopolymer that is widely used to manufacture substitute sustainable counterparts to plastic bags. Starch-based resins are biodegradable and have considerable thermomechanical and rheological properties. The main advantage of starch is its easy processing properties. It has been used for many years in plastic manufacturing as fillers, but adding starch reduces the mechanical properties of starch-based composites. This is one of the main reasons to keep starch content below 40% in any composite [121, 123]. Apart from natural biodegradable polymers, other commercially used synthetic resins like polybutylene succinate (PBS), polybutylene succinate adipate (PBSA) and poly (butyleneadipate-co-terephthalate) are popularly used as biodegradable polymers [124]. Biodegradable polymers do not show a competitive advantage over synthetic polymers due to their low mechanical properties. Keeping the biodegradability factor in the count, natural cellulose is blended with natural polymer instead of integrating these natural polymers and synthetic polymers, making it more mechanically advanced yet 100% natural. Microorganisms play a significant role in the faster degradation of composites. The degradation study includes bacteria and fungi with the highest hydrolytic activity. Studies showed soil bacteria degrade the material faster than do isolated bacteria away from soil or compost [122]. Different hydrolase enzymes like lysozyme glycosides are responsible for faster degradation of materials upon soil burial or composting. The degradation of the composite materials depends on which bacteria or fungal growth can penetrate into the surface faster [125]. In a case study of chitosan and tannic acid–based composite, it has been observed that it is very difficult for most of the microorganisms to grow in it due to antimicrobial and antifungal properties. The SEM analysis showed bacterial growth occurs on the composites' surface as a thin film. Further study shows that composite has 99% antibacterial shielding against *Staphylococcus aureus*, but only 28.42% shielding against *Escherichia coli* [108, 126]. Oxidation is the key parameter for all microorganisms and fungi to access the material surface quickly. Oxidation bio-assimilates the molecular structure and breaks the hydrophilic oxidants, thus facilitating the biodegradation process [108, 126]. Synthetic plastics blended with starch create a hydrophilic property that helps to catalyse the surface with amylase enzymes. These catalysed surfaces can further be easily attacked and removed by microorganisms. Plenty of microorganisms found in soil are responsible for faster material surface degradation, including yeast, *Streptomyces* strains, *Mucor rouxii* NRRL 1835 fungi, *Aspergillus flavus* and *Penicillium simplicissimum* YK, to name a few [101, 111, 113, 127].

1.6 DEVELOPMENTS AND APPLICATIONS

All-natural composites are increasingly used to solve many problems in our society. The field of application primarily varies according to the end user. In such an approach, all-natural composites made of either industrial waste fibres or natural fibres are replacing disposable plastics [2, 128]. Disposable plastics are used worldwide for takeaway-based restaurants, eateries and hotels. Streetside food vendors, grocery stores, flea markets, and vegetable and fish markets require disposable plastic packets or containers. This generates a massive amount of non-recyclable waste worldwide, which has become a prime concern for all environmentalists [36]. Microplastic pollution has reached the level where it has been found in the body fluids of unborn babies and in unexplored deep trenches in the ocean where human interference has not yet occurred. The only solution to avoid this problem is to reduce plastic usage and use sustainable practices wherever possible. Many researchers and upcoming business startups globally promote and produce sustainable and rapidly biodegradable cutlery and packaging units. The products are primarily developed using naturally available biodegradable fibre and polymer protein. Several researchers have developed biodegradable counterparts of plastic disposables using bagasse and starch-based composites. Bamboo fibre, textile waste, banana fibre, rice or wheat husk and straws have also been used widely to develop natural composites. In a similar approach, we have developed all-natural composites based on silk fibre waste and wheat gluten. Apart from these, other organic constituents are used to develop products. Natural lemon extract is used as a crosslinker, and castor oil is used as a natural plasticiser [129].

For our research, all the samples were prepared following the same process, where the wheat gluten (WG) and chopped filature silk fibre of 20 ($\mp$ 1.5) mm length were mixed in a ratio of 50:50 (w/w) for making the composite sheets. The process requires hot water at 60 °C, in which the wheat gluten was poured slowly and mixed gently by stirring the water at a fixed temperature of 60 °C with 10% castor oil as a plasticiser and 5% freshly extracted lemon juice as the crosslinker. The fibres were dipped into hot water for 30 minutes until they were soft enough to be added to the mixture. After 30 minutes, the water was drained out gently from the fibres, the wheat gluten mixture was poured onto it and proper mixing was done with the hand. When the mixture became homogeneous, a thin sheet was formed using the hand-lay-up method and kept inside a hot-air oven at 70 °C until it was dried completely. The dried sample sheets had a thickness of approximately 1 mm each and were stacked in layers inside a stainless steel mould to achieve a sample dimension of 160 mm in length, 105 mm in width and 2.5 mm in depth. The mould was then placed in the hot compression mould and compressed under a pressure of 120 bar for 10 minutes at 120 °C, followed by a 15-minute compression cycle without applying any heat. The prepared samples were then cooled inside the mould and taken out for in-depth analysis.

The thermomechanical analysis and biodegradation test further showed the viability of the manufactured sheets to be used as biodegradable counterparts of plastic. The impressive mechanical strength provides the necessary strength for the structure. The enhanced flexibility due to the addition of natural plasticiser, and the crosslinker helped the composite deform into different shapes under the hot press. The thermogravimetric analysis showed impressive stability of the developed composite under high temperatures. A natural crosslinker and plasticiser helped the wheat gluten form disulfide bonds, which enhanced the polymeric reaction and formed more extended chains, as shown in Figure 1.1. The better crosslinking helped the composite exhibit more strength under mechanical testing and better thermal stability.

The rapid (two- to three-week) biodegradability of the developed composite is also impressive as shown by the soil degradation test. The SEM images also showed good adhesion between matrix and fibre in the presence of crosslinker and plasticiser, which naturally improves the mechanical properties of the composite. The SEM fractography in Figure 1.2 clearly indicated the promising adhesion between the fibre and matrix under the influence of plasticiser and crosslinker.

The tensile test showed that the prepared composites notably improved in tensile strength and percentage elongation compared to the matrix material. Though

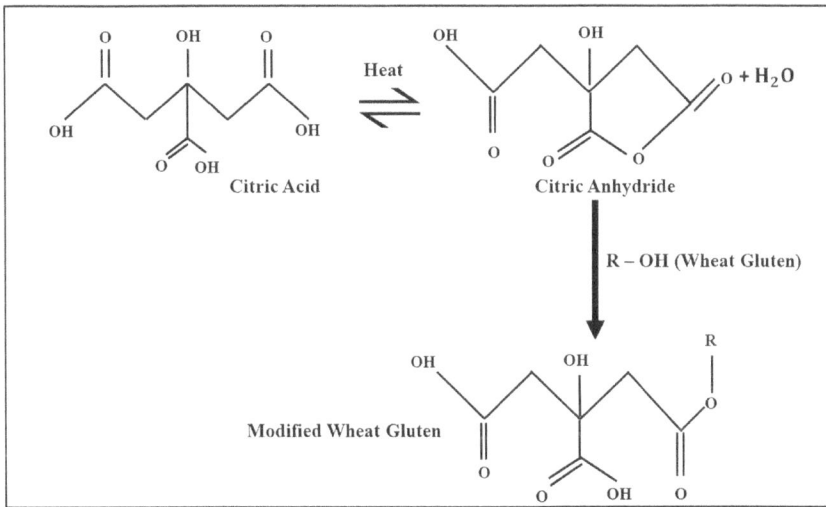

Figure 1.1 Chemical reaction between the wheat gluten and crosslinker to form a longer polymeric chain.

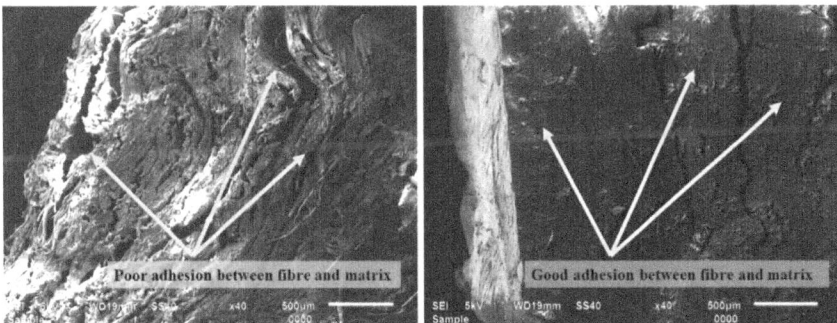

Figure 1.2 SEM fractographs showing the fibre matrix adhesion of the wheat gluten and filature silk composite with and without natural plasticiser and crosslinker.

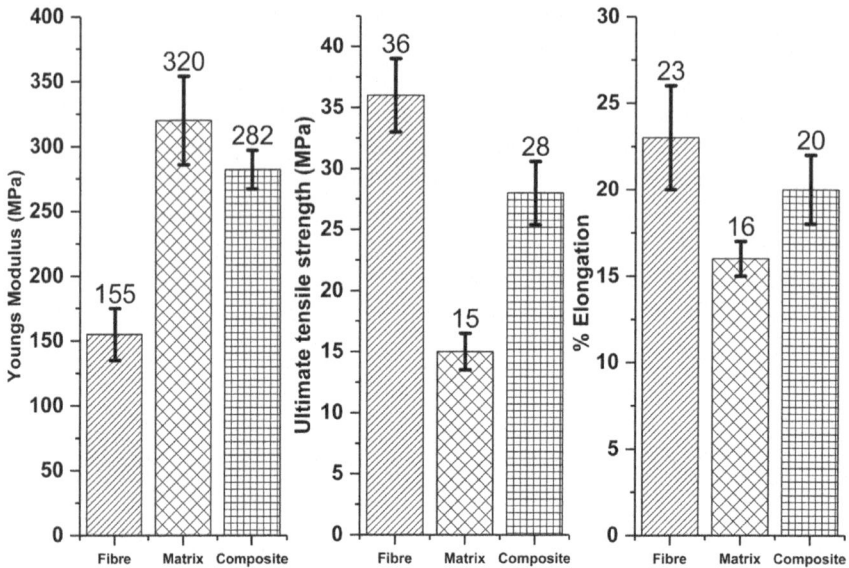

Figure 1.3 Comparison of different mechanical properties of filature silk, wheat gluten and wheat gluten and filature silk composite.

the addition of fibres enhances the strength of the composite, it also increases the brittleness of the composite. To reduce the overall brittleness, castor oil was incorporated into the composite as a plasticiser to make it more flexible. As a result, the final composite showed better elongation than the matrix material. The tensile test was performed on five samples, and the average outcome was calculated and compared, as shown in Figure 1.3. The composites showed an average ultimate tensile strength of 28 MPa, Young's modulus of 282 MPa and elongation of up to 20%. The strength and elongation of the prepared composite satisfy the conditions for low-pressure moulding. Low-pressure moulding ensures shaping the biocomposite sheets into different forms as per the requirement, with significantly lower applied pressure and no breakage.

1.7 CONCLUSION

Sustainable, green composites are needed urgently as they have the potential to replace plastic counterparts on a large scale. Natural composites exhibit advantageous thermomechanical properties. In addition, they are renewable, available in ample amounts and less expensive. Biocomposites have increasingly been perceived for their unmatchable biodegradability and low toxicity. Plenty of scientific research has already been conducted on the advantages of biocomposites. The outcome of our research demonstrated the immense potential for developing and utilising these composites according to market needs and size. The current necessity is to minimize the use of plastic products to the greatest extent possible. Developing newer materials and designs for biocomposites to ensure application-based performance capabilities should also be emphasised for future studies. As these biocomposites will be used for different end applications, enhancing the overall appearance and the performance of the biocomposites should also be studied widely for better market acceptance. Ultimately, the market should recognize the significance of replacing plastics with biocomposites, and regulations should be imposed to mandate using biocomposite-based products. If practised together, the problem of plastic waste can easily be countered with the newly developed biocomposites.

REFERENCES

[1] Ramteke S, Sahu BL (2020) Novel coronavirus disease 2019 (COVID-19) pandemic: Considerations for the biomedical waste sector in India. Case Stud Chem Environ Eng 2:100029. https://doi.org/10.1016/j.cscee.2020.100029
[2] Nielsen TD, Hasselbalch J, Holmberg K, Stripple J (2020) Politics and the plastic crisis: A review throughout the plastic life cycle. Wiley Interdiscip Rev Energy Environ 9:1–18. https://doi.org/10.1002/wene.360
[3] La Mantia FP, Morreale M (2011) Green composites: A brief review. Compos Part A Appl Sci Manuf 42:579–588. https://doi.org/10.1016/j.compositesa.2011.01.017
[4] Dunmade I (2013) Mechanical properties of renewable materials: A study on alpaca fibre. Int J Eng Sci Invent 2:56–62.
[5] Abdul Khalil HPS, Bhat AH, Ireana Yusra AF (2012) Green composites from sustainable cellulose nanofibrils: A review. Carbohydr Polym 87:963–979. https://doi.org/10.1016/j.carbpol.2011.08.078
[6] Mann GS, Singh LP, Kumar P, Singh S (2020) Green composites: A review of processing technologies and recent applications. J Thermoplast Compos Mater 33:1145–1171. https://doi.org/10.1177/0892705718816354
[7] Lee MW, Han SO, Seo YB (2008) Red algae fibre/poly(butylene succinate) biocomposites: The effect of fibre content on their mechanical and thermal properties. Compos Sci Technol 68:1266–1272. https://doi.org/10.1016/j.compscitech.2007.12.016
[8] Manshor MR, Anuar H, Nur Aimi MN, et al. (2014) Mechanical, thermal and morphological properties of durian skin fibre reinforced PLA biocomposites. Mater Des 59:279–286. https://doi.org/10.1016/j.matdes.2014.02.062
[9] Zhou M, Yan J, Li Y, et al. (2013) Interfacial strength and mechanical properties of biocomposites based on ramie fibers and poly(butylene succinate). RSC Adv 3:26418–26426. https://doi.org/10.1039/c3ra43713b
[10] Shibata M, Ozawa K, Teramoto N, et al. (2003) Biocomposites made from short abaca fiber and biodegradable polyesters. Macromol Mater Eng 288:35–43. https://doi.org/10.1002/mame.200290031
[11] Kim Y, Kwon OH, Park WH, Cho D (2013) Thermomechanical and flexural properties of chopped silk fiber-reinforced poly(butylene succinate) green composites: Effect of electron beam treatment of worm silk. Adv Compos Mater 22:437–449. https://doi.org/10.1080/09243046.2013.843830

[12] Studies A, District S, District M, et al. (2013) 2013-Ganesh-A comparative study on tensile behaviour of plant and animal fiber reinforced composites. Int J Innov Appl Stud 2:645–648.

[13] Koçak D, Tasdemir M, Usta I, et al. (2008) Mechanical, thermal, and microstructure analysis of silk- and cotton-waste-fiber-reinforced high-density polyethylene composites. Polym—Plast Technol Eng 47:502–507. https://doi.org/10.1080/03602550801977919

[14] Kim NK, Bruna FG, Das O, et al. (2020) Fire-retardancy and mechanical performance of protein-based natural fibre-biopolymer composites. Compos Part C Open Access 1:100011. https://doi.org/10.1016/j.jcomc.2020.100011

[15] Sharma A, Rao NN, Krupashankara MS (2018) Development of eco-friendly and biodegradable bio composites. Mater Today Proc 5:20987–20995. https://doi.org/10.1016/j.matpr.2018.6.490

[16] Iber BT, Okomoda VT, Rozaimah SA, Kasan NA (2021) Eco-friendly approaches to aquaculture wastewater treatment: Assessment of natural coagulants vis-a-vis chitosan. Bioresour Technol Reports 15:100702. https://doi.org/10.1016/j.biteb.2021.100702

[17] Shih YF, Cai JX, Kuan CS, Hsieh CF (2012) Plant fibers and wasted fiber/epoxy green composites. Compos Part B Eng 43:2817–2821. https://doi.org/10.1016/j.compositesb.2012.04.044

[18] Nine MJ, Kabiri S, Sumona AK, et al. (2020) Superhydrophobic/superoleophilic natural fibres for continuous oil-water separation and interfacial dye-adsorption. Sep Purif Technol 233:116062. https://doi.org/10.1016/j.seppur.2019.116062

[19] Sepe R, Bollino F, Boccarusso L, Caputo F (2018) Influence of chemical treatments on mechanical properties of hemp fiber reinforced composites. Compos Part B Eng 133:210–217. https://doi.org/10.1016/j.compositesb.2017.09.030

[20] Song Y, Zheng Q (2008) Improved tensile strength of glycerol-plasticised gluten bioplastic containing hydrophobic liquids. Bioresour Technol 99:7665–7671. https://doi.org/10.1016/j.biortech.2008.01.075

[21] Sun S, Song Y, Zheng Q (2007) Morphologies and properties of thermo-molded biodegradable plastics based on glycerol-plasticised wheat gluten. Food Hydrocoll 21:1005–1013. https://doi.org/10.1016/j.foodhyd.2006.03.004

[22] Rahman M, Brazel CS (2004) The plasticiser market: An assessment of traditional plasticisers and research trends to meet new challenges. Prog Polym Sci 29:1223–1248.

[23] Reddy N, Reddy R, Jiang Q (2015) Crosslinking biopolymers for biomedical applications. Trends Biotechnol 33:362–369. https://doi.org/10.1016/j.tibtech.2015.03.008

[24] Hemsri S, Grieco K, Asandei AD, Parnas RS (2012) Wheat gluten composites reinforced with coconut fiber. Compos Part A Appl Sci Manuf 43:1160–1168. https://doi.org/10.1016/j.compositesa.2012.02.011

[25] Emin MA, Quevedo M, Wilhelm M, Karbstein HP (2017) Analysis of the reaction behavior of highly concentrated plant proteins in extrusion-like conditions. Innov Food Sci Emerg Technol 44:15–20. https://doi.org/10.1016/j.ifset.2017.09.013

[26] McGauran T, Harris M, Dunne N, et al. (2021) Development and optimisation of extruded bio-based polymers from poultry feathers. Eur Polym J 158:110678. https://doi.org/10.1016/j.eurpolymj.2021.110678

[27] Jansson H, Swenson J (2010) The protein glass transition as measured by dielectric spectroscopy and differential scanning calorimetry. Biochim Biophys Acta—Proteins Proteomics 1804:20–26. https://doi.org/10.1016/j.bbapap.2009.06.026

[28] Liu Y, Cai Z, Sheng L, et al. (2019) Structure-property of crosslinked chitosan/silica composite films modified by genipin and glutaraldehyde under alkaline conditions. Carbohydr Polym 215:348–357. https://doi.org/10.1016/j.carbpol.2019.04.001

[29] Liu W, Qiu J, Chen T, et al. (2019) Regulating tannic acid-crosslinked epoxidised soybean oil oligomers for strengthening and toughening bamboo fibers-reinforced poly(lactic acid) biocomposites. Compos Sci Technol 181:107709. https://doi.org/10.1016/j.compscitech.2019.107709

[30] Nataraj D, Sakkara S, Meenakshi, HN, Reddy N (2018) Properties and applications of citric acid crosslinked banana fibre-wheat gluten films. Ind Crops Prod 124:265–272. https://doi.org/10.1016/j.indcrop.2018.07.076

[31] Reddy N, Tan Y, Li Y, Yang Y (2008) Effect of glutaraldehyde crosslinking conditions on the strength and water stability of wheat gluten fibers. Macromol Mater Eng 293:614–620. https://doi.org/10.1002/mame.200800031

[32] Yeng CM, Husseinsyah S, Ting SS (2015) Effect of cross-linking agent on tensile properties of chitosan/corn cob biocomposite films. Polym Plast Technol Eng 54:270–275. https://doi.org/10.1080/03602559.2014.977090

[33] Patil NV, Netravali AN (2016) Microfibrillated cellulose-reinforced nonedible starch-based thermoset biocomposites. J Appl Polym Sci 133. https://doi.org/10.1002/app.43803

[34] Windsor FM, Durance I, Horton AA, et al. (2019) A catchment-scale perspective of plastic pollution. Glob Chang Biol 25:1207–1221. https://doi.org/10.1111/gcb.14572

[35] Singh P, Sharma VP (2016) Integrated plastic waste management: Environmental and improved health approaches. Procedia Environ Sci 35:692–700. https://doi.org/10.1016/j.proenv.2016.07.068

[36] Joshi R, Ahmed S (2016) Status and challenges of municipal solid waste management in India: A review. Cogent Environ Sci 2. https://doi.org/10.1080/23311843.2016.1139434

[37] Xue C, Gao H, Hu Y, Hu G (2019) Hyperelastic characteristics of graphene natural rubber composites and reinforcement and toughening mechanisms at multi-scale. Compos Struct 228:111365. https://doi.org/10.1016/j.compstruct.2019.111365

[38] Krishna V, Kate KH, Satyavolu J, Singh P (2019) Additive manufacturing of natural fiber reinforced polymer composites: Processing and prospects. Compos Part B 174:106956. https://doi.org/10.1016/j.compositesb.2019.106956

[39] Chandramohan D, Presin Kumar AJ (2017) Experimental data on the properties of natural fiber particle reinforced polymer composite material. Data Br 13:460–468. https://doi.org/10.1016/j.dib.2017.06.020

[40] Tasdemir M, Akalin M, Kocak D, et al. (2010) Investigation of properties of polymer/textile fiber composites. Int J Polym Mater Polym Biomater 59:200–214. https://doi.org/10.1080/00914030903231415

[41] Chee SS, Jawaid M, Sultan M, et al. (2019) Thermomechanical and dynamic mechanical properties of bamboo/woven kenaf mat reinforced epoxy hybrid composites. Compos Part B Eng 163:165–174. https://doi.org/10.1016/j.compositesb.2018.11.039

[42] Shubhra QTH, Alam AKMM (2011) Effect of gamma radiation on the mechanical properties of natural silk fiber and synthetic E-glass fiber reinforced polypropylene composites: A comparative study. Radiat Phys Chem 80:1228–1232. https://doi.org/10.1016/j.radphyschem.2011.04.010

[43] Shubhra QTH, Alam AKMM, Beg MDH, et al. (2011) Mechanical and degradation characteristics of natural silk and synthetic phosphate glass fiber reinforced polypropylene composites. J Compos Mater 45:1305–1313. https://doi.org/10.1177/0021998310380290

[44] Kumar S, Manna A, Dang R (2022) A review on applications of natural fiber-reinforced composites (NFRCs). Mater Today Proc 50:1632–1636. https://doi.org/10.1016/J.MATPR.2021.09.131

[45] Roper H, Koch H (1990) The role of starch in biodegradable thermoplastic materials. Starch - Stärke 42:123–130. https://doi.org/10.1002/star.19900420402

[46] Ashori A (2008) Wood-plastic composites as promising green-composites for automotive industries! Bioresour Technol 99:4661–4667. https://doi.org/10.1016/j.biortech.2007.09.043

[47] Rasheed F, Plivelic TS, Kuktaite R, et al. (2018) Unraveling the structural puzzle of the giant glutenin polymer—An interplay between protein polymerization, nanomorphology, and functional properties in bioplastic films. ACS Omega 3:5584–5592. https://doi.org/10.1021/acsomega.7b02081

[48] Cao N, Fu Y, He J (2007) Preparation and physical properties of soy protein isolate and gelatin composite films. Food Hydrocoll 21:1153–1162. https://doi.org/10.1016/j.foodhyd.2006.09.001

[49] Lodha P, Netravali AN (2002) Characterisation of interfacial and mechanical properties of "green" composites with soy protein isolate and ramie fiber. J Mater Sci 37:3657–3665. https://doi.org/10.1023/A:1016557124372

[50] Chang YP, Abd Karim A, Seow CC (2006) Interactive plasticizing-antiplasticizing effects of water and glycerol on the tensile properties of tapioca starch films. Food Hydrocoll 20:1–8. https://doi.org/10.1016/j.foodhyd.2005.02.004

[51] Breuninger WF, Piyachomkwan K, Sriroth K (2009) Tapioca/cassava starch: Production and use. In Starch 541–568. https://doi.org/10.1016/B978-0-12-746275-2.00012-4

[52] Van Soest JJG, Hulleman SHD, De Wit D, Vliegenthart JFG (1996) Changes in the mechanical properties of thermoplastic potato starch in relation with changes in B-type crystallinity. Carbohydr Polym 29:225–232. https://doi.org/10.1016/0144-8617(96)00011-2

[53] Podshivalov A, Zakharova M, Glazacheva E, Uspenskaya M (2017) Gelatin/potato starch edible biocomposite films: Correlation between morphology and physical properties. Carbohydr Polym 157:1162–1172. https://doi.org/10.1016/J.CARBPOL.2016.10.079

[54] Rodríguez-Castellanos W, Martínez-Bustos F, Rodrigue D, Trujillo-Barragán M (2015) Extrusion blow molding of a starch–gelatin polymer matrix reinforced with cellulose. Eur Polym J 73:335–343. https://doi.org/10.1016/J.EURPOLYMJ.2015.10.029

[55] Peña C, de la Caba K, Eceiza A, et al. (2010) Enhancing water repellence and mechanical properties of gelatin films by tannin addition. Bioresour Technol 101:6836–6842. https://doi.org/10.1016/J.BIORTECH.2010.03.112

[56] Davidović S, Lazić V, Miljković M, et al. (2019) Antibacterial ability of immobilised silver nanoparticles in agar-agar films co-doped with magnesium ions. Carbohydr Polym 224:115187. https://doi.org/10.1016/J.CARBPOL.2019.115187

[57] Nogueira GF, Soares CT, Cavasini R, et al. (2019) Bioactive films of arrowroot starch and blackberry pulp: Physical, mechanical and barrier properties and stability to pH and sterilisation. Food Chem 275:417–425. https://doi.org/10.1016/J.FOODCHEM.2018.09.054

[58] Tarique J, Sapuan SM, Khalina A (2021) Effect of glycerol plasticiser loading on the physical, mechanical, thermal, and barrier properties of arrowroot (Maranta arundinacea) starch biopolymers. Sci Reports 11:13900. https://doi.org/10.1038/s41598-021-93094-y

[59] Grégoire M, Bar M, De Luycker E, et al. (2021) Comparing flax and hemp fibres yield and mechanical properties after scutching/hackling processing. Ind Crops Prod 172:114045. https://doi.org/10.1016/J.INDCROP.2021.114045

[60] Munawar SS, Umemura K, Kawai S (2007) Characterisation of the morphological, physical, and mechanical properties of seven nonwood plant fiber bundles. J Wood Sci 53:108–113. https://doi.org/10.1007/S10086-006-0836-X/METRICS

[61] Saba N, Paridah MT, Jawaid M (2015) Mechanical properties of kenaf fibre reinforced polymer composite: A review. Constr Build Mater 76:87–96. https://doi.org/10.1016/J.CONBUILDMAT.2014.11.043

[62] Defoirdt N, Biswas S, De Vriese L, et al. (2010) Assessment of the tensile properties of coir, bamboo and jute fibre. Compos Part A Appl Sci Manuf 41:588–595. https://doi.org/10.1016/J.COMPOSITESA.2010.01.005

[63] Yao W, Li Z (2003) Flexural behavior of bamboo–fiber-reinforced mortar laminates. Cem Concr Res 33:15–19. https://doi.org/10.1016/S0008-8846(02)00909-2

[64] Chen Z, Xu Y, Shivkumar S (2018) Microstructure and tensile properties of various varieties of rice husk. J Sci Food Agric 98:1061–1070. https://doi.org/10.1002/JSFA.8556

[65] Haameem M, Abdul Majid MS, Afendi M, et al. (2016) Mechanical properties of Napier grass fibre/polyester composites. Compos Struct 136:1–10. https://doi.org/10.1016/J.COMPSTRUCT.2015.09.051

[66] Chandrasekar M, Senthilkumar K, Senthil Muthu Kumar T, et al. (2021) Effect of adding sisal fiber on the sliding wear behavior of the coconut sheath fiber-reinforced composite. Tribol Polym Compos 115–125. https://doi.org/10.1016/B978-0-12-819767-7.00006-2

[67] Saragih SW, Lubis R, Wirjosentono B, Eddyanto (2018) Characteristic of abaca (Musa textilis) fiber from Aceh Timur as bioplastic. AIP Conf Proc 2049. https://doi.org/10.1063/1.5082463

[68] Senthilkumar K, Siva I, Rajini N, et al. (2018) Mechanical characteristics of tri-layer eco-friendly polymer composites for interior parts of aerospace application. Sustain Compos Aerosp Appl 35–53. https://doi.org/10.1016/B978-0-08-102131-6.00003-7

[69] Kavitha C, Sureshkumar K, Srividhya N, et al. (2022) Experimental investigation of tensile properties in banana fibre composites. Mater Today Proc. https://doi.org/10.1016/J.MATPR.2022.01.451

[70] Yusriah L, Sapuan SM, Zainudin ES, Mariatti M (2012) Exploring the potential of betel nut husk fiber as reinforcement in polymer composites: Effect of fiber maturity. Procedia Chem 4:87–94. https://doi.org/10.1016/J.PROCHE.2012.06.013

[71] Yu Y, Yang W, Wang B, Meyers MA (2017) Structure and mechanical behavior of human hair. Mater Sci Eng C 73:152–163. https://doi.org/10.1016/j.msec.2016.12.008

[72] Farran L, Ennos AR, Starkie M, Eichhorn SJ (2009) Tensile and shear properties of fingernails as a function of a changing humidity environment. J Biomech 42:1230–1235. https://doi.org/10.1016/j.jbiomech.2009.03.020

[73] Zhan M, Wool RP (2011) Mechanical properties of chicken feather fibers. Polym Compos 32:937–944. https://doi.org/10.1002/PC.21112

[74] Chen S, Liu M, Huang H, et al. (2019) Mechanical properties of Bombyx mori silkworm silk fibre and its corresponding silk fibroin filament: A comparative study. Mater Des 181:108077. https://doi.org/10.1016/J.MATDES.2019.108077

[75] Araya-Letelier G, Antico FC, Carrasco M, et al. (2017) Effectiveness of new natural fibers on damage-mechanical performance of mortar. Constr Build Mater 152:672–682. https://doi.org/10.1016/J.CONBUILDMAT.2017.07.072

[76] Mohan NH, Debnath S, Mahapatra RK, et al. (2014) Tensile properties of hair fibres obtained from different breeds of pigs. Biosyst Eng 119:35–43. https://doi.org/10.1016/J.BIOSYSTEMSENG.2014.01.003

[77] McGregor BA (2018) Physical, chemical, and tensile properties of cashmere, mohair, alpaca, and other rare animal fibers. Handb Prop Text Tech Fibres 105–136. https://doi.org/10.1016/B978-0-08-101272-7.00004-3

[78] Harizi T, Msahli S, Sakli F, Khorchani T (2010) Evaluation of physical and mechanical properties of Tunisian camel hair. http://dx.doi.org/101533/joti20050165 98:15–21. https://doi.org/10.1533/JOTI.2005.0165

[79] Gosline JM, DeMont ME, Denny MW (1986) The structure and properties of spider silk. Endeavour 10:37–43. https://doi.org/10.1016/0160-9327(86)90049-9

[80] Sharma AK, Bhandari R, Aherwar A, et al. (2020) A study of advancement in application opportunities of aluminum metal matrix composites. Mater Today Proc 26:2419–2424. https://doi.org/10.1016/J.MATPR.2020.02.516

[81] Vijaya Ramnath B, Elanchezhian C, Jaivignesh M, et al. (2014) Evaluation of mechanical properties of aluminium alloy–alumina–boron carbide metal matrix composites. Mater Des 58:332–338. https://doi.org/10.1016/J.MATDES.2014.01.068

[82] Ralph B, Yuen HC, Lee WB (1997) The processing of metal matrix composites—an overview. J Mater Process Technol 63:339–353. https://doi.org/10.1016/S0924-0136(96)02645-3

[83] Ellis MBD (2013) Joining of aluminium based metal matrix composites. Int Mater Rev 41:41–58. https://doi.org/10.1179/IMR.1996.41.2.41

[84] Lukkassen D, Meidell A (2007) Advanced materials and structures and their fabrication processes. B manuscript, Narvik Univ Coll HiN 2:1–14.

[85] Du J, Zhang H, Geng Y, et al. (2019) A review on machining of carbon fiber reinforced ceramic matrix composites. Ceram Int 45:18155–18166. https://doi.org/10.1016/J.CERAMINT.2019.06.112

[86] Chaiwong W, Samoh N, Eksomtramage T, Kaewtatip K (2019) Surface-treated oil palm empty fruit bunch fiber improved tensile strength and water resistance of wheat gluten-based bioplastic. Compos Part B 176:107331. https://doi.org/10.1016/j.compositesb.2019.107331

[87] Chen T, Wu Y, Qiu J, et al. (2020) Interfacial compatibilisation via in-situ polymerisation of epoxidised soybean oil for bamboo fibers reinforced poly(lactic acid) biocomposites. Compos Part A Appl Sci Manuf 138:106066. https://doi.org/10.1016/j.compositesa.2020.106066

[88] Ramesh M, Palanikumar K, Reddy KH (2013) Mechanical property evaluation of sisal-jute-glass fiber reinforced polyester composites. Compos Part B Eng 48:1–9. https://doi.org/10.1016/j.compositesb.2012.12.004

[89] Ku H, Wang H, Pattarachaiyakoop N, Trada M (2011) A review on the tensile properties of natural fiber reinforced polymer composites. Comp B Eng 42:856–873. https://doi.org/10.1016/j.compositesb.2011.01.010

[90] Haddadi SA, Ghaderi S, Sadeghi M, et al. (2021) Enhanced active/barrier corrosion protective properties of epoxy coatings containing eco-friendly green inorganic/organic hybrid pigments based on zinc cations/Ferula Asafoetida leaves. J Mol Liq 323:114584. https://doi.org/10.1016/j.molliq.2020.114584

[91] Gassan J, Gutowski VS, Bledzki AK. About the surface characteristics of natural fibres. https://doi.org/10.1002/1439-2054(20001101)283:1<132::AID-MAME132>3.0.CO;2-B

[92] Lacroix M, Vu KD (2013) Edible Coating and Film Materials: Proteins. Elsevier Ltd.

[93] Mokhothu TH, John MJ (2017) Bio-based coatings for reducing water sorption in natural fibre reinforced composites. Sci Rep 7:1–8. https://doi.org/10.1038/s41598-017-13859-2

[94] Tian Z, Zhang R, Liu Y, et al. (2021) Hemicellulose-based nanocomposites coating delays lignification of green asparagus by introducing AKD as a hydrophobic modifier. Renew Energy 178:1097–1105. https://doi.org/10.1016/j.renene.2021.06.096

[95] Ghaderi S, Ramazani SAA, Haddadi SA (2019) Applications of highly salt and highly temperature resistance terpolymer of acrylamide/styrene/maleic anhydride monomers as a rheological modifier: Rheological and corrosion protection properties studies. J Mol Liq 294:111635. https://doi.org/10.1016/j.molliq.2019.111635

[96] Van Nieuwenhove I, Salamon A, Peters K, et al. (2016) Gelatin- and starch-based hydrogels. Part A: Hydrogel development, characterisation and coating. Carbohydr Polym 152:129–139. https://doi.org/10.1016/j.carbpol.2016.06.098

[97] Pérez-Gago MB, Rhim JW (2013) Edible coating and film materials: Lipid bilayers and lipid emulsions. In Innovations in Food Packaging. 2nd ed. 325–350. https://doi.org/10.1016/B978-0-12-394601-0.00013-8

[98] Cruz AIC, Costa M da C, Mafra JF, et al. (2021) A sodium alginate bilayer coating incorporated with green propolis extract as a powerful tool to extend Colossoma macropomum fillet shelf-life. Food Chem 355. https://doi.org/10.1016/j.foodchem.2021.129610

[99] Gabriela da Silva Pires P, Bavaresco C, Daniela da Silva Pires P, et al. (2021) Development of an innovative green coating to reduce egg losses. Clean Eng Technol 2:100065. https://doi.org/10.1016/j.clet.2021.100065

[100] Darroman E, Durand N, Boutevin B, Caillol S (2015) New cardanol/sucrose epoxy blends for biobased coatings. Prog Org Coatings 83:47–54. https://doi.org/10.1016/j.porgcoat.2015.02.002

[101] Revuelta MV., Bogdan S, Gámez-Espinosa E, et al. (2021) Green antifungal waterborne coating based on essential oil microcapsules. Prog Org Coatings 151. https://doi.org/10.1016/j.porgcoat.2020.106101

[102] Tsupphayakorn-aek P, Suwan A, Tulyapitak T, et al. (2022) A novel UV-curable waterborne polyurethane-acrylate coating based on green polyol from hydroxyl telechelic natural rubber. Prog Org Coatings 163:106585. https://doi.org/10.1016/j.porgcoat.2021.106585

[103] Huang C, Zhang N, Wang Q, et al. (2021) Development of hydrophilic anti-crease finishing method for cotton fabric using alpha-lipoic acid without causing strength loss and formaldehyde release problem. Prog Org Coatings 151:106042. https://doi.org/10.1016/j.porgcoat.2020.106042

[104] Prakashkumar N, Vignesh M, Brindhadevi K, et al. (2021) Enhanced antimicrobial, antibiofilm and anticancer activities of biocompatible neem gum coated palladium nanoparticles. Prog Org Coat 151:106098. https://doi.org/10.1016/j.porgcoat.2020.106098

[105] Bortoletto-Santos R, Guimarães GGF, Roncato Junior V, Cruz DFD, Polito WL, Ribeiro C (2019) Biodegradable oil-based polymeric coatings on urea fertilizer: N release kinetic transformations of urea in soil. Sci Agric 77:e20180033.

[106] Asgari S, Jahanshahi M, Rahimpour A (2014) Cost-effective nanoporous agar-agar polymer/nickel powder composite particle for effective bio-products adsorption by expanded bed chromatography. J Chromatogr A 1361:191–202. https://doi.org/10.1016/j.chroma.2014.08.016

[107] Wang X, Liu Y, Liu X, et al. (2021) Degradable gelatin-based supramolecular coating for green paper sizing. ACS Appl Mater Interfaces 13:1367–1376. https://doi.org/10.1021/acsami.0c16758

[108] Nguyen TT, Huynh Nguyen TT, Tran Pham BT, et al. (2021) Development of poly (vinyl alcohol)/agar/maltodextrin coating containing silver nanoparticles for banana (Musa acuminate) preservation. Food Packag Shelf Life 29:100740. https://doi.org/10.1016/j.fpsl.2021.100740

[109] Sousa AMM, Sereno AM, Hilliou L, Gonçalves MP (2010) Biodegradable agar extracted from Gracilaria vermiculophylla: Film properties and application to edible coating. Mater Sci Forum 636–637:739–744. https://doi.org/10.4028/www.scientific.net/MSF.636-637.739

[110] Shrestha AK, Arcot J, Paterson JL (2003) Edible coating materials—their properties and use in the fortification of rice with folic acid. Food Res Int 36:921–928. https://doi.org/10.1016/S0963-9969(03)00101-7

[111] Geraldine RM, Soares NFF, Botrel DA, de Almeida Gonçalves L (2008) Characterisation and effect of edible coatings on minimally processed garlic quality. Carbohydr Polym 72:403–409. https://doi.org/10.1016/j.carbpol.2007.09.012

[112] Friedrich JCC, Silva OA, Faria MGI, et al. (2020) Improved antioxidant activity of a starch and gelatin-based biodegradable coating containing Tetradenia riparia extract. Int J Biol Macromol 165:1038–1046. https://doi.org/10.1016/j.ijbiomac.2020.09.143

[113] Aitboulahsen M, Zantar S, Laglaoui A, et al. (2018) Gelatin-based edible coating combined with mentha pulegium essential oil as bioactive packaging for strawberries. J Food Qual 2018. https://doi.org/10.1155/2018/8408915

[114] Cardoso GP, Dutra MP, Fontes PR, et al. (2016) Selection of a chitosan gelatin-based edible coating for color preservation of beef in retail display. Meat Sci 114:85–94. https://doi.org/10.1016/j.meatsci.2015.12.012

[115] Kingwascharapong P, Arisa K, Karnjanapratum S, et al. (2020) Effect of gelatin-based coating containing frog skin oil on the quality of persimmon and its characteristics. Sci Hortic (Amsterdam) 260:108864. https://doi.org/10.1016/j.scienta.2019.108864

[116] Zhang H, Liang Y, Li X, Kang H (2020) Effect of chitosan-gelatin coating containing nano-encapsulated tarragon essential oil on the preservation of pork slices. Meat Sci 166:108137. https://doi.org/10.1016/j.meatsci.2020.108137

[117] Kim HS, Kim HJ, Lee JW, Choi IG (2006) Biodegradability of bio-flour filled biodegradable poly(butylene succinate) bio-composites in natural and compost soil. Polym Degrad Stab 91:1117–1127. https://doi.org/10.1016/j.polymdegradstab.2005.07.002

[118] Liu L, Yu J, Cheng L, Yang X (2009) Biodegradability of poly(butylene succinate) (PBS) composite reinforced with jute fibre. Polym Degrad Stab 94:90–94. https://doi.org/10.1016/j.polymdegradstab.2008.10.013

[119] Kalka S, Huber T, Steinberg J, et al. (2014) Biodegradability of all-cellulose composite laminates. Compos Part A Appl Sci Manuf 59:37–44. https://doi.org/10.1016/j.compositesa.2013.12.012

[120] Muniyasamy S, Anstey A, Reddy MM, et al. (2013) Biodegradability and compostability of lignocellulosic based composite materials. J Renew Mater 1:253–272. https://doi.org/10.7569/JRM.2013.634117

[121] Olaiya NG, Surya I, Oke PK, et al. (2019) Properties and characterisation of a PLA-chitin-starch biodegradable polymer composite. Polymers (Basel) 11. https://doi.org/10.3390/polym11101656

[122] Kaczmarek-Szczepańska B, Sionkowska MM, Mazur O, et al. (2021) The role of microorganisms in biodegradation of chitosan/tannic acid materials. Int J Biol Macromol 184:584–592. https://doi.org/10.1016/j.ijbiomac.2021.06.133

[123] Wu CS (2003) Physical properties and biodegradability of maleated-polycaprolactone/starch composite. Polym Degrad Stab 80:127–134. https://doi.org/10.1016/S0141-3910(02)00393-2

[124] Hamad K, Kaseem M, Ko YG, Deri F (2014) Biodegradable polymer blends and composites: An overview. Polym Sci—Ser A 56:812–829. https://doi.org/10.1134/S0965545X14060054

[125] Kim IH, Yang HJ, Noh BS, et al. (2012) Development of a defatted mustard meal-based composite film and its application to smoked salmon to retard lipid oxidation. Food Chem 133:1501–1509. https://doi.org/10.1016/j.foodchem.2012.02.040

[126] Hasheminya SM, Mokarram RR, Ghanbarzadeh B, et al. (2019) Development and characterisation of biocomposite films made from kefiran, carboxymethyl cellulose and Satureja Khuzestanica essential oil. Food Chem 289:443–452. https://doi.org/10.1016/j.foodchem.2019.03.076

[127] Nissa RC, Fikriyyah AK, Abdullah AHD, Pudjiraharti S (2019) Preliminary study of biodegradability of starch-based bioplastics using ASTM G21–70, dip-hanging, and soil burial test methods. IOP Conf Ser Earth Environ Sci 277:012007. https://doi.org/10.1088/1755-1315/277/1/012007

[128] Tokiwa Y, Calabia BP, Ugwu CU, Aiba S (2009) Biodegradability of plastics. Int J Mol Sci 10:3722–3742. https://doi.org/10.3390/ijms10093722

[129] Bhowmik P, Kant R, Nair R, Singh H (2021) Influence of natural crosslinker and fibre weightage on waste kibisu fibre reinforced wheatgluten biocomposite. J Polym Res 28:106. https://doi.org/10.1007/s10965-021-02470-9

Chapter 2

Metallic functionally graded materials

Krishan Kumar, Khushboo Rakha,
Shahriar Reza and Harpreet Singh

2.1 INTRODUCTION

Materials development originates from monolithic materials and progresses into alloys, composites, and advanced composites such as functionally graded materials through various techniques. It is impossible to have conflicting properties in a single object with monolithic materials, while alloying is limited by thermodynamic behavior and degree of miscibility of their constituents. The traditional composites fail under a harsh working environment as the sharp interface (i.e., mismatch in thermal and mechanical properties of adjoining materials) acts as a debilitating stress concentration site causing crack initiation and propagation to eventual failure of the object by delamination [1]. This sharp interface of conventional composites is systematically eliminated in novel composites (i.e., functionally graded materials) through a gradually changing interface; consequently, stress concentration reduces [2]. FGMs are advanced engineering materials with spatial gradients in either elastic, plastic, or both elastic and plastic properties, used to fabricate multi-functional parts by achieving conflicting performance goals in the same part. These property gradients are achieved through a seamless transition in porosity, material composition, and/or microstructure in their spatial dimensions. Engineering continuously turns to nature to find solutions to distinct engineering problems. FGMs are biologically inspired materials that mimic the ubiquitous FGMs in the palm tree, bamboo, and spongy trabecular structure of animal bones, teeth, and tissue variation in muscle and wood [3]. Nature has designed these products to meet various performance requirements. For example, the outer part of a human tooth is made of wear-resistant enamel with a ductile inner structure that absorbs impact load to optimize the fatigue life of teeth. High surface hardness and ductile cores achieve similar performance requirements of resisting the surface wear and fatigue failure in bearing applications [4]. In a true sense, for the multi-functional status of a component, there should be both intralayer and interlayer variation in properties [5]. FGM promotes artifact design by customizing various regions of a part with different material properties. Artificial FGMs have been investigated for decades after their first use by Japan in 1984 in space shuttle bodies as a thermal barrier material to withstand temperature differences of 1000°C [6]. The leverage of spatial

DOI: 10.1201/9781032703046-2

property gradient is a swiftly advancing research area, especially given the evolution of additive manufacturing (AM). Figure 2.1 demonstrates diverse application streams of FGMs to fabricate products with heterogeneous properties like toughness and wear resistance in engine nozzles, automotive valve stems and turbine blades under thermo-mechanical stresses, gears under impact and abrasion wear, and compliant hydrostatic bearing and cutting tools in distinct engineering applications. Graded implants and dental restorations with better biocompatibility and mechanical functionality have been ubiquitous in biomedical applications [1, 7–9].

Based on heterogeneity, materials with strategically controlled porosity/ composition/microstructure or a combination of them across their volume are referred to as FGMs. A potential classification of FGMs is exemplified in Figure 2.2: (a) heterogeneous structure materials comprising variable densification or porosity within a homogeneous composition; (b) heterogeneous composition materials having multiple materials combined with smooth and a seamless transition in composition; and (c) heterogeneous microstructure in any of (a) and (b) using heat treatment processes [10].

The fabricated FGMs are often post-processed to enhance tribological behavior through microstructural gradients [11]. The heterogeneous structure is also known as density cellular structure, porous structure, and functionally graded cellular microstructure. It comprises a lattice structure instead of solid material for higher strength-to-weight ratio to fabricate lighter yet stronger components. The geometrical characteristics of pores, lattice cells, and interconnectivity are linked to the properties of this kind of FGMs. The ratio of the void to the solid volume of a material (i.e., porosity) is engineered for superior acoustic, optical, thermal, and mechanical behavior and reduced weight per unit volume.

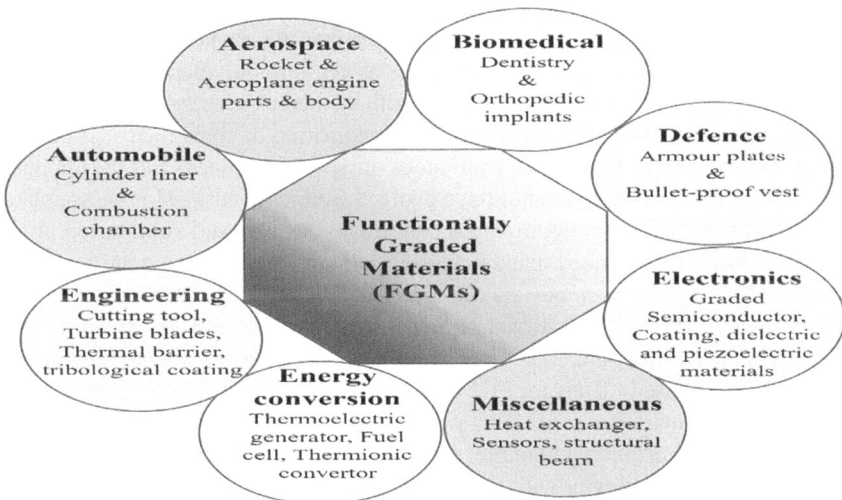

Figure 2.1 Potential applications of FGMs.

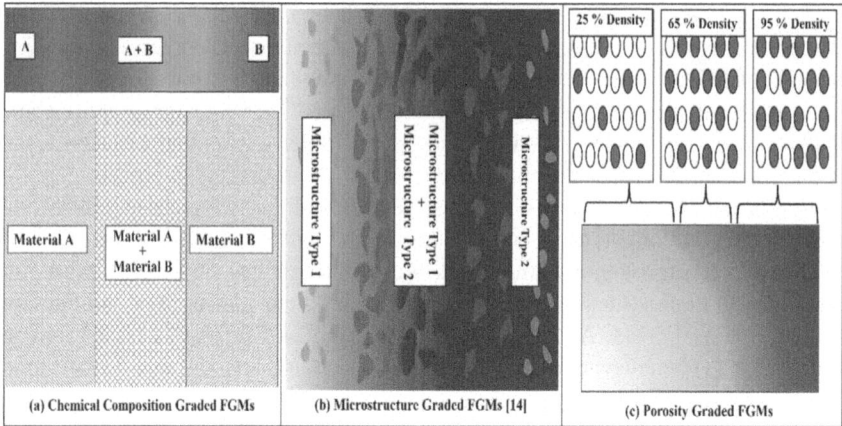

Figure 2.2 Classification of functionally graded materials.

Graded porosity or material density gradient can be obtained through variation in porosity density or pore size and shape. This kind of FGM has been ubiquitously used for surgical implants as porosity helps integrate the implant with surrounding tissues and blood circulation [12]. Chemical composition graded FGMs have multi-materiality aspects with dynamically composed gradients in the volume of bulk material. The materials may have single-phase or complex morphology with multiple phases. The distribution of materials may follow uniformity, steps, or other particular patterns in objects [3, 12]. In microstructure gradient FGMs, distinct microstructures are strategically modulated through various methods such as composition variation, post-processing like case hardening, quenching, and other heat treatment processes to obtain the desired property variation in material [13]. FGMs are fabricated using a combination of ceramic-polymer, ceramic-metal, ceramic-ceramic, and metal-metal. Along with the heterogeneity-based classification, FGMs are classified according to the combination of constituents, structure, nature of gradient (i.e., stepped, continuous, or specific pattern) and the application [14]. Fabrication approaches have evolved from conventional processes like gas-based thermal spray, liquid-based centrifugal casting, and solid-based powder metallurgy to advanced manufacturing methods like additive manufacturing (AM). AM technology empowers the direct fabrication of bespoke components by accurately placing the material at predefined positions in a design space [3]. Tungsten, nickel, titanium, aluminum, copper, and magnesium are reported as the most frequently used metals for FGM fabrication [15].

Several advantageous aspects of AM techniques include innovation through rapid mass customization and unprecedented design freedom, the ability to create high geometric complexity with short lead time and part count reduction (PCR), producing lightweight components by using lattice structures in place of solid material, minimizing waste generation or even eliminating it, embedding different functional component like motors, bearings, or electronic components into a single part through part consolidation, and enabling multiple materials additive manufacturing (MMAM) [7, 16]. These factors bring AM to the forefront of FGM fabrication techniques. However, there are also a few restrictive aspects of AM, such as a limited ability to build overhanging features, warpage and cracking due to thermal stresses, infill densities, stair-stepping, surface roughness, and microstructural anisotropy in the Z-direction [17]. Integrating the restrictive and opportunistic aspects of design for additive manufacturing (DfAM) allows for the production of complex geometries using multiple build materials while minimizing time, cost, material waste, and the risk of build failures [17, 18]. The AM processing variables such as beam current, beam focus, and beam speed enable the fabrication of multi-material parts with embedded components like multi-material airless tires, robot fingers, articulated camera probe, and articulated mechanisms [19]. Prabhu et al. [20, 21] explored the commensurate need for skilled workers and the importance of introducing AM educational intervention by institutions, academics, and professionals to improve the technical excellence of students' design outcomes. Recent innovations have evolved additive manufacturing from rapid prototyping (RPT) to direct digital manufacturing (DDM) of customized geometry with functional material properties. Researchers have developed several discretization mathematical models for composition control at every voxel using function-based models through piecewise blending or global function in MMAM. Then, techniques like topology optimization and material composition optimization are used to enhance fabricated parts' performance [22]. Metal-based FGMs have garnered much attention from both academia and industrial communities. Several review papers have been published, each addressing various aspects of FGMs. However, none of them comprehensively covers all facets of FGMs, including their origination, types, design techniques and tools, fabrication methods, post-processing requirements, frequently used techniques, and applications. This review is a compilation of all developments including materials, modeling, simulation, optimization, fabrication techniques, scanning patterns, delivery mechanism of multiple materials, and applications of FGMs, together with future scope of research. Being a state-of-the-art and comprehensive review, it can be useful for both researchers and industry. For example, designers could retrieve information about available resources and their limitations to improve their decisional process. Software developers could get insights to improve the constraints while modeling and optimizing the FGMs. Finally, researchers can obtain a comprehensive overview of the topic as a guide for future research.

The remaining article is organized as follows: Section 2.2 surveys the literature on modeling and optimization of FGMs. Section 2.3 highlights different developments in FGM fabrication processes. Various auxiliary aspects like toolpath planning and scanning strategy, powder feed mechanisms, and support structure are elaborated in Section 2.4. Post-processing requirements and techniques are reviewed in Section 2.5. Section 2.6 covers various testing and characterization tools used in FGM part fabrication. Section 2.7 discusses various fabrication challenges and future directions of FGMs research. Finally, this chapter is summarized in Section 2.8.

2.2 MODELING AND OPTIMIZATION OF FGMs

Even though the merits of FGMs have already been unveiled, the modeling and optimization of these materials is an effortful task. The increased number of variables at the design and fabrication stages makes the optimization process overly complicated, especially in the case of multiple materials. Numerous simulation-based analyses on additive manufacturing of composites have been conducted. However, as FGM fabrication is a newer application of AM processes, only a limited number of computational models have been investigated [12]. Computer-aided design technologies have helped researchers generate a multitude of 3D CAD representations with boundary representations (B-rep), spatial decomposition, constructive solid geometry (CSG), and function representation (F-rep) as commonly used geometric representation techniques. B-rep and F-rep techniques represent 3D geometry only while CSG and spatial decomposition along with hierarchy-based parallel representation describe both geometry and material [3]. Before developing special modules for FGMs in commercial software like ANSYS, ABAQUS, and COMSOL Multiphysics, manufacturing simulation used to be done through NC programming using G-code [23]. Optimal process parameters for minimizing the undesirable residual stress and part distortion can be determined using experimental trial and engineering judgment. However, the experimental trial becomes exorbitantly expensive for multi-material parts of complex geometries. The computational model can provide an accurate relation between process parameters and the magnitude of induced residual stress and part distortion. Furthermore, incorporating a feedback loop and real-time process parameter modification can prompt optimization. Computational tools enable optimization of the structure, composition, topology, or a combination of multiple variables using algorithms. Computational fluid dynamics (CFD) and fast Fourier transform (FFT) combined with finite element (FE), finite volume (FV), and finite difference (FD) models have been used for thermal-mechanical and thermo-fluid modeling [24, 25]. The modeling of the heterogeneous objects, that is, FGMs, enfold geometric modeling with varying material composition and heterogeneous lattice structure. Topology optimization (TO) has arisen as a powerful design tool and has been exploited to optimize the material or weight of an object, which further reduces fabrication time and material cost. The prominent TO methods for FGM fabrication include homogenization, solid isotropic material with penalization (SIMP), geometry projection method, level set, Heaviside projection method, phase-field method, moving morphable component (MMC) method, and evolutionary approach which were first developed for single material (SIMO) and then extended to multi-materials (MMTO) [26]. The modeling and optimization techniques used for microstructurally, structurally, and compositionally graded materials are outlined in Sections 2.2.1, 2.2.2, and 2.2.3, respectively. Various software programs and their application and salient features are presented in Table 2.1.

Table 2.1 Summary of design and optimization tools of FGMs

Compositionally graded material

Software/Tool	Salient features	Application/ Material	Reference
COMSOL Multiphysics	FEA results show good compatibility with that of the analytical method	Compliant hydrostatic bearing	[9]
	To understand the relation between operating parameters in coating process	Nickel alloy coated	[25]
	Topologically optimized, porous structure with weight saving	Photopolymer liquid resin	[27]
Thermo-Calc and Pandant software	Undesirable phase formation in FGMs, their identification, and avoidance	For any multiple materials FGM	[28–32]
	Optimum property gradients using the map-based planner		
	Computationally predicting phase transformations	Ti-6Al-4V/Inconel 718 FGM	[31]
ABAQUS	CFD simulation with variable operating parameters user-defined functions	FGM ofTi-6Al-4V	[24]
	Design elastic FEM model	Any Material	[33]
MATLAB, and modeFRONTIER®	A mathematical model for multi-objective optimization in DED	Inconel 718 and Ti-6Al-4V	[8]
MATLAB	Optimization of macro periodic structure using substructures	Any material	[34]
Autodesk, and GrabCAD Voxel Print	Voxel-based modeling software for multi-material 3D printing to simulate the design	Any material combination	[3]
Rpdata, and Netfabb	Slicing of the support structure and design structure	Any material combination	[35]
SolidWorks	3D geometric modeling and FEA simulations	TB+,VW+	[36]
Structurally Graded Material			
3D Multigrid Solver	A mathematical model for numerical analysis of substrate/coating with varying elastic properties	Any substrate/ coating combination	[37]
MATLAB, and OptiStruct software	MATLAB coding, gradient cellular structures design and optimization, mechanical and thermal performance analysis	Elastic properties-based optimization	[38]
	MATLAB topology optimization to accommodate the orientation dependence of material properties	Messerschmitt-Bolkow-Blohm Beam	[39]

(Continued)

Table 2.1 (Continued)

Compositionally graded material

Software/Tool	Salient features	Application/ Material	Reference
ANSYS, Rhinoceros	FEA for mechanical properties in ANSYS and 3D lattice distribution in Rhinoceros software	Hip implant	[40]
Unigraphics NX 10.0 and 3-Matic software	Design the graded porous scaffolds and optimization of same	Mimic gradient structure of bone for implant	[41]
3-Matic and Magics	Design of functionally graded lattice structures with cubic and honeycomb unit cell	Modeling the energy absorption structure	[42]
Magics and Slicer	Fix the meshed STL files and generate the toolpath	Any material combination	[43]

2.2.1 Microstructurally graded material

The graded microstructure can be induced in both homogeneous and heterogeneous composition materials through predetermined process parameters. The concept and fabrication technique of microstructurally graded FGMs using fine- and coarse-grained microstructure is demonstrated in reference [13]. In the case of homogeneous composition materials, gradation in microstructure is obtained by varying heat input to fabricate fine-grained and coarse-grained regions in a part. In the case of heterogeneous composition material, gradation in microstructure can be a function of composition or heat input during fabrication or a combination of both. Such types of FGMs are frequently fabricated using powder bed fusion-based AM techniques. No unique modeling and optimization are needed as simple parameter variation can fabricate such gradation. A graded microstructure can be obtained by selectively adding thermal energy to allow grain growth, and multiple scanning can result in tailored microstructural morphology comparable to wrought microstructure [17].

2.2.2 Structurally graded material

Geometric modeling with heterogeneous lattice structures is concerned with the geometric features of the objects. These FGMs came into existence earlier than compositionally graded materials and have been extensively researched. They involve lattice structures and have been a popular design application for fabricating lightweight and advanced energy-absorption structures. Each lattice structure comprises unit cells with solid and void regions [42]. The main challenge in lattice structure–based FGM design is bridging the designed computational models and complex material properties realized in fabrication. The orientation-dependent properties of metallic parts, such as horizontal features which have a rougher surface than do vertical ones, make the task even more complex [39]. These materials are further classified as strut-based lattice structures and triply periodic level surface (TPLS)-based cellular structures. Functionally graded cellular structures (FGCS) have objectively controlled lattice structures in different volume regions using software or the Voronoi algorithm [38]. Porous scaffold designs are optimized for mechanical and biological properties and for osseointegration [41]. Zhang et al. [44] fabricated various functionally graded triply periodic minimal surface structures (TPMS) using stereolithography and compared them for mechanical properties and deformation mechanisms. These structures offer smooth surface and pore interconnectivity to meet the biological and mechanical requirements of dental and other orthopedic tissues. Cetin and Baykasoğlu [45] developed graded lattice structure filled tubes (GLSFTs) for energy absorption under high-speed in-plane crushing.

Finite element methods (FEM) combined with fast Fourier transform have been widely used for numerical analysis, but they require massive CPU and memory. Boffy et al. [37] demonstrated an efficient numerical model using multigrid techniques within a finite-difference framework that offers accelerated convergence through a local refinement strategy such that large-scale problems can be solved with moderate memory and CPU capacity. Li et al. [38] developed a generative design and optimization technique for TPLS-based FGMs, and the results show better density distribution control that enhances structural and thermal properties. Wang et al. [40] demonstrated an evolutionary algorithm for graded cellular material by combining structural optimization with multi-scale analysis. Recently, Fu et al. [34] put forth an efficacious topology optimization method that substructures the global periodic structure into a series of the condensed linear system before optimizing the same. Smith and Norato [46] extended the geometry projection method of topology optimization to optimize multi-material structures.

2.2.3 Compositionally graded material

Applications involving extreme environments, such as nuclear power generation and aerospace engineering, require components capable of functioning at high-temperature gradients within the same part. Consequently, multiple properties like toughness, wear resistance, corrosion and oxidation resistance, lightweight, strength, and reasonable cost become essential in distinct regions. Nevertheless, because contrasting properties are rare in a single material, multi-material FGMs with tailored composition gradients and stable interfacial bonds between dissimilar materials are used [3]. The varying material modeling produces composition distribution for greater diversity in properties than structurally graded FGMs. This class of FGMs has become more prevalent with the development of the AM technique. However, these materials often face secondary phases with deleterious properties and cracks [28]. Kaufman and Ågren [29] came out with a novel technique of predicting the formation of such phases with the help of calculation of phase diagram (CALPHAD) software to manipulate composition accordingly.

The biggest drawback, however, was that the software could not visualize more than three elements at once. Hofmann et al. and Kirk et al. [30, 47] computationally predicted brittle intermetallic phases' free gradient paths and proposed that these undesirable phases could be avoided by adding an alloying element, or a discontinuous jump in composition. The methodology successfully produced deleterious phase–free FGMs, but it lacks an optimum gradient for properties.

Authors achieved the same in their other work to produce a composition path with monotonic property gradients [28, 31, 48]. Yan et al. [8] designed a multi-objective optimization algorithm with injection nozzle diameters, injection velocities, nozzle angles, laser power, and scanning speed variables for the FGM fabrication. Xu et al. [49] proposed the SIMP method for multi-material thermoelectric generators (TEG) to achieve high efficiency. Another innovation in CAD modeling of multi-material FGMs is algorithmically generated voxel models where the material deposition is controlled in each voxel [50].

2.3 MANUFACTURING TECHNIQUES OF FGM OBJECTS

FGMs can be a thin coating applied over a substrate or a bulk volume material for harsh operating conditions. A thin coating of the target material is used for applications demanding surface properties different from that of the bulk volume of material [51]. Bulk FGMs have varying properties across their volume. Fabrication methods of both types of FGMs are conventional and advanced (Figure 2.3), based on their ability to process complex shapes. Each technique possesses unique features for a type of material, complexity of shapes, dimension control, and volume of the part. For example, centrifugal casting holds the edge over other methods for fabrication of bulky parts, powder metallurgy-based methods suit the moderately complex part, and AM techniques excel in fabricating small complex parts [52]. Distinct types of fabricating methods falling under conventional and advanced techniques are outlined in Sections 2.3.1 and 2.3.2, respectively. An overview of various fabrication techniques, their features, and materials processed are summarized in Table 2.2.

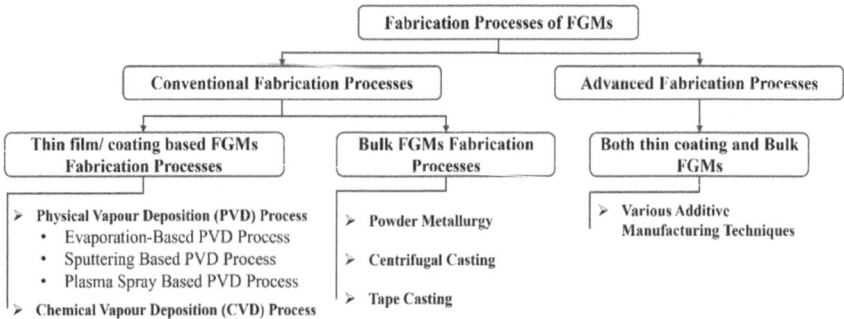

Figure 2.3 Classification of FGM fabrication methods.

Table 2.2 Overview of FGM fabrication methods and their salient features

Processing technique	Material	Features of product/fabrication method	Application	Reference
		Conventional fabrication processes		
Vacuum plasma spray	Ferritic martensitic high chromium steel EUROFER97 and Tungsten	Excellent quality and crack-free microstructure	Functionally graded coating	[53]
Magnetron sputtering process	Ferritic martensitic high chromium steel EUROFER97 and Tungsten	Poor interfacial strength as bonding between particles is only formed by mechanical clamping		
Powder metallurgy	Fine Mg-Zn particles Al2024/SiC FGMs	Specimen for microstructural and mechanical characterization, density measurement	Aerospace, transportation, biocompatible, and electronic	[16]
Centrifugal casting method	Rubber-like polymer for support, titanium for slipper	Compliant hydrostatic bearing for non-constant curvature of counter surface	Pump, motor, machine tool, civil engineering applications	[9]
	Functionally graded Al/B4C, Al/SiC, Al/Al2O3, and Al/TiB2	Hollow cylindrical shell	Wear resistance behavior study of FGMs	[54]
Tape casting method	Lanthanum strontium cobalt ferrite (LSCF)	Functionally graded cathode using LSCF	Solid oxide fuel cells	[55]
	Acicular hydrogen processed using freeze tape casting	Functionally graded acicular hydrogen electrode	Fuel cell	[56]
	Magnetocaloric materials	Specific magnetocaloric properties	Magnetic refrigeration (MR) technology	[57]
		Advanced fabrication processes		
Directed energy deposition (DED), laser engineered net shaping (LENS), or direct laser metal deposition (LMD)	304L stainless steel (SS304L) and Inconel 625 (IN625)	IN625 provides strength and corrosion resistance at elevated temperatures, and SS304L reduces the cost	Automobile engine valve stems, and corrosion-resistant coatings	[75]

(Continued)

Table 2.2 (Continued)

Processing technique	Conventional fabrication processes			
	Material	Features of product/fabrication method	Application	Reference
Laser powder bed fusion (L-PBF), electron beam powder bed fusion (E-PBF), selective laser melting (SLM), selective laser sintering (SLS)	Stainless steel (SS410), Ti6Al4V (Ti64) alloy, and Ni-Cr bond layer	Intermediate Ni-Cr bond layer in bimetallic structure reduces thermal and residual stresses	Cheaper bimetallic component with hot hardness & corrosion resistance	[77]
	Ti-6Al-4V/Inconel 718	Increasing microhardness with IN718 content	FGMs coating	[42]
	Inconel 718 (IN718)/SS 316L	Selective compositional range exclusion to produce a defect-free FGM	FGM fabrication and characterization	[44], [74]
	SS316L, IN718, Pure Cu	Intermediate layers fetch defect-free coating	Coating of pure Cu on SS316L substrate	[77]
	Copper-nickel (Cu-Ni)	Defect-free and high-density FGM	High energy storage capacity	[76]
	Ti6Al4V to Invar 36	Intermetallic phases FeTi, Fe2Ti, Ni3Ti, NiTi2	Improve the understanding, design	[95]
	316L stainless steel and Cu10Sn	6-channel ultrasonic selective powder delivery	Turbine disk, ring and blade	[5]
	Ti6Al4V	Rationally designed porous scaffolds for high strength and toughness	Load-bearing orthopedic applications like cortical bone	[41]
		Structurally graded FGMs	Advanced energy absorption device	[42]
	Inconel 718	Microstructurally graded FGMs	Aircraft engine and nuclear applications	[14]
		Characterization for defects and compared with computed tomography	The mock leading-edge segment of the turbine blade	[33, 58]

	IN738LC	Correlated scanning strategy with obtained microstructure	FGM development with tailored mechanical properties	[59]
	Soda-lime glass, Al_2O_3, Cu, polymer	Ultrasonic vibration-assisted feeding mechanism	Shoe sole, turbine blade, bearing	[7]
	Ti47Al2Cr2Nb	Compositionally graded FGM using raw powder	Demonstration of complex FGM	[60]
Stereolithography or vat photopolymerization	Photopolymer liquid resin	The topologically optimized porous structure	Structurally graded FGM fabrication	[27]
	Al_2O_3-Si_3N_4 mixed with photocurable resin to form a paste	Printed using optimized parameters like layer thickness and paste viscosity	Complicated FGM fabrication	[61]
	3DSR ultra-high resolution (UHR) resin, isobornyl acrylate, and iron oxide (Fe_3O_4)	Variation in the magnetic field can fetch different gradients of magnetic particles concentration	Compositionally graded polymer composites	[62]
Binder jetting (BJT)	Polyvinyl alcohol powder, graphene oxide ink	Compositionally graded FGM fabrication	High-porosity FGM for supercapacitor	[35]
	Vero cyan (VC) and tango black+ (TB+)	Voxel-based design of fatigue test sample	Analyzing stepwise and continuous grade	[63]
Sheet lamination (SHL)	Stainless steel, Al, and Cu foils	Optimum welding parameter selection	FGM fabrication using SHL	[64]
Material extrusion (MEX)	Al, and carbon nanotubes (CNT)	FGM with superior mechanical and thermal properties and a high weight-to-strength ratio	Transportation equipment parts and air-hydro cylinder tube body	[65]
Hybrid AM technology (SLM + cold spray)	Ti6Al4V, Al, and Al_2O_3	Thick, dense, and machinable FGMs free from intermetallic phases, improved hardness is achieved	Ti6Al4V part is manufactured using SLM and cold sprayed with Al and Al+Al_2O_3	[56]

2.3.1 Conventional fabrication processes

These methods are limited by their capabilities to fabricate only simpler shape objects. Therefore, these processes are further subclassed based on their application for fabricating thin coating and bulk volume-based FGMs, as elucidated in Figure 2.3.

2.3.1.1 Thin film/coating-based FGM fabrication processes

Thin coating-based FGMs are fabricated using physical and chemical vapor deposition methods. The physical vapor deposition (PVD) processes share their root with the electroplating process but have better process control with improved outcomes. In PVD processes, coating material (target material) is vaporized or atomized before being deposited over the substrate to be coated. Based on atomizing techniques, PVD processes are further categorized as evaporation based, sputtering based, and plasma spray based. Various kinds of heat sources like electrical resistance, laser, electron beam, ion beam, and several others are used for melting and evaporating the target material in a vacuum chamber. Thermal evaporation and electron beam-based PVDs are depicted in Figure 2.4 (a) and (b). The sputtering-based PVD process involves the collision of high-velocity

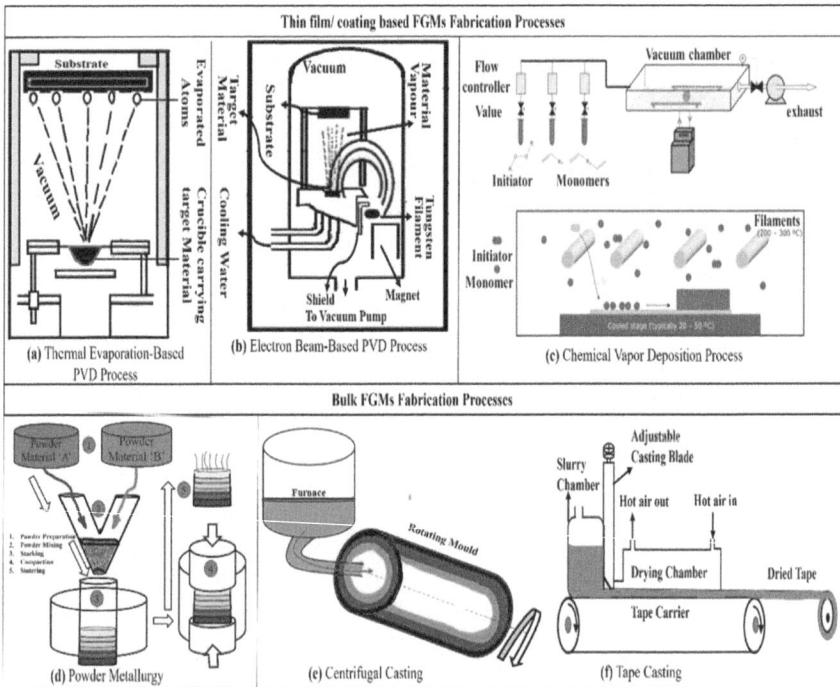

Figure 2.4 Conventional methods of fabricating FGMs.

plasma with atoms of target material and sputters them before depositing them on the substrate surface. Plasma spray-based PVD is a hybrid deposition process comprising traditional low-pressure plasma spraying (LPPS) for atomizing the target material and the PVD technique for depositing the same. In contrast to PVD processes, chemical vapor deposition (CVD) uses heat sources to atomize the atoms from the target material in a lesser vacuum chamber. The deposition of vaporized atoms at the substrate surface is achieved through a chemical reaction with precursor gases in the chamber, as shown in Figure 2.4 (c) [2]. CVD has the advantage over PVD in terms of lower power consumption, higher deposition rate, and coating of the underside of the substrate, but the chemical reaction's by-product can be toxic, explosive, and corrosive [52]. Tang et al. [66] synthesized W-Cu based FGM using spark plasma sintering after homogeneously mixing the powders of distinct compositions in a high-energy planetary ball milling system.

2.3.1.2 Bulk FGM fabrication processes

The conventional methods for fabricating the bulk FGMs include powder metallurgy, tape casting, and centrifugal casting [52]. Powder metallurgy involves depositing the predefined powder composition layer by layer followed by compaction and sintering, as illustrated in Figure 2.4 (d) [10]. Centrifugal casting involves pouring the molten material comprising multiple constituents into a rotating mold, as depicted in Figure 2.4 (e). The centrifugal force causes the constituents to segregate into layers, with the denser material at the outermost side and the next most dense coated inside it, rendering a high-density product [67]. Saleh et al. [54] reviewed the literature on the fabrication of functionally graded metal matrix composites (FGMMCs) using centrifugal casting and synthesized process parameters for distinct materials. Tape casting is a popular method of fabricating large-scale multilayer ceramic substrates by casting the amalgamation of ceramic powder into a solvent, with dispersants, plasticizers, and binders followed by drying and sintering. The process involves spreading the slurry of materials over a moving belt by a blade edge and stacking tapes of varying compositions, as demonstrated in Figure 2.4(f) [68, 69]. Another approach for FGM fabrication is adjacent co-flow of slurries (side-by-side tape casting) [70]. Metal-ceramic reinforced FGMs combine the best properties of both materials, like the mechanical properties of metals and the hot hardness of ceramic [1]. For instance, TiB/Ti-based FGMs for fabricating armor are ductile and tough with excellent anti-penetration properties. The acoustic impedance matching principle can optimize the anti-penetration performance and secondary strike ability of such FGMs. Their high strength-to-weight ratio with high ballistic resistance makes them the leading candidate for defense applications to improve maneuver capacity and battlefield survivability [57]. Another application of tape casting is fabricating graded electrodes [55]. Recent advancements have demonstrated the use of hybrid AM processes to reinforce hard particles using cold spray into a selectively laser-sintered soft substrate; decreasing the concentration from the surface into the matrix gives graded surface hardness with a tough core [56].

2.3.2 Advanced fabrication processes

Advanced fabrication methods hold an edge over conventional methods in terms of their capability to fabricate complex shape parts easily and without any special jigs or fixtures [68]. Functionally graded additive manufacturing (FGAM) techniques provide a time-efficient and economical way to produce complicated geometries, especially low-volume, customized, personalized products with advanced material directly from the 3D CAD model with a drastically compressed supply chain [71]. The manufacturing procedures for FGAM begin with a 3D CAD model, slicing, converting CAD file into an appropriate data exchange file format, optimal orientation determination, toolpath definition, support generation, fabrication, and post-processing (if needed) in the same sequence. Potential data exchange formats are STL file and OBJ file (describing geometric features only), AMF file, FAV file, SVX file, and 3MF file (describes both geometric and material-related features). CAD software such as SolidWorks, CATIA, AutoCAD 2022, Inventor, Rhino, and Mesh Mixer can generate these file formats [72]. After its inception in the mid-1980s, AM has blossomed and unfolded into a plethora of processes, including several techniques to fabricate FGMs. The fabrication of FGMs using AM techniques involves one extra step of material description within the object before fabrication. Furthermore, growing competition among AM machine manufacturers has reduced the entry threshold for investment. By gradually varying the spatial distribution of various constituents, objects with graded magnetic, hysteresis, and mechanical properties can be fabricated [3]. All commercially available AM techniques (classification according to ASTM F2792 standards) – stereolithography (SLA), material jetting (MJT), fused filament fabrication (FFF), binder jetting (BJT), powder bed fusion (PBF), and sheet lamination and directed energy deposition (DED) – are being used to fabricate multiple material parts with tailored properties [15, 72]. The electron beam and laser-based AM processes – DED and PBF – offer far more design freedom and structure customization at the microscale. Moreover, open-source firmware, slicers, printer setup, control software, and hardware design have reduced the cost of 3D printers significantly [40]. Selective laser sintering (SLS) for fabricating polymer-based products and direct metal laser sintering (DMLS) or selective laser melting (SLM) for metallic parts are subsumed under the classification of L-PBF technologies [73]. SLM shows enormous potential to fabricate geometrically complex parts of multiple materials with fully dense structures and material saving using topology optimization. One crucial drawback of AM parts is a rough surface, which needs post-processing such as abrasive flow machining (AFM) and magnetic abrasive finishing [74]. A generic overview of the FGAM process is illustrated in Figure 2.5.

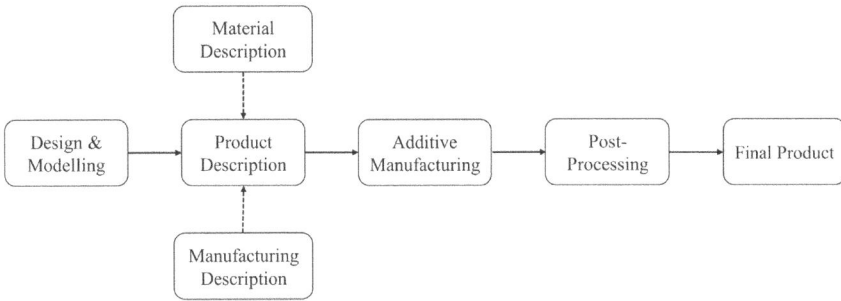

Figure 2.5 Generic functionally graded additive manufacturing process flow [3].

2.3.2.1 Powder bed fusion (PBF) AM process

Laser powder bed fusion (L-PBF) and electron beam powder bed fusion (E-PBF) are superior techniques to fabricate fully dense complex 3D components with functionally graded properties, high resolution, and surface finish [33]. The selection of heat sources depends on multiple factors – type of material, desired functional properties, and other user-defined functions. A few researchers have also conducted a comparative analysis to provide an essential guide for selecting the appropriate heat source. The E-PBF process is conducted in a vacuum, and L-PBF uses inert gas to avoid oxidation of the melt pool. The electron beam has higher absorptivity and lower heat dissipation due to reduced convection in a vacuum [75]. Mechanical properties of additively manufactured objects depend on feature orientation relative to the build plane [39]. As an advancement from the conventional E-PBF process, Zhou et al. [60] demonstrated an E-PBF based novel process, called selective evaporation, for fabricating FGMs from a homogeneous composition powder by varying irradiating energy density onto distinct regions in a powder bed; for instance, by varying the quantity of aluminum being evaporated from Ti47Al2Cr2Nb powder. Ghorbanpour et al. [58] systematically examine the effect of manufacturing parameters on grain size, crystallographic textures, precipitate and lave phases, unmelted particles, and porosity in L-PBF fabricated parts. In similar studies, Tan et al. [76] and Mao et al. [77] examined the effect of the power of heat source, hatching pattern, and scanning speed on interfacial characterization and mechanical properties of build part. Material characteristics of the substrate such as thermal conductivity, melting point, and solubility with coating material play a vital role, as significant differences in thermal conductivity and coefficient of thermal expansion cause misfit strain in the interface, and thermal gradient affects residual stress.

Pure titanium (Ti) and its alloys, especially Ti6Al4V, also abbreviated as Ti64, have the highest strength-to-weight ratio, making it the most versatile alloy. Ti64's excellent mechanical properties, good corrosion tolerance, and favorable osteo-integration make it the most used alloy for orthopedics [72]. El-Sayed et al. [78] and Caiazzo et al. [79] fabricated porosity-graded implants in cancellous bone and trabecular bone of the human body. However, its poor wear resistance, low thermal conductivity, high coefficient of friction, great affinity toward other materials, and reaction with cutting tool material make it difficult to machine material. Consequently, AM provided a fantastic opportunity for exploiting the properties of Ti alloys for various applications. Popovich et al. [13] fabricated a microstructurally graded cylindrical rod of nickel-based superalloy, using the SLM process with coarse grains at the center and fine grains at the outer shell, allowing a trade-off between creep and fatigue performance. For iron, nickel, and titanium-based materials, a high -power source renders columnar coarse-grained, and more refined equiaxed grained microstructure is obtained by using a low power source [80–82]. Similarly, continuous-wave (CW) laser results in coarser grain, and pulsed-wave laser fabricates fine grain microstructure [83]. The reason in both cases is that high power, and continuous-wave laser, result in a larger heat affected zone (HAZ) or bigger melt pool such that grains have more extended time for growth before solidification completes, and the opposite is the case with low power and pulsed-wave laser. Coarse-grain microstructure has better creep resistance, and fine grains have longer fatigue life. So, distinct properties can be achieved in different regions of the same part [80–83]. Recent works by Zhang et al. [59] and Wei et al. [5, 61] demonstrate the fabrication of compositionally graded FGMs using the SLM through point-by-point selective powder deposition using a dual ultrasonic vibrational powder feeding system.

2.3.2.2 Directed energy deposition AM process

This class of AM process uses multi-coaxial nozzles to deliver materials in varying proportions to fabricate FGMs. The supplied materials are melted by using a directed laser at the intersection region of exiting materials [72]. Ji et al. [31] fabricated Ti-6Al-4V/Inconel 718 based FGM with graded microstructure and increased microhardness with Inconel 718 content as a secondary phase precipitate. Kim et al. [48] and Su et al. [84] fabricated compositionally graded 316L stainless steel and Inconel 718 material and performed characterization for variation in microstructural, chemical composition, and mechanical properties. Zhang et al. [85] fabricated compositionally graded FGMs of 316L stainless steel and Inconel 625 using DED and evaluated them for elemental composition, phases, microstructure, and mechanical properties. DED has also been used for fabricating defect-free Cu-Ni-based FGMs for energy storage applications [86]. Thermal and residual stresses induced when coating the stainless steel substrate with Ti6Al4V (Ti64) alloy can be reduced by depositing intermediate Ni-Cr bond layers [87] or Cu layers [88].

2.3.2.3 Stereolithography AM process

Stereolithography can fabricate components of photocurable polymers with high accuracy and fewer defects, but it has been comparatively less exploited for FGM fabrication. Topologically optimized porous structures with contrasting mechanical properties using UV light curable photopolymer liquid resin (SPR6000 epoxy) have been fabricated using stereolithography [75]. In another study, compositionally graded Al2O3-Si3N4 mixed with photocurable resin was fabricated [27]. Recent work used novel magnetic-field-assisted digital light processing (DLP) stereolithography (M-SL) AM technique for printing compositionally graded polymer composites with embedded magnetic particles with the help of ultrasonic homogenizer [89, 90]. Valizadeh et al. [62] used grayscale-masked stereolithography (MSLA) to tailor orange tough resin-based hyperelastic FGMs such that the degree of photopolymerization is controlled by changing grayscale pixels.

2.3.2.4 Binder jetting (BJT) AM process

The BJT AM technique is frequently used for fabricating ceramic parts, but it has also been used for fabricating FGMs [91]. Shen et al. [92] reported a novel binder jetting approach for fabricating the multi-material FGMs using a switching nanoparticle binding ink. Although a sizeable number of articles demonstrate the application of BJT in fabricating FGMs, most of these are ceramic-based.

2.3.2.5 Material jetting (MJT) AM process

Another infrequently used FGAM technique is material jetting. It uses raster-based toolpaths to provide on and off the respective jets to obtain desired geometry and material composition. Salcedo et al. [36] developed a physical model of polymer-based FGM. Kaweesa and Meisel [35] presented a voxel-based design and compared stepwise gradients against continuous gradients. It was observed that FGM fabrication applications of MJT are restricted to polymer-based materials only. The hybrid AM process can fabricate significantly good-quality FGMs [63].

2.3.2.6 Sheet lamination (SHL) AM process

The sheet lamination technique is also infrequently used to fabricate metallic FGMs. It involves ultrasonically welding the thin sheets of dissimilar materials to replicate the CAD model [64].

2.3.2.7 Material extrusion (MEX) AM process

In this fabrication process, the fused filament is extruded through a moving nozzle along a predefined toolpath to fabricate components by adding layer upon layer. High melting temperature and oxidation issues restrict MEX in metallic FGM fabrication. Stoner et al. [43] demonstrated the TPMS method for structural gradation using Magic and Slic3r software to fix the meshed STL files and generate the toolpath. Kim et al. [65] fabricated tubular shape aluminum (Al) and carbon nanotubes (CNT)-based FGM by a hot extrusion process.

2.4 AUXILIARY ASPECTS OF FGM FABRICATIONS

2.4.1 Toolpath planning and scanning strategies for advanced fabrication processes

Directed material delivery-based AM processes (DED, MJT, and MEX) use tool-path planning to supply materials. In contrast, powder bed or liquid pool-based AM processes like PBF, BJT, and stereolithography use term-scanning strategies as a heat source to scan the bedded material for melting or curing. Research has proved that toolpath planning and scanning strategy play a vital role in obtaining texture and microstructure in the fabricated part. Different toolpath planning for single-composition structurally graded and compositionally graded FGMs are used [12]. Specific region-filling toolpaths for single material deposition are unidirectional linear, zigzag, and continuous contours. For multi-material FGMs, the toolpath for material deposition is described by functions that correlate the material composition with the position inside the part. Xiao and Joshi [93] proposed a global automatic toolpath planning strategy for the DED AM process using a slice file containing geometry and material boundary information. The FGM part can be represented as an assembly of sliced and pixelated STL files. The material function provides a material description for each voxel, and the algorithm prioritizes the scanning of different pixels. No experimental work is reported yet, but in principle, proper implementation toolpath planning or scanning strategy fetch quicker fabrication of superior quality products at less cost. In recent work, Chandrasekaran et al. [94] used a bidirectional (zigzag) scanning strategy to depose multiple layers while fabricating FGM using ER70S-6 and ER2209 coating on duplex stainless steel substructure using wire arc additive manufacturing (WAAM).

In the case of powder bed or liquid pool-based FGAM processes, striking a balance between heat source power and scanning speed plays a vital role in the producing high-quality parts. During additive manufacturing, local stresses (residual stresses) develop in the already built region. It is continuously subjected to non-uniform thermal contraction under different cooling rates at distinct locations and intermittent expansions caused by reheating as the heat source moves [95]. This emphasizes the importance of selecting the appropriate scanning strategy and post-processing techniques to minimize residual stresses, as these are critical parameters. Control of thermal energy supplied during the fabrication process allows spatial control of microstructural architecture [17]. Jhabvala et al. [96] proposed four different scanning strategies leading to homogeneous heating of parts and concluded that low scanning speed with a thermal gradient can lead to local solidification with cracks. High scanning speed needs higher power and may cause delamination and balling defects. Bobbio et al. [97] used a hatch angle of 90° to reduce thermal stresses during the fabrication of Ti6Al4V to Invar 36 FGMs. Geiger et al. [98] fabricated microstructurally graded IN738LC using three different scanning patterns in the SLM process and performed a comparative study. An overview of prominent scanning strategies is summarized in Table 2.3.

Table 2.3 Overview of scanning strategies for FGM fabrication and their outcomes

Strategy	Schematic diagram	Key features	Reference
Parallel scanning		• Easiest to generate from CAD file • Large temperature gradient orthogonal to scan direction for a single scan • Overheating causes thermal stress, consequently balling and cracks • Multiple pass scanning fetches better results	[96]
Spiral scanning		• No overheating near sides but severe overheating at the center of part, causing cracks and balling effect at center • Multiple pass scanning fetches high-quality product provided that deposited energy in each cycle is more prominent than heat loss in between	
Paintbrush scanning		• Similar to parallel scanning but has smaller scan width with reduced overheating and temperature gradient for both high and low conductivity powders • Dense part with reasonable accuracy but delamination for the tall part due to weak interlayer bonding	

(Continued)

Table 2.3 (Continued)

Strategy	Schematic diagram	Key features	Reference
Chessboard scanning		• The consolidation area divided into smaller cells • Parallel scanning in alternate directions can be seen as scaling down of standard parallel scanning • Modified thermodynamics renders reduced temperature inhomogeneity	
Scanning at 45° to layer deposition plane		Z-axis is taken as building direction, and each deposited layer is scanned at 45° between X and Y with chessboard scanning. Anisotropic microstructure using two heat sources of distinct capacity	[13, 99]
Hatch angle 90° or scan vector rotate 90°		Compositionally graded FGM fabrication using DED to improve the understanding of FGMs	[97]

2.4.2 Powder feeding system of PBF-based AM process

The DED process has outpaced the PBF-based processes for fabricating multi-material FGM components, as the latter have been limited to single-component microstructurally graded FGMs. Nevertheless, the development of an ultrasonic vibrational powder feeding system has paved the way for multi-material components fabrication using the SLM for high-end applications [59]. This novel approach combines point-by-point selective powder delivery with a conventional PBF-based AM process for printing multi-material components [5, 61]. Wei et al. [5] developed another multiple selective powder delivery array device that can be incorporated into a conventional SLM system and can deposit up to six dissimilar materials with point-by-point control. Most of the researchers have fabricated the FGMs with inter-layer composition gradient with constant composition within a layer. This limits the application of FGMs, as industries would need both horizontal and vertical property gradients for high-end applications. The efficiency of the powder-based AM fabrication process depends on powder flow characteristics such as powder deposition, which further affect the track geometry, molten pool volume, and capture efficiency. Katinas et al. present a computational fluid dynamics (CFD) model using ANSYS Fluent software for predicting powder capture efficiency and other process characteristics in four-nozzle injection systems in an LMD AM process [32].

2.4.3 Support structure in AM of FGMs

The support structure is vital in AM fabrication as it can act as (i) heat diffuser and rigidity enhancer in fabrication processes involving high thermal gradient causing shape distortion and residual stress; (ii) substrate for local deposition in MJT and DED AM processes; and (iii) support structure for overhanging parts [100]. Research has been conducted to optimize the support structures to reduce support material waste and print time. Optimization includes part orientation to eliminate or minimize the need for support [101–103], using cellular structures in place of solid [104–106] and use of sacrificial or soluble material (rare in the case of metallic fabrications) [107]. Critical thermal contraction and expansion coefficient, sufficient rigidness, high oxidation resistance at elevated temperatures, and suitable brittleness allow for easy removal through induced vibration or some mechanical means. This makes ceramic and cermet potential candidates for the support structure. Wei et al. [108] developed an easy to remove SiC-316L composite as a support structure material for the L-PBF process. Embedding the support structures into cellular structures improves mechanical performances [41]. Removing these structures is a tedious task and debilitates additive manufacturing as it needs post-processing such as cutting, breaking, and machining. Tailored brittleness can alleviate the removal process using mechanical vibration or machining. It is common to use the support structure material similar to the build material [109]. Nevertheless, removing the metallic support structure is difficult and time-consuming; so, using easy-to-remove composite materials is becoming popular. Support material dispenser is used to create a brittle composite that can be easily removed [110]. Self-terminating dissolvable support materials with chemical solutions can alleviate the tiresome removal process 109]. Weakening the support structure by adjusting its geometries and processing parameters can minimize post-processing. However, weak support structures can result in severe deformation in overhanging features of components [85].

2.5 POST-PROCESSING

AM technologies showed enormous potential to fabricate end-use products with tailored spatially varying properties and several other features, as detailed in previous sections. Despite the promising potential, AM fabricated parts are prone to defects like rough surfaces due to the stair-stepping/stair-casing effect, delamination, insufficient fusion, cracks, and pores, among many other defects that compromise quality [18]. Numerous research efforts have addressed these issues through part orientation, layer thickness, and build direction. However, post-processing is identified as a robust solution to improve these restrictive aspects of AM and impart the desired properties to the fabricated part. Surface finishing operations include conventional machining operations like CNC, chemical machining, turning and milling, abrasive flow machining, electroplating, laser operations for surface finishing, shot peening, abrasive blast (ceramic and grit), and optical polish [111]. The additively manufactured metallic parts have benchmarked amalgamation of heat-assisted post-processing steps – (a) stress relief to relieve residual stress, (b) hot isostatic pressing (HIP) for demoting porosity, and (c) aging for distribution of precipitate in the microstructure – before putting the part in service [112]. Weber et al. [53] normalized EUROFER97 and tungsten-based FGMs to obtain residual stress-free and martensitic microstructure. Popovich et al. [13] fabricated a microstructurally graded tensile test bar Inconel 718 using electrodischarge machining (EDM). Grinding and polishing are used to remove residue from the EDM before post-processing including HIP and HIP + homogenization + ageing (HIP + H/T) to get needle-like $Ni3Nb-\delta$ precipitate from existing brittle laves phase and reduce porosity [99]. Geiger et al. [98] used full heat treatment (full HT) comprising recrystallization above $\acute{\gamma}$ solvus temperature, HIP, post-HIP solution, and precipitation hardening, in this order, only to reduce the elastic anisotropy by reducing the degree of crystallographic texture through annealing twinning. Vibro-polishing following conventional polishing has also been used for sample preparation for characterization. Shen et al. [113] performed thermal-induced phase evolution (homogenization) to obtain FCC structure from residuals of BCC lattice. Elevated temperature annealing has also been used to change the microstructure for property enhancement through grain growth or martensitic decomposition [114, 115]. Ituarte et al. (2020) used ultrasonic vibration-assisted ball burnishing to improve surface finish and microhardness and reduce residual stress in additively manufactured tool steel. Various post-processing methods and their outcomes are summarized in Table 2.4.

Table 2.4 Overview of post-processing techniques and their effects

Post-processing	Objectives	Reference
Hot isostatic pressing (HIP)	Enhanced mechanical properties (ductility, strength, and fatigue behavior) by undesirable δ-phase and laves-phase dissolution and reduced porosity by plastic flow to close flaws	[99]
Annealing heat treatment (H/T)	Alleviates residual stresses induced during fabrication, precipitation hardening (depends on alloying elements)	[99, 112, 114]
Aging (often combined with H/T)	For desired precipitation distribution, dissolve impairing intermetallic compounds and carbides	[99, 116]
Homogenization	Elevated temperature annealing to dissolve undesired phases	[111]
Machining operations	Remove part from support or baseplate and obtain the desired geometry with a good surface finish	[13, 116]
Grinding and polishing	Surface finishing, dimensional control, and sample preparation	

2.6 TESTING AND CHARACTERIZATION

The growing research trend and ever-increasing application of AM processes for end-use products have intrigued post-processing, testing, and characterization researchers. The varied objectives determine serviceable properties before putting the product into application and anomaly detection for process control. Recent research shows the evolution from *ex situ* testing to decipher the ongoing process, and the efforts are to develop real-time process monitoring and control. Thermal imaging by placing a photodiode and a near-IR CMOS camera around melt pools has been used for generating process signatures, which are further used to form a basis for process monitoring and control. The deviation from benchmark reference conditions is being monitored. A feedback control method has also been proposed in different studies where the experimental process variables control the laser power input [115]. Clijsters et al. [114] developed a high-speed, real-time melt pool monitoring system using optical sensors and a field-programmable gate array (FPGA) for *in situ* quality control in SLM process. The microstructure and electrochemical properties of tape-casted porosity-graded cathodes have been characterized using SEM and electrochemical impedance spectroscopy. Crystallographic preferred orientations (texture) can be measured by SEM-based electron backscatter diffraction (EBSD) for evaluating the correlation between anisotropy of Young's modulus and texture [98]. Popovich et al. [13] fabricated microstructurally graded FGMs of nickel-based superalloy, Inconel 718, using SLM-based AM process, and they post-processed the same [99] followed by fatigue and creep test analysis at operating conditions analogous to actual operational conditions of turbine blades [117]. In recent work, Hussain et al. [118] presented a proportional–integral–derivative (PID) and classical feedback controller to circumvent the intertrack disturbance in the melt pool region. Grasso et al. [119] proposed a principal component analysis (PCA) using a fast image capturing camera with high-resolution integrated with data analyses and a machine vision system for detecting local overheating in L-PBF. Mahmoudi et al. [73] adopted a dual-wavelength imaging pyrometer for *in situ* anomaly detection.

Even in recent literature, *in situ* monitoring methods are confined to a particular defect [120]. However, advancements in sensors and hardware systems have made original equipment manufacturers (OEMs) integrate the multi-component *in situ* sensing modules in the AM system itself (like EOSTATE melt pool monitoring system) for evaluating process variables like melt pool characteristics, powder integrity, and chamber temperature [121]. Zhang and Coddet [122] observed the melting mechanism and densification behavior of iron in SLM under various physical descriptors. Ghorbanpour et al. [58] investigated the fatigue crack growth behavior under microstructural anisotropy-graded Inconel 718 structure. Various characterization techniques/tools and corresponding characterized features are summarized in Table 2.5.

Table 2.5 Overview of characterized features, tools used in different characterizations, and observations

Characterized feature	Tools/process used	Observation	Reference
Microstructure and interfacial defects	Optical microscopy and SEM	Micron-sized particles/precipitates, carbide and secondary phase, grain morphology and size, microscale morphology, porosity, fractured section, composition	[13, 27, 31, 41, 42, 48, 57, 58, 63, 65, 66, 70, 76, 77, 84, 90, 97]
	Transmission electron microscopy (TEM)	Presence of ceramics, segregants, and precipitates, elemental composition using TEM	
	X-ray diffraction	XRD plot of a compositionally graded sample, and phase identification	
	Electron backscattered diffraction (EBSD) technique in SEM	Crystallographic maps to study texture and microstructure at fatigue crack using EBSD	
Accurate physical properties	Micro-CT system	3D morphology is evaluated for optimization	[41]
Chemical composition	EDX spectroscopy in SEM	Graded composition	[31, 58, 66, 90, 65, 97, 113]
	Auger electron spectroscopy (AES)	Present phase identification	
	X-ray diffraction	Qualitative distribution of constituent materials	

Characterized feature	Tools/process used	Observation	Reference
Local mechanical properties Microhardness and hardness	Nano indenter XP device, triboindenter, contact clip gauge (extensometer) Vickers indenter, pin-on-disk wear machine, durometer	Varying properties with graded chemical composition	[3, 13, 48, 53, 58, 63, 65, 90]
Transmissivity of laser	Calorimeter	Transmissivity varies with particle size and layer thickness	[122]
Surface roughness	3D profilometer	The smoothest surface obtained with a scanning velocity of 0.05m/s and 110W laser	
Surface modification	Age hardening, ball burnishing, Taguchi method, ANOVA test, and XRD analysis	Enhanced average surface roughness (Ra), microhardness, beneficial residual stress, and ANOVA of process parameter optimization	[86, 92, 118]
Tensile strength	Tensile testing machine	Variation in wear resistance, tensile strength, and elongation	[77, 90]
Strain map	Video extensometer, Istra 4D software	Evaluate strain evolution with crosshead displacement through digital image	[13]
Compression strength	Universal mechanical testing machine	Compression strength, strain rate, and deformation measurement	[42]
Interface analysis	Vibrating sample magnetometer	Isothermal entropy change was measured	[70]
Porosity	Mercury intrusion porosimetry (MIP), optical microscope (OM)	The theoretical pore diameter is evaluated, and geometric properties like strut size and pore size of the scaffolds by OM images	[41, 42]
Particle distribution	Nano-computed tomography and OM	Grayscale analysis using OM	
Three-point bending test	Universal materials testing machine	Bending strength variation is measured with composition gradation	[92]
		Measure fatigue crack growth rate and critical stress intensity factor	[58]
Inverse pole figure (IPF)	TSL-OIM analysis software	Maps for microstructural analyses	[58]
Fatigue testing	Servo hydraulic MTS machine	Fatigue behavior also depends on scanning direction	
Energy storage capacity	Single-cycle automated ball indentation	2.25 times increase in energy storage capacity with the increase in Cu percentage	[86]

(Continued)

Table 2.5 (Continued)

Characterized feature	Tools/process used	Observation	Reference
Viscosity characteristics	Rotational rheometer	Viscosity measurement of formed paste	[27, 63]
Thermal analysis of resin	Fourier-transform infrared spectroscopy (FTIR), differential scanning calorimetry (DSC), thermogravimetric analysis (TG)	Thermal analysis of formed paste	
Creep and TMF testing	Servo hydraulic MTS 858 tabletop system	Analyze the creep and thermomechanical fatigue behavior of the FGM part	[13]

2.7 CHALLENGES AND FUTURE DIRECTIONS

This study put forth a conceptual understanding of design, modeling and optimization, fabrication techniques, post-processing, and applications of FGMs. It is observed in the literature that even after various technological advancements, a few technical challenges impede the tapping of the full potential of FGMs as advanced materials for novel design with tailored properties. Design and optimizing software developments are underway for process parameters, but various physical and technical factors can introduce potential deviations between predicted and manufactured objects. This section discusses the challenges mentioned in previously published papers, identifying research gaps and future directions in metallic FGMs.

2.7.1 Modeling and design

    i. Defining the optimal material distribution function for a multi-material FGM requires extensive knowledge of chemical composition, material properties, and the limitations of manufacturing processes. The lack of design guidelines on material compatibility and thermophysical properties has circumscribed material selection. Furthermore, a framework for optimal distribution of property by tailoring the gradient and transition phase distribution in design space is lacking [3]. Consequently, accurate predictive process controls like scanning strategies and heat source power are complicated. Recent research has shown the importance of thermodynamic modeling to model phase transformations and thermophysical properties, but more work for computational development is currently needed.

    ii. The CAD software lacks in-built computer representation for automatically controlling the toolpath for FGM fabrication. Voxel-based modeling allows assigning material by transferring mesh data into voxels.

2.7.2 Alloy incompatibility and insolubility

i. Databases of material performance information provide some insight into the "processing-structure-property" relationship [123, 124]. Nevertheless, for multi-material FGMs, cracks and delamination defects are common.

ii. Alloy incompatibility can lead to the formation of insoluble brittle phases, resulting in residual stresses that may initiate and propagate cracks.

iii. Mixing dissimilar materials with different melting temperatures may result in preferential vaporization of constituents with lower melting temperatures while elements with high melting temperatures may remain intact. Moreover, mixing the constituents with significant differences in liquid surface tension and density may render dimensional inaccuracy, segregation, and a porous structure.

2.7.3 Process modeling

i. Real-time melt pool monitoring and *in situ* quality control of FGM fabrication is in the earliest stage, and more research for monitoring and feedback mechanism is needed for higher-quality fabrication.

ii. At present, fabrication machines lack in-built *in situ* monitoring of the process, and published research demonstrates monitoring of only one process parameter such as temperature, the morphology of melt pool, heat affected zone, and porosity. Simultaneous monitoring and control of multiple process parameters are needed.

2.7.4 Advanced characterization

Material characterization has evolved significantly over the past few decades with methods to differentiate features at the nanoscale or even near-atomic scale. Two of these techniques, 3D atom probe tomography (APT) and transmission electron microscopy (TEM), are employed in characterizing and designing advanced materials like semiconductors, aerospace materials, and composites. These techniques would prove to be an excellent tool for narrowing the gap between the properties of predicted and manufactured objects. Especially in metallic FGMs, both APT and TEM can resolve the compositional gradation and interlayer interaction for equipping researchers to design and optimize frontier FGMs. In addition, these techniques provide 3D nanoscale characterization along with sub-nanometer scale spatial resolution [125].

Based on the literature survey, the future directions for inclusive growth of FGMs are (Figure 2.6): (i) material, (ii) process, (iii) modeling and optimization, (iv) mass production and development of new manufacturing techniques, and (v) post-processing and testing. These trends aspire to develop and improve FGMs by using one or more new materials, new manufacturing methods, numerical simulation, or mathematical models for fabricating a graded layer or region. Based on

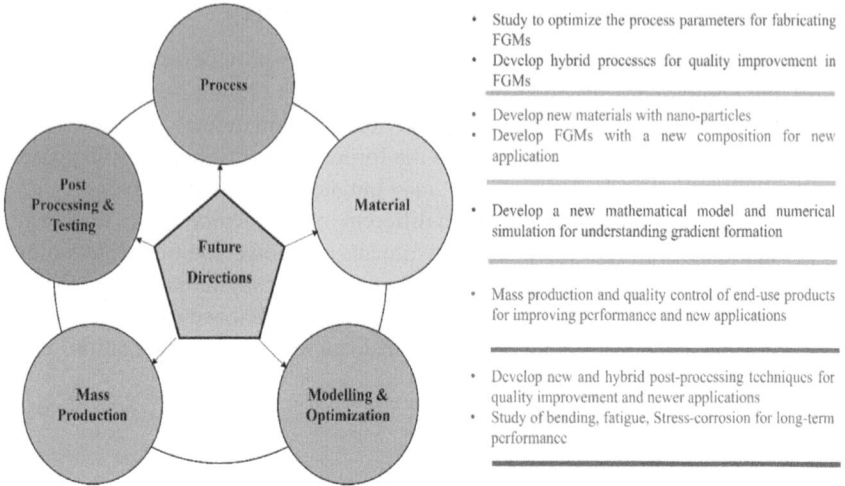

- Study to optimize the process parameters for fabricating FGMs
- Develop hybrid processes for quality improvement in FGMs

- Develop new materials with nano-particles
- Develop FGMs with a new composition for new application

- Develop a new mathematical model and numerical simulation for understanding gradient formation

- Mass production and quality control of end-use products for improving performance and new applications

- Develop new and hybrid post-processing techniques for quality improvement and newer applications
- Study of bending, fatigue, Stress-corrosion for long-term performance

Figure 2.6 Potential future research trends in FGMs.

published research, mathematical modeling is a paramount research direction for crucial future developments like producing FGMs with high performance and at a lower cost [2, 126]. Based on the summaries of different modeling, simulation, and optimization methods, it becomes evident that further research with more university-industry collaboration and technology transfer is needed. The forums for sharing ideas and disseminating information about the frontiers of FGM research, including the current state-of-the-art, educational aspects, identifying future potential and gaps in the present scenario, and promoting technology transfer through university-industry collaboration should be developed and advertised.

2.8 SUMMARY AND CONCLUDING REMARKS

FGMs fundamentally change the paradigm of part fabrication as material properties become a parameter in process design. Parts are fabricated with tailored regions utilizing varied materials or structures for imparting required properties. Several engineering operations require components with different properties at different spatial positions. Although varying properties can be achieved through traditional composites, their limitation under extreme environments makes FGMs a potential solution to meet these operational requirements. This review objectively assessed the diverse aspects of FGMs, beginning with material developments and followed by the evolution in fabrication methods, modeling, simulation, and optimization techniques. Substantial progress in each of these aspects has been observed in the domain of FGMs. Additive manufacturing has been developed from rapid prototyping to rapid manufacturing, enabling on-demand manufacturing of high-end products, especially in the biomedical and aerospace industries, with tremendous potential for future applications in many more domains. Multiple prominent features of FGMs have made them one of the key materials for a future role in sustainable development. Therefore, researchers strive for inclusive development of FGMs to expand their scope of utilization to befit the fourth industrial revolution.

This chapter provides academicians, researchers, and professional experts with a dependable platform to understand the fundamental concepts plus complexities of FGM fabrication processes, the sequence of their development, and several crucial parameters to obtain the required gradation of properties. Furthermore, it delivers extensive work exploration and research directions for new researchers in this domain. Therefore, researchers and practitioners will benefit from the present study's highlights and save effort and time to grasp critical aspects of functionally graded materials.

2.8.1 Statements and declarations

2.8.1.1 Acknowledgement

The authors would like to thank Indian Institute of Technology Ropar for the research facilities.

2.8.2 Competing interest

The authors have no relevant financial or non-financial interests to disclose.

2.8.3 Funding

The authors declare that no funds, grants, or other support were received during the preparation of this manuscript.

REFERENCES

[1] B. Zhang, P. Jaiswal, R. Rai, and S. Nelaturi, "Additive Manufacturing of Functionally Graded Material Objects: A Review," J. Comput. Inf. Sci. Eng., vol. 18, no. 4, 2018, doi: 10.1115/1.4039683.

[2] B. Saleh et al., "30 Years of Functionally Graded Materials: An Overview of Manufacturing Methods, Applications and Future Challenges," Compos. Part B Eng., vol. 201, 2020, doi: 10.1016/j.compositesb.2020.108376.

[3] G. H. Loh, E. Pei, D. Harrison, and M. D. Monzón, "An Overview of Functionally Graded Additive Manufacturing," Addit. Manuf., vol. 23, pp. 34–44, 2018, doi: 10.1016/j.addma.2018.06.023.

[4] M. A. Klecka, G. Subhash, and N. K. Arakere, "Determination of Subsurface Hardness Gradients in Plastically Graded Materials via Surface Indentation," J. Tribol., vol. 133, no. 3, 2011, doi: 10.1115/1.4003859.

[5]   C. Wei, Z. Sun, Q. Chen, Z. Liu, and L. Li, "Additive Manufacturing of Horizontal and 3D Functionally Graded 316L/Cu10Sn Components via Multiple Material Selective Laser Melting," J. Manuf. Sci. Eng. Trans. ASME, vol. 141, no. 8, 2019, doi: 10.1115/1.4043983.

[6]   M. Naebe and K. Shirvanimoghaddam, "Functionally Graded Materials: A Review of Fabrication and Properties," Appl. Mater. Today, vol. 5, pp. 223–245, 2016, doi: 10.1016/j.apmt.2016.10.001.

[7]   Y. H. Chueh, X. Zhang, C. Wei, Z. Sun, and L. Li, "Additive Manufacturing of Polymer-Metal/Ceramic Functionally Graded Composite Components via Multiple Material Laser Powder Bed Fusion," J. Manuf. Sci. Eng. Trans. ASME, vol. 142, no. 5, 2020, doi: 10.1115/1.4046594.

[8]   J. Yan, I. Battiato, and G. M. Fadel, "A Mathematical Model-Based Optimization Method for Direct Metal Deposition of Multimaterials," J. Manuf. Sci. Eng. Trans. ASME, vol. 139, no. 8, 2017, doi: 10.1115/1.4036424.

[9]   J. P. A. Nijssen and R. A. J. Van Ostayen, "Compliant Hydrostatic Bearings Utilizing Functionally Graded Materials," J. Tribol., vol. 142, no. 11, 2020, doi: 10.1115/1.4047299.

[10] R. M. Mahamood and E. T. Akinlabi, "Types of Functionally Graded Materials and Their Areas of Application," Top. Mining, Metall. Mater. Eng., pp. 9–21, 2017, doi: 10.1007/978-3-319-53756-6_2.

[11] Y. Watanabe, Y. Hattori, and H. Sato, "Distribution of Microstructure and Cooling Rate in Al-Al2Cu Functionally Graded Materials Fabricated by a Centrifugal Method," J. Mater. Process. Technol., vol. 221, pp. 197–204, 2015, doi: 10.1016/j.jmatprotec.2015.01.028.

[12] Y. Liu, G. Zheng, N. Letov, and Y. F. Zhao, "A Survey of Modeling and Optimization Methods for Multi-Scale Heterogeneous Lattice Structures," J. Mech. Des. Trans. ASME, vol. 143, no. 4, 2021, doi: 10.1115/1.4047917.

[13] V. A. Popovich, E. V. Borisov, A. A. Popovich, V. S. Sufiiarov, D. V. Masaylo, and L. Alzina, "Functionally Graded Inconel 718 Processed by Additive Manufacturing: Crystallographic Texture, Anisotropy of Microstructure and Mechanical Properties," Mater. Des., vol. 114, pp. 441–449, 2017, doi: 10.1016/j.matdes.2016.10.075.

[14] I. M. El-Galy, B. I. Saleh, and M. H. Ahmed, "Functionally Graded Materials Classifications and Development Trends from Industrial Point of View," SN Appl. Sci., vol. 1, no. 11, Nov. 2019, doi: 10.1007/s42452-019-1413-4.

[15] S. F. Hassan, O. Siddiqui, M. F. Ahmed, and A. I. Al Nawwah, "Development of Gradient Concentrated Single-Phase Fine Mg-Zn Particles and Effect on Structure and Mechanical Properties," J. Eng. Mater. Technol. Trans. ASME, vol. 141, no. 2, 2019, doi: 10.1115/1.4041865.

[16] I. Gibson, D. Rosen, B. Stucker, and M. Khorasani, "Additive Manufacturing Technologies," Addit. Manuf. Technol., 2021, doi: 10.1007/978-3-030-56127-7.

[17] P. A. Morton, J. Mireles, H. Mendoza, P. M. Cordero, M. Benedict, and R. B. Wicker, "Enhancement of Low-Cycle Fatigue Performance from Tailored Microstructures Enabled by Electron Beam Melting Additive Manufacturing Technology," J. Mech. Des. Trans. ASME, vol. 137, no. 11, 2015, doi: 10.1115/1.4031057.

[18] M. Ansari, E. Jabari, and E. Toyserkani, "Opportunities and Challenges in Additive Manufacturing of Functionally Graded Metallic Materials via Powder-Fed Laser Directed Energy Deposition: A Review," J. Mater. Process. Technol., vol. 294, 2021, doi: 10.1016/j.jmatprotec.2021.117117.

[19] R. R. Ma, J. T. Belter, and A. M. Dollar, "Hybrid Deposition Manufacturing: Design Strategies for Multimaterial Mechanisms via Three- Dimensional Printing and Material Deposition," J. Mech. Robot., vol. 7, no. 2, pp. 1–10, 2015, doi: 10.1115/1.4029400.

[20] R. Prabhu, S. R. Miller, T. W. Simpson, and N. A. Meisel, "Complex Solutions for Complex Problems? Exploring the Role of Design Task Choice on Learning, Design for Additive Manufacturing Use, and Creativity," J. Mech. Des. Trans. ASME, vol. 142, no. 3, 2020, doi: 10.1115/1.4045127.

[21] R. Prabhu, S. R. Miller, T. W. Simpson, and N. A. Meisel, "Exploring the Effects of Additive Manufacturing Education on Students' Engineering Design Process and its Outcomes," J. Mech. Des. Trans. ASME, vol. 142, no. 4, 2020, doi: 10.1115/1.4044324.

[22] I. T. Ozbolat and A. K. M. B. Khoda, "Design of a New Parametric Path Plan for Additive Manufacturing of Hollow Porous Structures With Functionally Graded Materials," J. Comput. Inf. Sci. Eng., vol. 14, no. 4, 2014, doi: 10.1115/1.4028418.

[23] P. Muller, P. Mognol, and J. Y. Hascoet, "Functionally Graded Material (FGM) Parts: From Design to the Manufacturing Simulation," ASME 2012 11th Bienn. Conf. Eng. Syst. Des. Anal. ESDA 2012, vol. 4, pp. 123–131, 2012, doi: 10.1115/ESDA2012-82586.

[24] B. Cheng, S. Price, J. Lydon, K. Cooper, and K. Chou, "On Process Temperature in Powder-Bed Electron Beam Additive Manufacturing: Model Development and Validation," J. Manuf. Sci. Eng. Trans. ASME, vol. 136, no. 6, 2014, doi: 10.1115/1.4028484.

[25] F. Wirth and K. Wegener, "A Physical Modeling and Predictive Simulation of the Laser Cladding Process," Addit. Manuf., vol. 22, pp. 307–319, 2018, doi: 10.1016/j.addma.2018.05.017.

[26] A. M. Mirzendehdel and K. Suresh, "A Pareto-Optimal Approach to Multimaterial Topology Optimization," J. Mech. Des. Trans. ASME, vol. 137, no. 10, 2015, doi: 10.1115/1.4031088.

[27] H. Xing, B. Zou, X. Liu, X. Wang, C. Huang, and Y. Hu, "Fabrication Strategy of Complicated Al2O3-Si3N4 Functionally Graded Materials by Stereolithography 3D Printing," J. Eur. Ceram. Soc., vol. 40, no. 15, pp. 5797–5809, 2020, doi: 10.1016/j.jeurceramsoc.2020.05.022.

[28] T. Kirk, R. Malak, and R. Arroyave, "Computational Design of Compositionally Graded Alloys for Property Monotonicity," J. Mech. Des. Trans. ASME, vol. 143, no. 3, 2021, doi: 10.1115/1.4048627.

[29] L. Kaufman and J. Ågren, "CALPHAD, First and Second Generation—Birth of the Materials Genome," Scr. Mater., vol. 70, no. 1, pp. 3–6, 2014, doi: 10.1016/j.scriptamat.2012.12.003.

[30] T. Kirk, E. Galvan, R. Malak, and R. Arroyave, "Computational Design of Gradient Paths in Additively Manufactured Functionally Graded Materials," J. Mech. Des. Trans. ASME, vol. 140, no. 11, 2018, doi: 10.1115/1.4040816.

[31] S. Ji, Z. Sun, W. Zhang, X. Chen, G. Xie, and H. Chang, "Microstructural Evolution and High Temperature Resistance of Functionally Graded Material Ti-6Al-4V/Inconel 718 Coated by Directed Energy Deposition-Laser," J. Alloys Compd., vol. 848, 2020, doi: 10.1016/j.jallcom.2020.156255.

[32] C. Katinas, W. Shang, Y. C. Shin, and J. Chen, "Modeling Particle Spray and Capture Efficiency for Direct Laser Deposition Using a Four Nozzle Powder Injection System," J. Manuf. Sci. Eng. Trans. ASME, vol. 140, no. 4, 2018, doi: 10.1115/1.4038997.

[33] B. Ealy et al., "Characterization of Laser Additive Manufacturing-Fabricated Porous Superalloys for Turbine Components," J. Eng. Gas Turbines Power, vol. 139, no. 10, 2017, doi: 10.1115/1.4035560.

[34] J. Fu, L. Xia, L. Gao, M. Xiao, and H. Li, "Topology Optimization of Periodic Structures With Substructuring," J. Mech. Des. Trans. ASME, vol. 141, no. 7, 2019, doi: 10.1115/1.4042616.

[35] D. V. Kaweesa and N. A. Meisel, "Quantifying Fatigue Property Changes in Material Jetted Parts Due to Functionally Graded Material Interface Design," Addit. Manuf., vol. 21, pp. 141–149, 2018, doi: 10.1016/j.addma.2018.03.011.

[36] E. Salcedo, D. Baek, A. Berndt, and J. E. Ryu, "Simulation and Validation of Three Dimension Functionally Graded Materials by Material Jetting," Addit. Manuf., vol. 22, pp. 351–359, 2018, doi: 10.1016/j.addma.2018.05.027.

[37] H. Boffy, M. C. Baietto, P. Sainsot, and A. A. Lubrecht, "An Efficient 3d Model of Heterogeneous Materials for Elastic Contact Applications Using Multigrid Methods," J. Tribol., vol. 134, no. 2, 2012, doi: 10.1115/1.4006296.

[38] D. Li, N. Dai, Y. Tang, G. Dong, and Y. F. Zhao, "Design and Optimization of Graded Cellular Structures With Triply Periodic Level Surface-Based Topological Shapes," J. Mech. Des. Trans. ASME, vol. 141, no. 7, 2019, doi: 10.1115/1.4042617.

[39] C. Sharpe and C. C. Seepersad, "Lattice Structure Optimization With Orientation-Dependent Material Properties," J. Mech. Des. Trans. ASME, vol. 143, no. 9, 2021, doi: 10.1115/1.4050299.

[40] Y. Wang, S. Arabnejad, M. Tanzer, and D. Pasini, "Hip Implant Design With Three-Dimensional Porous Architecture of Optimized Graded Density," J. Mech. Des. Trans. ASME, vol. 140, no. 11, 2018, doi: 10.1115/1.4041208.

[41] Y. Z. Xiong, R. N. Gao, H. Zhang, L. L. Dong, J. T. Li, and X. Li, "Rationally Designed Functionally Graded Porous Ti6Al4V Scaffolds With High Strength and Toughness Built via Selective Laser Melting for Load-Bearing Orthopedic Applications," J. Mech. Behav. Biomed. Mater., vol. 104, 2020, doi: 10.1016/j.jmbbm.2020.103673.

[42] S. Y. Choy, C. N. Sun, K. F. Leong, and J. Wei, "Compressive Properties of Functionally Graded Lattice Structures Manufactured by Selective Laser Melting," Mater. Des., vol. 131, pp. 112–120, 2017, doi: 10.1016/j.matdes.2017.06.006.

[43] B. Stoner, J. Bartolai, D. V. Kaweesa, N. A. Meisel, and T. W. Simpson, "Achieving Functionally Graded Material Composition Through Bicontinuous Mesostructural Geometry in Material Extrusion Additive Manufacturing," JOM, vol. 70, no. 3, pp. 413–418, Mar. 2018, doi: 10.1007/s11837-017-2669-z.

[44] C. Zhang et al., "Mechanical Characteristics and Deformation Mechanism of Functionally Graded Triply Periodic Minimal Surface Structures Fabricated Using Stereolithography," Int. J. Mech. Sci., vol. 208, 2021, doi: 10.1016/j.ijmecsci.2021.106679.

[45] E. Cetin and C. Baykasoğlu, "Crashworthiness of Graded Lattice Structure Filled Thin-Walled Tubes Under Multiple Impact Loadings," Thin-Walled Struct., vol. 154, 2020, doi: 10.1016/j.tws.2020.106849.

[46] H. Smith and J. A. Norato, "Topology Optimization With Discrete Geometric Components Made of Composite Materials," Comput. Methods Appl. Mech. Eng., vol. 376, 2021, doi: 10.1016/j.cma.2020.113582.

[47] D. C. Hofmann et al., "Developing Gradient Metal Alloys Through Radial Deposition Additive Manufacturing," Sci. Rep., vol. 4, 2014, doi: 10.1038/srep05357.

[48] S. H. Kim et al., "Selective Compositional Range Exclusion via Directed Energy Deposition to Produce a Defect-Free Inconel 718/SS 316L Functionally Graded Material," Addit. Manuf., vol. 47, 2021, doi: 10.1016/j.addma.2021.102288.

[49] X. Xu, Y. Wu, L. Zuo, and S. Chen, "Topology Optimization of Multimaterial Thermoelectric Structures," J. Mech. Des. Trans. ASME, vol. 143, no. 1, 2021, doi: 10.1115/1.4047435.

[50] I. F. Ituarte, N. Boddeti, V. Hassani, M. L. Dunn, and D. W. Rosen, "Design and Additive Manufacture of Functionally Graded Structures Based on Digital Materials," Addit. Manuf., vol. 30, 2019, doi: 10.1016/j.addma.2019.100839.

[51] R. Fathi et al., "Past and present of functionally graded coatings: Advancements and future challenges," Elsevier, Accessed: Apr. 23, 2022. [Online]. Available: www.sciencedirect.com/science/article/pii/S2352940722000129

[52] R. M. Mahamood and E. T. Akinlabi, "Processing Methods of Functionally Graded Materials," Top. Mining, Metall. Mater. Eng., pp. 23–45, 2017, doi: 10.1007/978-3-319-53756-6_3.

[53] T. Weber et al., "Functionally Graded Vacuum Plasma Sprayed and Magnetron Sputtered Tungsten/EUROFER97 Interlayers for Joints in Helium-Cooled Divertor Components," J. Nucl. Mater., vol. 436, no. 1–3, pp. 29–39, 2013, doi: 10.1016/j.jnucmat.2013.01.286.

[54] B. Saleh, J. Jiang, A. Ma, D. Song, and D. Yang, "Effect of Main Parameters on the Mechanical and Wear Behaviour of Functionally Graded Materials by Centrifugal Casting: A Review," Met. Mater. Int., vol. 25, no. 6, pp. 1395–1409, Nov. 2019, doi: 10.1007/s12540-019-00273-8.

[55] Y. Chen, J. Bunch, T. Li, Z. Mao, and F. Chen, "Novel Functionally Graded Acicular Electrode for Solid Oxide Cells Fabricated by the Freeze-Tape-Casting Process," J. Power Sources, vol. 213, pp. 93–99, 2012, doi: 10.1016/j.jpowsour.2012.03.109.

[56] S. Yin, X. Yan, C. Chen, R. Jenkins, M. Liu, and R. Lupoi, "Hybrid Additive Manufacturing of Al-Ti6Al4V Functionally Graded Materials with Selective Laser Melting and Cold Spraying," J. Mater. Process. Technol., vol. 255, pp. 650–655, 2018, doi: 10.1016/j.jmatprotec.2018.01.015.

[57] Z. Zhong et al., "Design and Anti-Penetration Performance of TiB/Ti System Functionally Graded Material Armor Fabricated by SPS Combined With Tape Casting," Ceram. Int., vol. 46, no. 18, pp. 28244–28249, 2020, doi: 10.1016/j.ceramint.2020.07.325.

[58] S. Ghorbanpour et al., "Effect of Microstructure Induced Anisotropy on Fatigue Behaviour of Functionally Graded Inconel 718 Fabricated by Additive Manufacturing," Mater. Charact., vol. 179, 2021, doi: 10.1016/j.matchar.2021.111350.

[59] X. Zhang, C. Wei, Y. H. Chueh, and L. Li, "An Integrated Dual Ultrasonic Selective Powder Dispensing Platform for Three-Dimensional Printing of Multiple Material Metal/Glass Objects in Selective Laser Melting," J. Manuf. Sci. Eng. Trans. ASME, vol. 141, no. 1, 2019, doi: 10.1115/1.4041427.

[60] J. Zhou et al., "Fabrication of Functionally Graded Materials From a Single Material by Selective Evaporation in Electron Beam Powder Bed Fusion," Mater. Sci. Eng. A, vol. 793, 2020, doi: 10.1016/j.msea.2020.139827.

[61] C. Wei, L. Li, X. Zhang, and Y. H. Chueh, "3D Printing of Multiple Metallic Materials via Modified Selective Laser Melting," CIRP Ann., vol. 67, no. 1, pp. 245–248, 2018, doi: 10.1016/j.cirp.2018.04.096.

[62] I. Valizadeh, A. Al Aboud, E. Dörsam, and O. Weeger, "Tailoring of Functionally Graded Hyperelastic Materials via Grayscale Mask Stereolithography 3D Printing," Addit. Manuf., vol. 47, 2021, doi: 10.1016/j.addma.2021.102108.

[63] F. Liravi and M. Vlasea, "Powder Bed Binder Jetting Additive Manufacturing of Silicone Structures," Addit. Manuf., vol. 21, pp. 112–124, 2018, doi: 10.1016/j.addma.2018.02.017.

[64] S. Kumar, "Development of Functionally Graded Materials by Ultrasonic Consolidation," CIRP J. Manuf. Sci. Technol., vol. 3, no. 1, pp. 85–87, 2010, doi: 10.1016/j.cirpj.2010.07.006.

[65] D. Kim et al., "Carbon Nanotubes-Reinforced Aluminum Alloy Functionally Graded Materials Fabricated by Powder Extrusion Process," Mater. Sci. Eng. A, vol. 745, pp. 379–389, 2019, doi: 10.1016/j.msea.2018.12.128.

[66] X. Tang et al., "Fabrication of W-Cu Functionally Graded Material by Spark Plasma Sintering Method," Int. J. Refract. Met. Hard Mater., vol. 42, pp. 193–199, 2014, doi: 10.1016/j.ijrmhm.2013.09.005.

[67] P. Jaiswal et al., "Build Orientation Optimization for Additive Manufacturing of Functionally Graded Material Objects," Springer, vol. 96, no. 1–4, pp. 223–235, Apr. 2018, doi: 10.1007/s00170-018-1586-9.

[68] J. C. Wang, H. Dommati, and S. J. Hsieh, "Review of Additive Manufacturing Methods for High-Performance Ceramic Materials," Int. J. Adv. Manuf. Technol., vol. 103, no. 5–8, pp. 2627–2647, Aug. 2019, doi: 10.1007/S00170-019-03669-3.

[69] R. S. Parihar, S. G. Setti, and R. K. Sahu, "Recent Advances in the Manufacturing Processes of Functionally Graded Materials: A Review," IEEE J. Sel. Top. Quantum Electron., vol. 25, no. 2, pp. 309–336, Mar. 2018, doi: 10.1515/secm-2015-0395.

[70] R. Bulatova, C. Bahl, K. Andersen, L. T. Kuhn, and N. Pryds, "Functionally Graded Ceramics Fabricated with Side-by-Side Tape Casting for Use in Magnetic Refrigeration," Int. J. Appl. Ceram. Technol., vol. 12, no. 4, pp. 891–898, Jul. 2015, doi: 10.1111/ijac.12298.

[71] D. Ding, Z. Pan, D. Cuiuri, and H. Li, "Wire-Feed Additive Manufacturing of Metal Components: Technologies, Developments and Future Interests," Int. J. Adv. Manuf. Technol., vol. 81, no. 1–4, pp. 465–481, Oct. 2015, doi: 10.1007/S00170-015-7077-3.

[72] I. Gibson, D. Rosen, and B. Stucker, "Additive manufacturing technologies: 3D printing, rapid prototyping, and direct digital manufacturing," 2nd ed., pp. 1–498, Jan. 2015, doi: 10.1007/978-1-4939-2113-3.

[73] M. Mahmoudi, A. A. Ezzat, and A. Elwany, "Layerwise Anomaly Detection in Laser Powder-Bed Fusion Metal Additive Manufacturing," J. Manuf. Sci. Eng. Trans. ASME, vol. 141, no. 3, 2019, doi: 10.1115/1.4042108.

[74] J. Guo et al., "Internal Surface Quality Enhancement of Selective Laser Melted Inconel 718 by Abrasive Flow Machining," J. Manuf. Sci. Eng. Trans. ASME, vol. 142, no. 10, 2020, doi: 10.1115/1.4047141.

[75] H. Bikas, P. Stavropoulos, and G. Chryssolouris, "Additive Manufacturing Methods and Modeling Approaches: A Critical Review," Int. J. Adv. Manuf. Technol., vol. 83, no. 1–4, pp. 389–405, Mar. 2016, doi: 10.1007/s00170-015-7576-2.

[76] C. Tan, K. Zhou, and T. Kuang, "Selective Laser Melting of Tungsten-Copper Functionally Graded Material," Mater. Lett., vol. 237, pp. 328–331, 2019, doi: 10.1016/j.matlet.2018.11.127.

[77] S. Mao, D. Z. Zhang, Z. Ren, G. Fu, and X. Ma, "Effects of Process Parameters on Interfacial Characterization and Mechanical Properties of 316L/CuCrZr Functionally Graded Material by Selective Laser Melting," J. Alloys Compd., vol. 899, 2022, doi: 10.1016/j.jallcom.2021.163256.

[78] M. A. El-Sayed, K. Essa, M. Ghazy, and H. Hassanin, "Design Optimization of Additively Manufactured Titanium Lattice Structures for Biomedical Implants," Int. J. Adv. Manuf. Technol., vol. 110, no. 9–10, pp. 2257–2268, Oct. 2020, doi: 10.1007/S00170-020-05982-8.

[79] F. Caiazzo, V. Alfieri, and B. D. Bujazha, "Additive Manufacturing of Biomorphic Scaffolds for Bone Tissue Engineering," Int. J. Adv. Manuf. Technol., vol. 113, no. 9–10, pp. 2909–2923, Apr. 2021, doi: 10.1007/S00170-021-06773-5.

[80] H. Choo et al., "Effect of Laser Power on Defect, Texture, and Microstructure of a Laser Powder Bed Fusion Processed 316L Stainless Steel," Mater. Des., vol. 164, 2019, doi: 10.1016/j.matdes.2018.12.006.

[81] R. Shi, S. A. Khairallah, T. T. Roehling, T. W. Heo, J. T. McKeown, and M. J. Matthews, "Microstructural Control in Metal Laser Powder Bed Fusion Additive Manufacturing Using Laser Beam Shaping Strategy," Acta Mater., vol. 184, pp. 284–305, 2020, doi: 10.1016/j.actamat.2019.11.053.

[82] B. Attard, S. Cruchley, C. Beetz, M. Megahed, Y. L. Chiu, and M. M. Attallah, "Microstructural Control During Laser Powder Fusion to Create Graded Microstructure Ni-Superalloy Components," Addit. Manuf., vol. 36, 2020, doi: 10.1016/j.addma.2020.101432.

[83] C. Guo et al., "A Comparing Study of Defect Generation in IN738LC Superalloy Fabricated by Laser Powder Bed Fusion: Continuous-Wave Mode Versus Pulsed-Wave Mode," J. Mater. Sci. Technol., vol. 90, pp. 45–57, 2021, doi: 10.1016/j.jmst.2021.03.006.

[84] Y. Su, B. Chen, C. Tan, X. Song, and J. Feng, "Influence of Composition Gradient Variation on the Microstructure and Mechanical Properties of 316L/Inconel718 Functionally Graded Material Fabricated by Laser Additive Manufacturing," J. Mater. Process. Technol., vol. 283, 2020, doi: 10.1016/j.jmatprotec.2020.116702.

[85] X. Zhang, Y. Chen, and F. Liou, "Fabrication of SS316L-IN625 Functionally Graded Materials by Powder-Fed Directed Energy Deposition," Sci. Technol. Weld. Join., vol. 24, no. 5, pp. 504–516, Jul. 2019, doi: 10.1080/13621718.2019.1589086.

[86] S. Yadav, A. N. Jinoop, N. Sinha, C. P. Paul, and K. S. Bindra, "Parametric Investigation and Characterization of Laser Directed Energy Deposited Copper-Nickel Graded Layers," Int. J. Adv. Manuf. Technol., vol. 108, no. 11–12, pp. 3779–3791, Jun. 2020, doi: 10.1007/s00170-020-05644-9.

[87] H. Sahasrabudhe, R. Harrison, C. Carpenter, and A. Bandyopadhyay, "Stainless Steel to Titanium Bimetallic Structure Using LENSTM," Addit. Manuf., vol. 5, pp. 1–8, 2015, doi: 10.1016/j.addma.2014.10.002.

[88] L. Li, X. Zhang, W. Cui, F. Liou, W. Deng, and W. Li, "Temperature and Residual Stress Distribution of FGM Parts by DED Process: Modeling and Experimental Validation," Int. J. Adv. Manuf. Technol., vol. 109, no. 1–2, pp. 451–462, Jul. 2020, doi: 10.1007/s00170-020-05673-4.

[89] E. B. Joyee and Y. Pan, "Investigation of a Magnetic-Field-Assisted Stereolithography Process for Printing Functional Part With Graded Materials," 2020 Int. Symp. Flex. Autom. ISFA 2020, 2020, doi: 10.1115/ISFA2020-9650.

[90] S. Safaee and R. K. Chen, "Development of a Design and Characterization Framework for Fabrication of Functionally Graded Materials Using Magnetic Field-Assisted Digital Light Processing Stereolithography," J. Manuf. Process., vol. 67, pp. 314–324, 2021, doi: 10.1016/j.jmapro.2021.04.058.

[91]   P. Badoniya, A. Yadav, M. Srivastava, P. K. Jain, and S. Rathee, "Fabrication of Functionally Graded Materials (FGMs) via Additive Manufacturing Route," pp. 191–213, 2022, doi: 10.1007/978-981-16-7377-1_9.

[92]   X. Shen, M. Chu, F. Hariri, G. Vedula, and H. E. Naguib, "Binder Jetting Fabrication of Highly Flexible and Electrically Conductive Graphene/PVOH Composites," Addit. Manuf., vol. 36, 2020, doi: 10.1016/j.addma.2020.101565.

[93]   X. Xiao and S. Joshi, "Automatic Toolpath Generation for Heterogeneous Objects Manufactured by Directed Energy Deposition Additive Manufacturing Process," J. Manuf. Sci. Eng. Trans. ASME, vol. 140, no. 7, 2018, doi: 10.1115/1.4039491.

[94]   S. Chandrasekaran, S. Hari, and M. Amirthalingam, "Wire Arc Additive Manufacturing of Functionally Graded Material for Marine Risers," Mater. Sci. Eng. A, vol. 792, 2020, doi: 10.1016/j.msea.2020.139530.

[95]   S. Jayanath and A. Achuthan, "A Computationally Efficient Finite Element Framework to Simulate Additive Manufacturing Processes," J. Manuf. Sci. Eng. Trans. ASME, vol. 140, no. 4, 2018, doi: 10.1115/1.4039092.

[96]   J. Jhabvala, E. Boillat, T. Antignac, and R. Glardon, "On the Effect of Scanning Strategies in the Selective Laser Melting Process," Virtual Phys. Prototyp., vol. 5, no. 2, pp. 99–109, Jun. 2010, doi: 10.1080/17452751003688368.

[97]   L. D. Bobbio et al., "Additive Manufacturing of a Functionally Graded Material From Ti-6Al-4V to Invar: Experimental Characterization and Thermodynamic Calculations," Acta Mater., vol. 127, pp. 133–142, 2017, doi: 10.1016/j.actamat.2016.12.070.

[98]   F. Geiger, K. Kunze, and T. Etter, "Tailoring the Texture of IN738LC Processed by Selective Laser Melting (SLM) by Specific Scanning Strategies," Mater. Sci. Eng. A, vol. 661, pp. 240–246, 2016, doi: 10.1016/j.msea.2016.03.036.

[99]   V. A. Popovich, E. V. Borisov, A. A. Popovich, V. S. Sufiiarov, D. V. Masaylo, and L. Alzina, "Impact of Heat Treatment on Mechanical Behaviour of Inconel 718 Processed With Tailored Microstructure by Selective Laser Melting," Mater. Des., vol. 131, pp. 12–22, 2017, doi: 10.1016/j.matdes.2017.05.065.

[100] J. Jiang, X. Xu, and J. Stringer, "Support Structures for Additive Manufacturing: A Review," J. Manuf. Mater. Process., vol. 2, no. 4, 2018, doi: 10.3390/jmmp2040064.

[101] P. Das, K. Mhapsekar, S. Chowdhury, R. Samant, and S. Anand, "Selection of Build Orientation for Optimal Support Structures and Minimum Part Errors in Additive Manufacturing," Comput. Aided. Des. Appl., vol. 14, pp. 1–13, Nov. 2017, doi: 10.1080/16864360.2017.1308074.

[102] R. Paul and S. Anand, "Optimization of Layered Manufacturing Process for Reducing Form Errors With Minimal Support Structures," J. Manuf. Syst., vol. 36, pp. 231–243, 2015, doi: 10.1016/j.jmsy.2014.06.014.

[103] Z. Hai-ming, H. Yong, F. Jian-zhong, and Q. Jing-jiang, "Inclined Layer Printing for Fused Deposition Modeling Without Assisted Supporting Structure," Robot. Comput. Integr. Manuf., vol. 51, pp. 1–13, 2018, doi: 10.1016/j.rcim.2017.11.011.

[104] G. Strano, L. Hao, R. M. Everson, and K. E. Evans, "A New Approach to the Design and Optimisation of Support Structures in Additive Manufacturing," Int. J. Adv. Manuf. Technol., vol. 66, no. 9–12, pp. 1247–1254, Jun. 2013, doi: 10.1007/s00170-012-4403-x.

[105] J. Vanek, J. A. G. Galicia, and B. Benes, "Clever Support: Efficient Support Structure Generation for Digital Fabrication," Eurographics Symp. Geom. Process., vol. 33, no. 5, pp. 117–125, 2014, doi: 10.1111/cgf.12437.

[106] G. Zhao, C. Zhou, and S. Das, "Solid Mechanics Based Design and Optimization for Support Structure Generation in Stereolithography Based Additive Manufacturing," Proceedings of the ASME 2015 International Design Engineering Technical Conferences and Computers and Information in Engineering Conference. Volume 1A: 35th Computers and Information in Engineering Conference, V01AT02A035, 2015, doi: 10.1115/DETC2015-47902.

[107] C. S. Lefky, B. Zucker, D. Wright, A. R. Nassar, T. W. Simpson, and O. J. Hildreth, "Dissolvable Supports in Powder Bed Fusion-Printed Stainless Steel," 3D Print. Addit. Manuf., vol. 4, no. 1, pp. 3–11, Mar. 2017, doi: 10.1089/3dp.2016.0043.

[108] C. Wei, Y. H. Chueh, X. Zhang, Y. Huang, Q. Chen, and L. Li, "Easy-to-Remove Composite Support Material and Procedure in Additive Manufacturing of Metallic Components Using Multiple Material Laser-Based Powder Bed Fusion," J. Manuf. Sci. Eng. Trans. ASME, vol. 141, no. 7, 2019, doi: 10.1115/1.4043536.

[109] I. Gibson, D. Rosen, B. Stucker, and M. Khorasani, "Materials for Additive Manufacturing," Addit. Manuf. Technol., pp. 379–428, 2021, doi: 10.1007/978-3-030-56127-7_14.

[110] T. D. Ngo, A. Kashani, G. Imbalzano, K. T. Q. Nguyen, and D. Hui, "Additive Manufacturing (3D Printing): A Review of Materials, Methods, Applications and Challenges," Compos. Part B Eng., vol. 143, pp. 172–196, 2018, doi: 10.1016/j.compositesb.2018.02.012.

[111] N. N. Kumbhar and A. V. Mulay, "Post Processing Methods Used to Improve Surface Finish of Products Which Are Manufactured by Additive Manufacturing Technologies: A Review," J. Inst. Eng. Ser. C, vol. 99, no. 4, pp. 481–487, Aug. 2018, doi: 10.1007/s40032-016-0340-z.

[112] W. Sames, "Additive manufacturing of Inconel 718 using electron beam melting: processing, post-processing, & mechanical properties," 2015, Accessed: Apr. 03, 2022. [Online]. Available: https://oaktrust.library.tamu.edu/handle/1969.1/155230.

[113] C. Shen et al., "Thermal Induced Phase Evolution of Fe–Fe3Ni Functionally Graded Material Fabricated Using the Wire-Arc Additive Manufacturing Process: An In-Situ Neutron Diffraction Study," J. Alloys Compd., vol. 826, 2020, doi: 10.1016/j.jallcom.2020.154097.

[114] S. Clijsters, T. Craeghs, S. Buls, K. Kempen, and J. P. Kruth, "In Situ Quality Control of the Selective Laser Melting Process Using a High-Speed, Real-Time Melt Pool Monitoring System," Int. J. Adv. Manuf. Technol., vol. 75, no. 5–8, pp. 1089–1101, Oct. 2014, doi: 10.1007/s00170-014-6214-8.

[115] P. Charalampous, I. Kostavelis, and D. Tzovaras, "Non-Destructive Quality Control Methods in Additive Manufacturing: A Survey," Rapid Prototyp. J., vol. 26, no. 4, pp. 777–790, May 2020, doi: 10.1108/RPJ-08-2019-0224.

[116] I. F. Ituarte et al., "Surface Modification of Additively Manufactured 18% Nickel Maraging Steel by Ultrasonic Vibration-Assisted Ball Burnishing," J. Manuf. Sci. Eng. Trans. ASME, vol. 142, no. 7, 2020, doi: 10.1115/1.4046903.

[117] V. A. Popovich, E. V. Borisov, V. Heurtebise, T. Riemslag, A. A. Popovich, and V. S. Sufiiarov, "Creep and Thermomechanical Fatigue of Functionally Graded Inconel 718 Produced by Additive Manufacturing," Miner. Met. Mater. Ser., vol. Part F12, pp. 85–97, 2018, doi: 10.1007/978-3-319-72526-0_9.

[118] S. Z. Hussain et al., "Feedback Control of Melt Pool Area in Selective Laser Melting Additive Manufacturing Process," Processes, vol. 9, no. 9, 2021, doi: 10.3390/pr9091547.

[119] M. Grasso, A. G. Demir, B. Previtali, and B. M. Colosimo, "In Situ Monitoring of Selective Laser Melting of Zinc Powder via Infrared Imaging of the Process Plume," Robot. Comput. Integr. Manuf., vol. 49, pp. 229–239, 2018, doi: 10.1016/j.rcim.2017.07.001.

[120] M. Grasso, V. Laguzza, Q. Semeraro, and B. M. Colosimo, "In-Process Monitoring of Selective Laser Melting: Spatial Detection of Defects via Image Data Analysis," J. Manuf. Sci. Eng. Trans. ASME, vol. 139, no. 5, 2017, doi: 10.1115/1.4034715.

[121] Q. Y. Lu and C. H. Wong, "Additive Manufacturing Process Monitoring and Control by Non-Destructive Testing Techniques: Challenges and In-Process Monitoring," Virtual Phys. Prototyp., vol. 13, no. 2, pp. 39–48, Apr. 2018, doi: 10.1080/17452759.2017.1351201.

[122] B. Zhang and C. Coddet, "Selective Laser Melting of Iron Powder: Observation of Melting Mechanism and Densification Behavior via Point-Track-Surface-Part Research," J. Manuf. Sci. Eng. Trans. ASME, vol. 138, no. 5, 2016, doi: 10.1115/1.4031366.

[123] E. L. Doubrovski, E. Y. Tsai, D. Dikovsky, J. M. P. Geraedts, H. Herr, and N. Oxman, "Voxel-Based Fabrication Through Material Property Mapping: A Design Method for Bitmap Printing," CAD Comput. Aided Des., vol. 60, pp. 3–13, 2015, doi: 10.1016/j.cad.2014.05.010.

[124] E. Mahamood, R. M. Akinlabi, "Functionally Graded Materials, Topics in Mining," Metallurgy and materials engineering. Springer International Publishing, Cham, 2017. https://scholar.google.co.in/scholar?hl=en&as_sdt=0%2C5&q=Functionally+graded+materials%2C+Topics+in+Mining&btnG= (accessed Mar. 31, 2022).

[125] A. Devaraj et al., "Three-Dimensional Nanoscale Characterisation of Materials by Atom Probe Tomography," Int. Mater. Rev., vol. 63, no. 2, pp. 68–101, Feb. 2018, doi: 10.1080/09506608.2016.1270728.

[126] M. Sireesha, J. Lee, A. S. Kranthi Kiran, V. J. Babu, B. B. T. Kee, and S. Ramakrishna, "A Review on Additive Manufacturing and Its Way Into the Oil and Gas Industry," RSC Adv., vol. 8, no. 40, pp. 22460–22468, 2018, doi: 10.1039/c8ra03194k.

Chapter 3

An overview of power factor correction techniques

Jagdeep Kaur Brar and Yadwinder Pal Sharma

3.1 INTRODUCTION

Electric machines, transformers, and fluorescent lamps, all inductive appliances, require a magnetic field to perform tasks. The magnetic field is required, but it does not produce any productive work. The utility must supply the energy required to complete the task. The charge usually needed by motor drives, lamps, and computer systems has been derived from a combination of instantaneous active and reactive elements.

Appliances such as a furnace only necessitate that actual value of current. Several applications, including an induction machine, require simultaneous real and reactive currents. The actual current is the proportion of the electrical current that is converted into useful work by the equipment, including heat generation by a heating chamber. The ampere (A) is a current measurement unit, while the watt (voltage real current) is a power standard unit (W). The component necessary to generate the flux required for induction device operation is reactive current. The current is measured in amperes (A) and the reactive power is measured in volt ampere reactive (VARs). The three categories of utility grid are resistive consumer load, inductive consumer load, and capacitive consumer loads. To work effectively, all inductive loads require two types of power. The work is done by active power (P), but the magnetic field is maintained by reactive power (Q). Figure 3.1 depicts the relationship between apparent power, useful power, and wasteful power.

$$\text{Apparent Power (S)} = \sqrt{P^2 + Q^2} \tag{3.1}$$

Power factor: The electric power factor is the ratio between usable and perceived power (S). Energy factor values vary from 0 to 1.00. Low wattage factor can be defined as less than 0.80. The power factor is denoted by PF or $\cos\theta$. The power factor is written as follows:

$$\text{PF (Power Factor)} = \frac{\text{Active Power}(P)}{\text{Apparent Power}(S)} \tag{3.2}$$

DOI: 10.1201/9781032703046-3

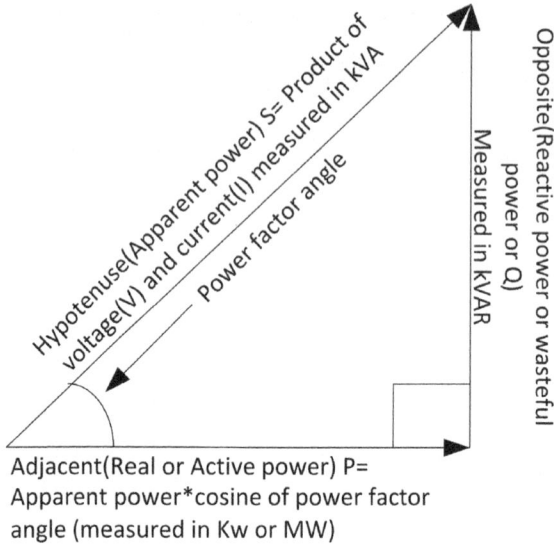

Adjacent(Real or Active power) P=
Apparent power*cosine of power factor
angle (measured in Kw or MW)

Figure 3.1 Triangle of electricity.

In AC circuits, the power factor is critical since it determines how much power is consumed.

$$P = VI \cos\theta \, (\text{for single-phase supply})$$

So, the value of current is given as: I (current) $= \dfrac{P}{V \cos\theta}$

Also, $P = \sqrt{3} \; VI\cos\theta$ (for three-phase supply):

$$I = \frac{P}{\sqrt{3}V \cos\theta}$$

For constant power voltage, the utility current is generally inversely proportional to $\cos\theta$, as mentioned earlier. As the power factor declines, the demand current rises, and vice versa.

3.2 DRAWBACKS OF LOW POWER FACTOR

Low power factor merely suggests inefficient use. Because $\cos\theta$ ranges from 0 to 1, a value of 0.9 to 1.0 is considered an excellent power factor, implying used power is almost equal. In view of the consumer's standpoint, it basically implies getting the most out of what they paid for. When the power factor is low, the utility must provide both productive active and non-productive reactive power. Larger generators, transformers, cables, and other system components generate higher capital and operating expenses for the utility, which they mainly permit to industrial users in the form of power factor penalties. As a result, a higher power factor aids in avoiding those penalties. The following drawbacks occur when the power factor is less than unity [1].

I. **Required large kVA rating of equipment**: To illustrate, kVA is often used to evaluate electric machines like transformers and switches. Considering $kVA = \dfrac{kW}{\cos\theta}$, it is evident that the kVA value of such an apparatus is inversely related to the power factor. A stronger transformer rating is achieved with a lower power factor. Whenever the power factor is insufficient, the transformer rating must be increased, resulting in a larger and more powerful apparatus.

II. **Requires larger size of conductor**: At a low power factor, the conductor must carry more current to transmit a set quantity of power at constant voltage. This requires the use of big conductors.

III. **High copper losses**: There are enhanced copper losses in all of the supplied system parts only when current is large at a lagging power factor. As a result, efficiency suffers.

IV. **Decrease in voltage regulation**: The electric charge at a poor trailing power factor causes severe power loss in electric machines, transformers, and transmission system and distribution companies. As a consequence, overall voltage supplied at the generator terminals is diminished, affecting system performance. To keep the terminal voltage energy within reasonable parameters, extra equipment is required.

V. **System handling capacity is reduced**: The trailing power factor affects the system's overall handling capacity. Because the current's reactive component limits maximum implementation of installed capacity, the system handling capacity is reduced.

3.3 LOW POWER FACTOR CAUSES

A poor power factor is troublesome from an economic perspective. The distribution system is a function load power factor that is usually below 0.8. A lagging power factor can occur for a number of reasons.

1. The bulk of electric drives include inductive motors with a reduced lag power factor under minimal load. The power factor of such machines

is comparatively low (0.2 to 0.3), but it improves to 0.8 or 0.9 during peak load.

2. Flash bulbs, electrical discharge machining light bulbs, and industrial heating burners each have a low trailing power factor.

3. The power system's demand fluctuates during the day, with peak load in the dawn and dusk and lesser demands at the other intervals. During down times, the source voltage rises, which boosts the magnetization flow of current. The power factor is weakened as a consequence [2].

3.4 IMPORTANCE OF INCREASING POWER FACTOR

As the following describes, power factor enhancement is crucial for both households and producers.

a. **Consumers**: Users pay for their total demand in kVA as well as the units used. When their power factor improves, their maximal kVA prices decrease, leading to yearly cost reductions according to required power expenditures. Despite the fact that power factor correction equipment requires an estimated annual outlay, achieving an ideal power factor results in total yearly savings for such customers.

b. **Generating stations**: Power factor improvements are just as important to a generating station as they are to the user. The generators of power plants are rated in kVA, while the result is evaluated in kW. As a result of the station's production $kVA = \dfrac{kW}{\cos\theta}$, like a function, the power factor measures the number of units it can supply. The higher the power factor of a generating station, the more kWh it provides to the structure. As just an outcome, a greater power factor boosts the generating potential of the generating station.

3.5 POWER FACTOR CORRECTION CALCULATION

In the power factor analysis, a voltmeter and an ammeter are used to measure the source voltage level or current drawn. A wattmeter is often used to measure real power. As active power $(P) = VI\cos\theta$ watt. Whereas, power factor $(\cos\theta) = \dfrac{P}{VI}$

So, reactive power is calculated by $Q = VI\sin\theta$. A shunt capacitor or capacitor bank is connected in the circuit to compensate the reactive power. A following formula is used to compute the value of a capacitor: $Q = \dfrac{V^2}{X_c}$

Whereas, $C = \dfrac{Q}{2\pi f V^2}$ farad.

3.6 IMPROVING THE ENERGY FACTOR TECHNIQUES

The overall energy factor of a large producing station is usually between 0.8 and 0.9. It is sometimes smaller; in such circumstances, it is usually advisable to take additional measures to boost the output power. Static capacitors, synchronous condensers, and phase advancers can all help with this.

3.7 STATIC CAPACITOR

Power factor can be enhanced by installing capacitor banks alongside equipment that has a trailing power factor. A capacitor draws a leading current to neutralize the trailing reactive component of the demand current, which boosts the power factor of the demand. A capacitor can be linked in a delta or star configuration for three-phase loads, as shown in Figure 3.2. In factories, static capacitors are almost always employed to enhance the power factor.

> **How capacitors handle a lower power factor**: Reduced power factor seems to be a concern that can be remedied by incorporating a PFI capacitor element into the industrial power grid. Capacitors work primarily as reactive current producers, providing the circuit's overall power flow (kVAR). By creating their own voltage level, industrial applications save money for the utilities. As a result, the utility's extent of power (kVA) will be dropped, culminating in proportionately lower expenses. Capacitors reduce the amount of current taken from the power grid, allowing it to run at full capacity.

Capacitors for PFI come in three primary varieties:

A. **Single capacitor units** are required for each inductive apparatus. Individual capacitors have several benefits:
   i. Enhanced system capacity: Real power adjustment attached to the demand enhances system capacity while lowering power dissipation.

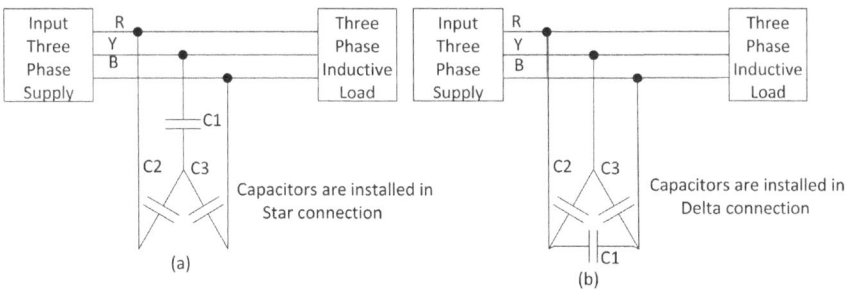

Figure 3.2 Static capacitor placed with the demand

ii. Relatively cool functionality: As voltages decline, flow rises, providing excessive heat problems due to the marginally higher flow of current. Whenever voltage dips are rectified nearer to the demand, these heating concerns are prevented at the outset.

iii. The control logic is streamlined, and there are fewer control parts, because the capacitor and motor can be switched on and off simultaneously. There is minor risk of overcompensating because the particular capacitor is sized to the exact load and switches concurrently with that load. Individual capacitor selection is simple and straightforward, and no complex computations are required.

B. **Units of capacitor banks**: A bank of capacitor units consists of multiple capacitors housed within a container and joined at a centralized spot in the power grid [3]. Fully automated capacitors include a capacitor in an identical container with a circuit or thyristor or silicon-controlled rectifier (SCR) controlled by a processor, whereas static capacitors contain a capacitor mounted in the identical yard with no switching. On larger inductive loads, a single capacitor is placed, and the bank is usually assembled at primary lines or switchgear, among many other places. Fixed or automatic bank systems have a number of advantages. Banks are much more cost-effective than standalone capacitor devices when the main purpose is to minimize utility power prices and/or reduce current in primary feeds from a significant generating station or substation. The ease of power factor adjustment is aided by the fact that just one installation is required. Instead of a number of separate capacitors next to every inductive load, a single fixed or programmable shunt capacitor device can save money.

C. **A mixture of the aforementioned**: A single capacitor is generally mounted on significant induction machines, and groups are mounted on significant feeders or switchboards, among other things. One of most effective banks are generally individualized and regulated. Single capacitors are employed on powerful motors, whereas banks are mounted on distribution systems. To determine the required overall power factor correction, it is necessary to review both the total kVAR requirement and the desired power factor [4].

3.8 POWER FACTOR CORRECTION BY PASSIVE CIRCUIT

Although a larger inductor is required for passive PFC than for active PFC, it is indeed cheaper. Although using a capacitor bank to address non-linearity in a load is a straightforward solution, it's not quite as successful as active PFC. Motor control centers, utility energy meters, substation transformers, and other sectors with numerous inductive loads use these power factor correction procedures. This filter suppresses the harmonic components, helping the non-linear instrument appear to be a linear demand. By adding capacitors or inductors as necessary, the power factor can now be brought approximately to unity. On the other hand, it requires high-current inductors that are complex and costly. In contrast to the converter, mainly passive components can be used in passive PFC to improve the shape of the line current. This kind of power factor technique can raise the power factor to a value of 0.7 to 0.8.

3.9 POWER FACTOR ENHANCEMENT BY ACTIVE CIRCUIT

A power electronic system that enhances the power factor by modifying the wave shape of current generated by a demand is known as an active power factor compensator (active PFC). The strategy is to create demand circuits with power quality that appears completely resistive (apparent power equal to real power). In this case, voltage and current are in phase, and energy is zero. This enables one of the most efficient deliveries of electricity from the utility to that same end user. An active PFC is an electronic power system that manages the amount of energy consumed by an application in return for a power factor as close to unity as possible. Every active filter design regulates the load's incoming current to produce a supply current that closely resembles the mains' output voltages (i.e. a sinusoidal wave). A mix of reactive components and active switches is used to improve the efficacy of source current shaping and generate an adjustable voltage output [5].

3.10 SYNCHRONOUS CONDENSERS

When a synchronous motor is over-excited, it absorbs a leading current and functions like the capacitors. An over-excited synchronous motor running at no loads is referred to as a synchronous condenser. When this machine is connected to the supply simultaneously, it generates a higher current that partially compensates for the load's lagging reactive component. As a result, the bad factor has improved. Figure 3.3 depicts the enhancement of the synchronous condenser method's power factor. A well-known way is to employ a synchronous motor as reactive power compensation. In comparison to a set of capacitors, if a synchronous motor is simply utilized for reactive power adjustment, the system will be inefficient and expensive [1]. Synchronous motors can have a power factor of unity, trailing, or going to lead, but there are significant drawbacks, such as the need for human control under varying loads.

3.11 PHASE ADVANCERS

Induction motors employ a phase advancer to rectify their power factor. Because the excitation windings of the stator have lagging current, the power factor of the machine is reduced; however, if this current is supplied from an external source, the power factor of the machine is improved and the phase advancer is reduced to

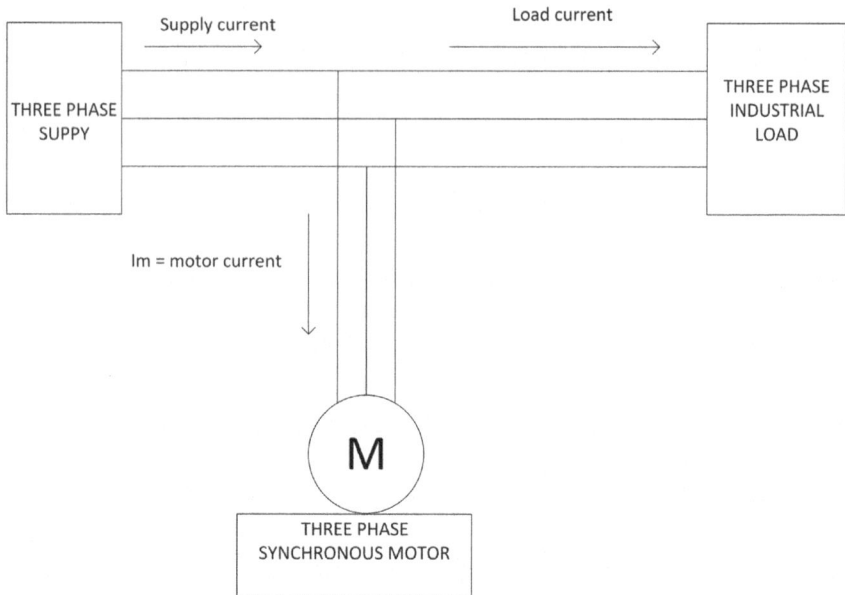

Figure 3.3 Power factor improvements by synchronous condenser.

a simple AC exciter. This approach can be employed in situations when using a synchronous condenser is not cost effective [6], providing a means to improve the power factor of induction motors. An inductive motor's stator winding draws an excited current that would be 90° behind voltage supply. For this reason, electric motors will have a poor power factor. If the excitation comes from somewhere else, the stator winding would be liberated or continued. As a result, the induction motor's power factor could be enhanced. Phase advancers provide this additional excitation. It's simply referred to as an AC exciter. It is positioned along the same axis as the primary motor and is connected to the rotor winding of the machine. It provides stimulating ampere turns at intermediate frequencies to the rotor winding. To provide more ampere turns as required, the induction motor could be structured to function under leading power factor, similar to an overexcited synchronous machine.

3.12 POWER FACTOR IMPROVEMENT BY USING ACTIVE FILTER

Active power filters minimize harmonic pollution in transmission lines. Over the last two decades, the proliferation of transmission losses, including stable converters, heaters, and other load variations, has resulted in a variation of unwanted phenomena in the functioning of energy systems, some of which are intractable with passive LC filtering. An LC circuit is a particular electrical circuit configuration that consists of an inductor, typically denoted by the symbol "L," and a capacitor, typically represented by the symbol "C." Active filters can adapt to randomly changing currents, which is the main difference between LC filters and active filters. Active filters have the capability to enhance power factor and simultaneously alleviate harmonic distortions, providing a more comprehensive approach to addressing power quality concerns. Managing a current source parallel connected to the power converter produces the required harmonic currents [7]. In this arrangement, the mains need to deliver the essential current that avoids pollution problems through the transmission line. In extreme cases, the load also requires reactive power, which can be created by the same source of current. This kind of filter can then pay for both harmonics and power factor concerns. In three-phase unbalanced loads, individual mains phase currents can also be redistributed and equalized as long as the overall extent of active power stays constant. In this design, the filter can address three issues instantly: (a) harmonic reduction, (b) modification of power factor, and (c) power reassignment that keeps systems stable. In the practical employment of this type of processing, force converter, current ampere controller, and voltage source converters are typically used. The superiority and performance of an active power filter are primarily determined by three factors: (a) the design of the power inverter; (b) the instrumentation technique often used to continue the current framework (leakage currents, triangular carriage, regular intervals representative sample); and (c) the active filter control strategy permits power factor correction to unity, harmonic elimination, and compensation for load unbalance. To build the reference template, a "Reference Current Calculator" schematic diagram has been created in Figure 3.4. I_S is the source current, I_F is the filter current, and I_L is the load current.

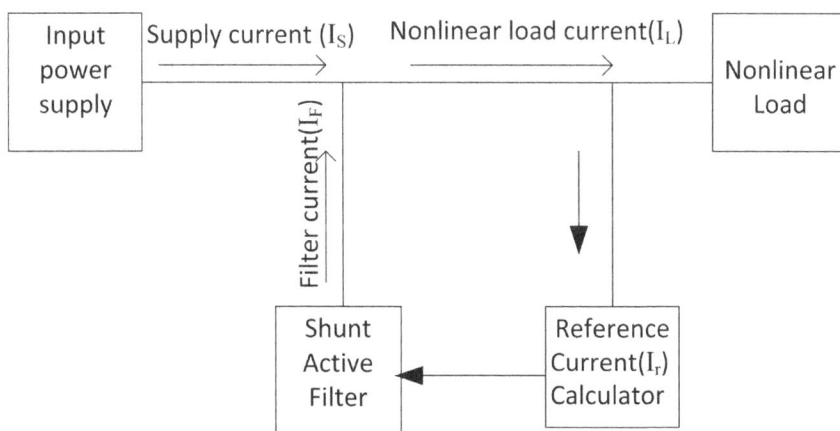

Figure 3.4 Parallel active power filter's working principle.

For determining the active component of the demand current, "Sample and Hold" circuits are used as passive components. Harmonic components, power factor, and non-linear load compensation patterns are created by the network. Most of the ways for developing the guide and help for power systems use the concept of instantaneous reactive power. This theory offers an extremely precise method for acquiring the reference template and allows for a clear difference between active and reactive power in the present. This technique has a few distinguishing features. First, separating the averaging active power over a certain duration is much more comfortable in many instances and from a mains perspective than separating the instantaneous active power, because the first method prevents flashes and lowers temporal issues from mains. Another challenge seems to be the difficulty of creating and modifying the electronic circuit necessary to acquire the reference, which comprises d-q conversions, splits, and a significant number of multiplications. This effort has produced a substantially simpler solution by incorporating sample and hold circuits (S&H), which avoids expensive conversions as well as mathematical operations such as multiplication and division [8].

3.13 POWER FACTOR IMPROVEMENT BY MPPT OR DC/DC CONVERTER

Due to the instability of solar energy, the large penetration of photovoltaic (PV) systems in the power grid now poses numerous issues. To address this issue, this chapter examines the power factor correction (PFC) management technique for supporting converters that are integrated with the maximum power point tracking (MPPT) algorithm using the existing power grid. After that, an experimental circuit model incorporating two control approaches will be built and tested under real-world settings. When PFC assistance is used, the output voltage response is more stable than when the simple MPPT algorithm is used [9]. Power distribution stability and system reliability are becoming key problems for all sorts of electronic power systems, including solar energy. Figure 3.5 depicts the proposed

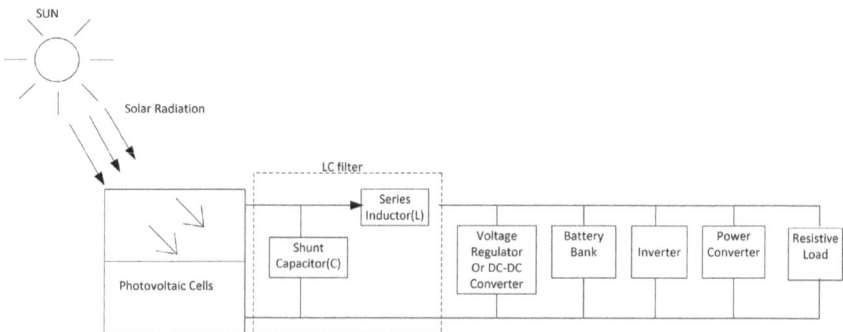

Figure 3.5 Power factor improvements in a photovoltaic scheme.

utility-connected PV system's power circuit. When linking a solar system to the utility, the PV system will meet both the harmonic and active power supply requirements. An inverter provides reactive current adjustment, harmonic removal, and active power supply, while a DC/DC converter handles maximum power tracking. In addition, the suggested system is connected to a utility line through a series inductor that provides for reactive power regulation.

An inductor-capacitor (L-C) filter is used to connect the load to the reactor and inverter at the common coupling point. When production of the solar system is inadequate, the power source delivers real energy to the system, and the PV control scheme introduces compensatory current to remove the harmonic content created by the power converter. As a result, the PV system solely supplies power flow to the utility in this mode, eliminating the harmonic content caused by the load. The photovoltaic system's primary function in normal operating conditions is to provide active power to the load or utility as well as performing necessary uninterruptible power supply (UPS) services for the load [9].

As a result, the suggested system offers the following benefits:

1. A power factor close to one.
2. Demand protection with a UPS service.
3. Capacity to balance the load.

3.14 THYRISTORIZED AUTOMATIC POWER FACTOR IMPROVEMENT UNIT

The following discusses using a thyristor in an autonomous power factor adjustment unit. For these types of designs, SCR or triode for alternating current (TRIAC) are commonly employed. Thyristors act as a switch/valve between the load circuit and capacitor bank. Assume a power factor enhancement system has 10 capacitor banks, because when the power factor goes below a specific threshold, additional capacitors must be connected to the load circuit; as the power factor approaches unity, only a few capacitors must be attached to the circuit [10]. As shown in Figure 3.6, reactive current is sensed by embedded systems, instrument transformer and relay systems that continuously evaluate the power factor of the entire power system. The embedded systems trigger the thyristors by sending signals to their gate connections depending on the power factor value or comparative power demand. Once the thyristor has been activated to adjust for power flow or preserve the power factor, the capacitor bank is attached to the load. The thyristor gate initiating pulses are generated by zero crossing across conductive firing procedures. The use of AC power at zero points (between the positive and negative half cycle) where the voltage is almost zero is referred to as "zero cross over". Gate pulses will only engage devices when the power across the thyristor is low. This approach is used to avoid nonlinearities, overvoltage, and distortions from generating during switching. Significant enhancements in energy at

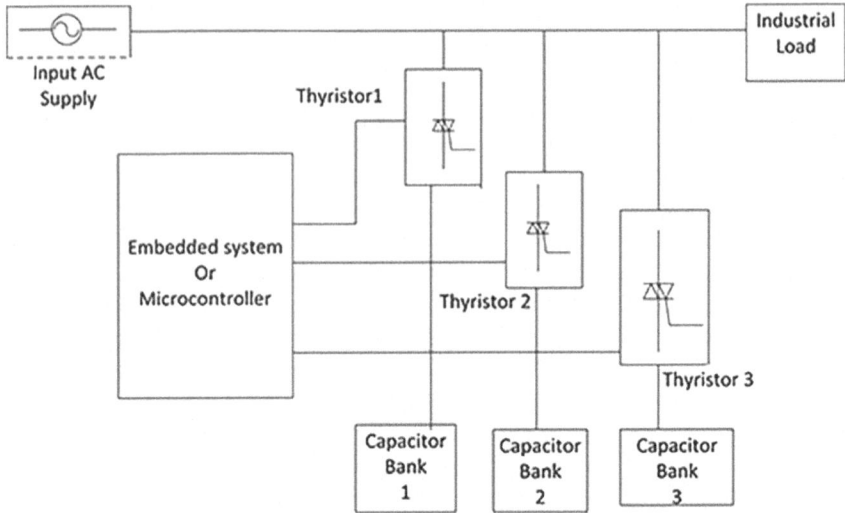

Figure 3.6 Automatic power factor adjustment device single line schematic diagram.

home and industrial loads are recommended in this technique. The properties of a thyristor toggling mechanism are that the rate of variation is immediate and the leakage current is protected. When switching operations, there seems to be no disturbance. The activation of thyristors through gate pulses occurs only when the voltage across them reaches zero. This method is employed during switching to prevent the occurrence of transients, voltage spikes, and harmonics. This approach yields notable improvements in power quality for both industrial and residential loads. The characteristics of the thyristor switching module include inrush current protection, instantaneous change, noise-free operation during switching, precise automatic zero recognition for zero voltage switching "ON" and zero active flipping "OFF," extended operational lifespan, over-temperature protection above 70 °C, dv/dt protection through an resistor capacitor (RC) snubber circuit, user-friendly operation, and the elimination of transients through zero flipping technology.

3.15 AUTOMATIC POWER FACTOR CONTROLLER

With relay/microprocessor logic, several forms of automatic power factor controllers (APFCs) are accessible [11]. The automatic power factor regulator enhances the power factor when it declines. The processor sends the signal to the control circuit, which triggers the switches when the power factor falls below a specified level. By turning on relays, the capacitor bank rapidly links to the

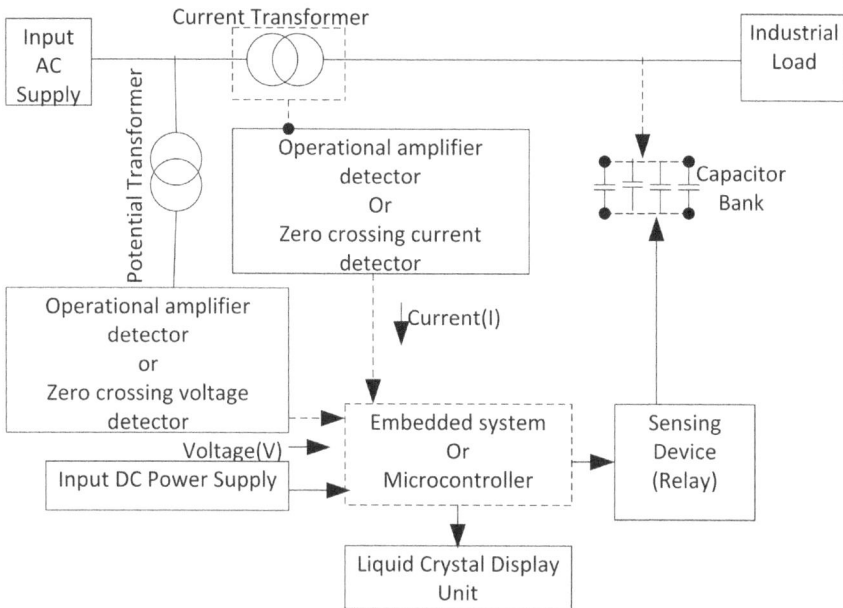

Figure 3.7 Schematic representation of automatic power factor regulator.

load and boosts power factor. An overview of the proposed method is illustrated in Figure 3.7.

Technical requirements: An electricity supply, the potential transformer, the current transformers (CTs), the zero crossing detector, a processor, a switch controller, a switch, a liquid crystal display (LCD) screen, and switching capacitors are the key components of this APFC system.

A. **Power supply**: The AC supply is acquired by the main of a step-down power transformer. The output of the transformer is sent to a rectifier. Controllers are supplied by a 5-volt DC power supply. A microcontroller intended to operate at a voltage greater than 5 volts will burn out and be ruined quickly.

B. **Potential transformer and current transformer**: The potential transformer (PT) decreases the supply voltage to the appropriate value for the circuit to operate. Microcontrollers rarely execute above 5 volts. If a microcontroller is designed to function at a voltage higher than 5 V, it will immediately burn out and be damaged. For measurement, control, and protection, the PT is used to step down voltage. A CT, on the other hand, is a device that converts large current into lesser current that can be used in electrical circuits. The current transformer can also be used to detect, monitor, and secure the network [12].

C. **Zero crossing sensor and microprocessor**: The outputs of the current and voltage transformers are sent into a zero crossing detector that transforms the sinusoidal waveform into a square waveform. The voltage level is zero in this situation. The output voltage waveform indicates when and in which direction and input signal crosses zero volts. The power output will travel more slowly through one saturation level to another if the voltage level is a reduced frequency signal. When there is disturbance between the two input terminals, the output may swing between positive and negative saturating potential Vsat. An output wave is produced and transmitted to the microcontroller whenever the zero crossing happens. The phase of the current and voltage signals is used to calculate the power factor. A microcontroller drives the APFC. The microcontroller performs calculations, takes action, and generates additional income. The phase difference between two outputs and, eventually, maximum power is calculated in a microcontroller by comparing the current and voltage from set reference detectors. The microcontroller will decide whether to switch the capacitor based on the current value of the power factor, which will be conveyed to the LCD display.

D. **Liquid crystal display (LCD) and relay driver**: A relay driver permits a reduced device to control signals, while an LCD seems to be a basic module that can be used in numerous systems for visual reasons. The switch driver is used in the circuit because the switches require a greater amount of current. The relays cannot be controlled directly by the microcontroller. A relay is a switch that is controlled by electricity when a relay is interfaced with a microcontroller, it can contribute to the enhancement of power factor within an electrical system. This collaborative process involves the microcontroller's continuous monitoring of power factor and the relay's pivotal role in controlling the connection and disconnection of power factor correction capacitors. The microcontroller diligently observes the power factor of the electrical system, utilizing appropriate sensors or meters for measurement. Upon detecting the necessity for power factor correction, the microcontroller transmits a control signal to the relay, directing it to close and thereby establish the connection between the power factor correction capacitors and the electrical system. The relay is commonly employed to govern the switching of these capacitors. With the relay in the closed position, the power factor correction capacitors are effectively linked in parallel with the system. These capacitors introduce leading reactive power to offset the lagging reactive power generated by inductive loads, leading to an improvement in the power factor. As a result, the power factor is brought closer to unity (1.0).

E. **Capacitor**: Capacitors boost the factor of energy enhancement devices by adding a reactive load to the circuit. The power factor of the circuit dictates the number of capacitors employed. The relay operates and provides a capacitor to the circuit whenever the power factor drops below a specified level [13]. If the power factor is much lower than the standard range, however, one capacitance is inserted. An additional capacitor is supplied if the power factor dips significantly below the predetermined value, and the system will continue in this fashion. The circuit will continue to add capacitors in parallel to the load until an acceptable power factor is attained [14].

3.16 SINGLE-PHASE POWER FACTOR CORRECTION USING PSO-BASED FIXED PWM

In the circuit shown in Figure 3.8, power factor adjustment was achieved using two loops: The inner loop, also known as the current loop, is used to control current, whereas the outer loop is used to control output voltage. The outer loop is connected to an outside proportional integral (PI) controller that uses particle swarm optimization (PSO) to keep the power output coordinated with the reference signal. The fixed pulse width modulated (PWM) current converter based on bang-bang/PSO is coupled to and controls the internal current loop. The current traveling through the inductor is then compared to the reference current (Iref). A constant frequency current converter based on bang-bang and PSO receives this prediction error signal. The constant current control converter, depending on bang-bang/PSO, also regulates the internal current loop. But after, the current traveling through the inductor is compared to the reference current (Iref) (I_{L1}). A fixed frequency current converter based on bang-bang and PSO receives the current error signal. The metal-oxide-semiconductor field-effect transistor (MOSFET) is triggered by the PWM pulses. A DC-DC converter with a larger output value than the voltage is the single-ended primary-inductor converter (SEPIC) DC-DC

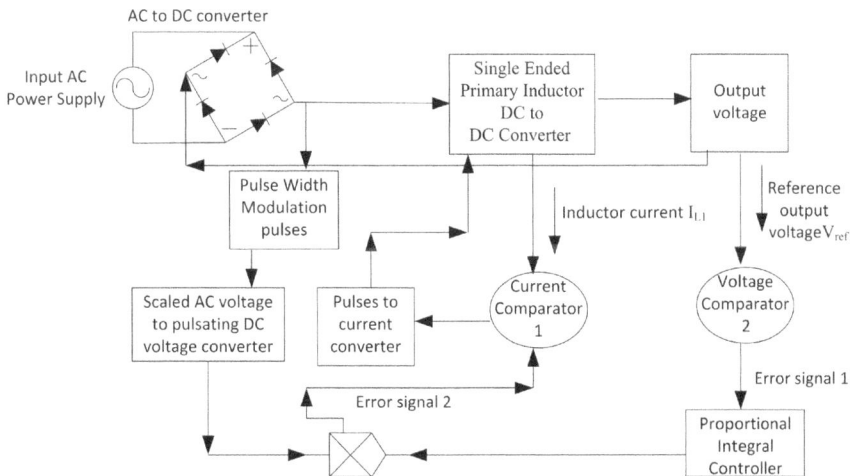

Figure 3.8 Power factor enhancements by use of a DC-DC SEPIC converter.

converters. In essence, although SEPIC DC-DC converters are not inherently engineered for the sole purpose of PFC, they can be integrated into power supply designs in conjunction with PFC techniques. This integration works to bolster power quality and efficiency within the system. The adaptability of SEPIC converters in managing fluctuating input voltages, preserving output voltage stability, and operating efficiently renders them suitable for scenarios where PFC is a consideration, particularly in settings characterized by variable input conditions and dynamic loads. Their utilization as part of power supply design should form part of a comprehensive approach to address power factor correction and voltage regulation concerns [15]. SEPIC DC-DC converters provide the following advantages:

1. They have a high-power density.
2. In comparison to other switching converters, they have a fast transient response.
3. They can execute both buck and boost operations, which is a major benefit.

3.17 POWER FACTOR IMPROVEMENT WITH HELP OF SEPIC DC-DC CONVERTER

Brushless direct current motors, PFC, and variable speed through a SEPIC are all explored. Proposed is an appropriate solution for driving applications, as well as a new approach for controlling the speed of the motor and PFC using SEPIC and regulating DC link values. The SEPIC conversion could be used in discontinuous conduction mode to perform reactive power compensation for variable speed across a wide range in power conversion circuits. The conventional converter proposed only examined DC-DC conversion because that was created to deliver a non-pulsating input current despite operating in continuous current mode between both inductors. The modified SEPIC converter is an AC-DC conversion technique that comprises a cascade of updated buck–boost converter and improved converters [16]. Figure 3.9 depicts the proposed system's outline.

The buck-boost converter employs continuous current while the boosted converter works with intermittent current. Because this method provides a significant

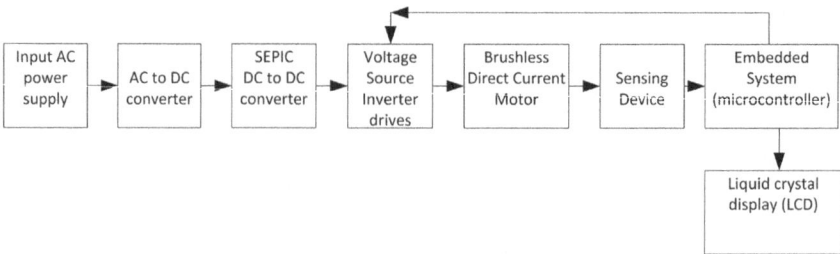

Figure 3.9 Block diagram of power factor correction by a SEPIC converter.

power factor, the output voltage must be adjusted using a simple feedback control mechanism. This proposed method aims to accomplish reactive power compensation without brushed motor drives using SEPIC converter configurations. Corrected power factor converters are essential for improving power quality (PQ). The voltage-source inverter (VSI) is predominantly employed for the conversion of a consistent DC voltage into an AC voltage characterized by adjustable magnitude and frequency. The diode bridge rectifier is removed to reduce conduction loss and component count. The approach of perturbation and observation can be used to obtain the highest power. The low pass filter can help to reduce ripples. A unique bidirectional bridgeless standalone SEPIC converter is employed to achieve a power factor close to unity.

3.18 THE BOOST POWER FACTOR CORRECTION CONVERTER

All connected equipment in an electricity distribution network frequently requires rectification. To eliminate ripples in the DC output, single-phase diode rectifiers with a high capacitor are employed, although this generates a nonlinear waveform of current. As a result, several power factor correction solutions are receiving a lot of attention. Among the different PFC topologies, boost topology is the most widely employed [17]. A boost converter could also help to both enhance the power factor and lower harmonic distortion. The rectifier circuit, depicted in Figure 3.10, receives the AC power supply and converts it to DC. The boost converter has two modes of operation: one in which it performs the function and the other in which it does not [18].

1. When the switch S1 is open, the inductor (L) de-energizes and the energy is transmitted to the capacitor (C) and load (RL).
2. In the reverse biased state, in which the inductor is powered by the AC supply via the rectifier, switch S2 is closed.

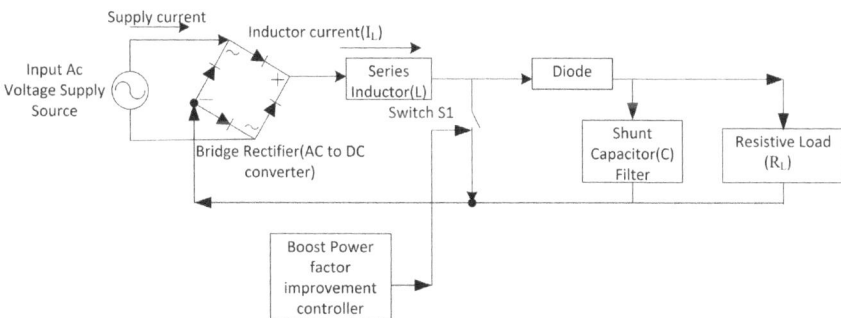

Figure 3.10 Boost power factor correction converters.

Because the inductor current falls in state 1 and rises in state 2, the inductor current falls in state 1 and rises in state 2. The duty cycle defines how often an inductor current decreases in comparison to how often it increases. The average inductor current is modified by adjusting the duty cycle, which results in an increase in power factor as well as reduction in total harmonic distortion (THD). A closed loop control maintains the output voltage while keeping the current in phase with it. The switch is connected to the boost PFC block, which generates a pulse width modulated signal that modulates the voltage.

3.19 PLC-BASED POWER FACTOR CORRECTION

A programmable logic controller (PLC) is an established manufacturing processor that controls the state of output devices by continuously monitoring the status of inputs and making decisions based on a programming [19]. After reading the power factor with the measuring unit, compare the present readings of the power factor displayed by the measuring unit to the planned power factor. The desired power factor is high enough to suit the needs of the user. There are then two scenarios to consider. The current reading and the target power factor are both the same in the first case. There is no need to make any changes to the system's current configuration. The two variables in the second scenario differ from one another. In this circumstance, power factor correction via system configuration modification is required. The next better pattern of inputs demands a change in setup, which adds more capacitance to the system circuit. If the next pattern meets the needed criteria, the circuit is configured using that pattern; otherwise, continue to the next better configuration pattern until the required criteria are met or the circuit has the maximum allowed capacitance. The status (ON or OFF) of the PLC's inputs, as well as the logic of the programming downloaded into it, determine the output signals [20].

When the motor is powered by a single-phase 230 V 50 Hz power supply, the total current drawn by an induction motor (IM) from the power supply results from the combination of two main components: the magnetizing current and the torque-producing component, therefore a current transformer is placed across one phase. This circuit uses a 1A/1mA current transformer, with the output going to a bridge rectifier, which converts the AC power to DC. The comparator receives DC values. The value of current recorded by the current transformer is converted to DC value. One of these input pairs is provided with 24 V and is decreased to specific values. The other input is provided through the output of the bridge rectifier, that is, the current value determined by using the current transformer. When the two values are the same, the respective output is turned on. The PLC's output is sent to relays, which then switch the capacitor bank on and off. The current transformer monitors the current of the load, and manual switching push buttons provide another input to the PLC. There are a total of eight inputs to the PLC, including four for manual operation and four for the CT. The output is then given to relays, which in turn switch the capacitors, according to the programming done in the PLC. The switched mode power supply (SMPS) is a device that supplies power to the PLC and relays. The PLC output is relay switching, which turns the condenser. If the power factor goes below a predetermined level, the motor will be turned off [21]. The AC supply is fed to the transformer, which is made up of three distinct rating transformers that act on the lagging power factor. Different rated transformers connected to the PLC's input port vary the power factor [22]. The relay coil becomes energized when the lever S is turned on, and SMPS sends 24 volts DC to the PLC. PLC also activates the capacitor bank, which provides leading current while compensating for reactive power. As a result, the power factor has improved.

3.20 IMPROVE POWER FACTOR THROUGH AN ON-DEMAND TAP CHANGER

Although vital for voltage regulation, the on-demand tap changer's voltage control philosophy is only suited for unidirectional load flow power systems. The traditional operation of the on-demand tap changer will be greatly impacted when distributed generators begin absorbing/injecting reactive power, resulting in high voltage values that exceed allowable standards, regular voltage magnitude variations, and an overall increased reactive power source [23]. The on-load tap changer (OLTC) is the most frequent and successful device for voltage regulation in power distribution networks. The OLTC's efficiency lies in its capability to manage the voltage of multiple feeds by keeping the secondary voltage of the power transformer constant as the load changes. As a result, the OLTC adjusts the transformer winding ratio to maintain a consistent transformer secondary voltage/bus-bar voltage as the network load changes.

After analyzing the voltage magnitude, the OLTC will modify its tap position to ensure that the voltage is comparable to the defined and fixed voltage reference value. Evaluating the voltage at a remote site is dependent on demand current and power factor. Because the technique still relies on a set OLTC reference voltage, it is unable to adjust to voltage problems posed by distributed generators.

3.21 POWER FACTOR IMPROVEMENT BY USING DYNAMIC VOLTAGE RESTORER (DVR)

Quality assurance is a pressing problem in present-day energy systems, with implications for both customers and producers. As a result of the amalgamation of renewable energy sources, micro-grid systems, and widespread use of power electronic devices, the present electric power system has encountered several challenges. Current-voltage harmonics, voltage sag, and swell can all damage sensitive electronics. Input voltage fluctuations in these devices can be caused by interference from other parts of the system. As a result, power quality is crucial for the power system's dependable and safe functioning in today's environment, where sensitive and expensive electronic equipment is becoming increasingly widespread [24]. A dynamic voltage restorer, the power quality device that injects both voltage and power into the system, can ameliorate voltage disturbances. DVRs commonly incorporate energy storage elements such as capacitors or batteries. When the DVR identifies a voltage sag, it taps into the stored energy within these components to rectify the situation. The DVR introduces supplementary voltage in series with the impaired load. It produces a compensatory voltage to counter the sag, guaranteeing that the load maintains the necessary voltage, even in the presence of a sag event. The precise quantity of voltage injection required for a DVR is contingent on various factors, encompassing the attributes of the voltage sag, the specifications of the load, and the governing algorithm of the DVR. The ability to correct the power factor at the source side's point of common coupling (PCC) uses a dynamic voltage restorer (DVR). Using the phase angle control (PAC) approach, the DVR compensating voltage will be injected in series with the transmission line with a specific phase angle and magnitude, resulting in a power factor angle shift of the resultant load voltage. As a consequence, the source voltage is always in phase with the source current under varied load situations, meaning that power factor correction is performed at the source side PCC. A laboratory prototype of the DVR is used to evaluate the suggested control algorithm. On the source side, a power factor of approximately unity may be maintained [25].

3.22   POWER FACTOR IMPROVEMENT BY LLC RESONANT CONVERTER

An AC and DC power converter assembly consists of a power factor correction module and an LLC resonant DC-DC converter module. This converter employs just two switches, three fewer diodes, and one fewer switch than the popular LLC resonant converter. The rectifier of interest offers a greater energy efficiency while being smaller than its traditional counterpart due to soft-switching in the LLC resonant converter. The single-stage double-switched (DS)-LLC rectifier has a wide output range and may attain unity power factor, indicating its efficacy and utility [26]. The following are the advantages of this topology over the typical architecture, which includes a pre-stage boost topology and an LLC resonant converter:

- A high-PF construction enables easier boost PFC.
- Because the PFC circuit stage shares a pair of switches with the LLC resonant converter stage, fewer power valves are required.
- High energy efficiency is obtained by using soft-switching in the LLC resonant stage.

3.23 LATEST TECHNOLOGY IN POWER FACTOR CORRECTION TECHNIQUE IS STATIC VAR GENERATORS

The sinexcel static VAR generator (SVG) is the most recent wall-mounted generation technology for power factor correction, and it offers more benefits than a capacitor bank solution. The SVG detects the load current, evaluates the reactance of the load, and injects the precise inverse reactive compensating current in real time. It can also maintain a power factor of 0.99 lagging or unity, and it can work at low voltages. The current transformer decreases the magnitude of current provided to the digital signal processing (DSP) and central processing unit (CPU), and the reactive power algorithm is used to distinguish between active and reactive power. It is delivered to the insulated-gate bipolar transistor (IGBT) control, which generates a 20 kHz frequency signal using pulse width modulation. Both an inductive (lagging) and capacitive (leading) load can be used with static VAR generators.

SVG's advantages are listed as follows.

*   Lowers a customer's electricity bill.
*   Customers obtain a return on their investment (ROI).
*   The SVG minimizes line load by lowering the electrical system's heat effect, resulting in longer life and lower maintenance costs.
*   SVG is a wall-mounting solution that frees up space on the floor.
*   The SVG minimizes maximum current by balancing the load across all three phases, lowering the maximum demand tariff (electricity costs) and the load on the circuit.
*   It reduces harmonic distortion.

3.24 REAL-TIME ENHANCEMENT OF REACTIVE POWER FOR AN INTELLIGENT SYSTEM USING A GENETIC ALGORITHM

This method of regulating the power factor in real time using multi-step capacitors is a novel process. Consequently, no optimization strategies were originally advocated for controlling this problem. The genetic algorithm (GA) is a meta-heuristic optimization approach for solving real-time optimization issues like the PF corrective problem. For real-time optimization of numerous power structure challenges, GA has shown promising outcomes. It's also suitable for issues involving binary choice variables. The power factor optimal solution uses the genetic method [27]. Genetic algorithms are built on the principles of natural selection and natural genetics. The natural genetics principles of replication, crossovers, and mutations are all used in the genetic search approach. The GA creates a randomly generated population of chromosomals, the size of which can be specified, before starting the process. After evaluating the fitness function for all created solutions as well as the model's constraints, the algorithm terminates and transmits the achieved solution to the controller by turning on/off the multi-step capacitors at appropriate stages. Otherwise, the preceding generation's parents are employed to generate a new population. Reproduction characterizes the selection of some of the chromosomes with the best fitness values. The crossover technique is determined by the encoding method, which takes a subset of chromosomes that are combined to generate a new pair of chromosomes. To avoid slipping below the local optimum, mutation is conducted after reproduction and crossover. This genetic algorithm provides a complete iteration in search of the optimum solution. The algorithm repeats these steps until it finds a solution that fulfills the stopping requirement. Ultimately, the controller turns on and off the capacitor bank and the detuned filter. The correction process is repeated after an interval to continuously evaluate and right the power factor [28].

3.25 POLYMER-BASED THERMOELECTRIC MATERIALS USED FOR IMPROVING POWER FACTOR

Because of their potential, thermoelectric (TE) materials are gaining prominence due to their ability to convert heat to electricity immediately. In order to increase global energy efficiency, they are utilized to capture electrical energy from waste heat. Polymer-based TE materials are mainly suitable for wearable and moveable devices because of their low density, great flexibility, and lower toxicity [28].

3.26 POWER FACTOR CORRECTION'S BENEFITS

Implementing power factor correction has a number of advantages. Among the benefits are lesser demand charges on the power system, which increase load-carrying capacity in existing circuits and lower overall power system losses. The advantages of power factor modification are not just financial; they also have substantial environmental advantages, which are described as follows.

1. **Decreased consumption charges**: Maximum electric utility suppliers charge for the highest metered demand dependent on the higher registered demand in kilowatts (KW meter) or a percentage of the higher registered demand in kilovolts (KVA) (KVA meter). The extent of the measured KVA will be significantly greater than the KW required if the power factor is inadequate. As a result, power factor correction reduces the demand charge, thereby lowering electricity costs.

2. **Existing circuits' harnessing capabilities**: Reactive current is required in case of loads that consume reactive power. When power factor correction capacitors are fixed near inductive loads at the end of existing circuits, current carried by each circuit is reduced. The upgraded power factor may

permit the circuit to carry new loads, saving dollars in unneeded upgrade expenses. Additionally, the decreased current flow lowers the circuit's resistive losses. Furthermore, power factor correction capacitors reduce line current while providing reactive current locally. When line current is lowered, transformers and feeder circuits utilize fewer kVA. Thermal overloads on transformers, transmission lines, generators, and cables are minimized as a result of the shunt capacitor correction.

3. **Improved voltage**: At a given load, larger current flow is due to a lower power factor. The voltage decreases when the line current increases in the wire, resulting in low voltage display on the device. a higher power factor reduces the voltage drop in the conductor, which advances the voltage at the equipment. Shunt capacitors minimize inductive current in an electric circuit, resulting in a better voltage profile. As the line current is lowered, the IR and IX voltage drops decrease and increase the voltage level of the system from the capacitor site to the source. Voltage of 0.95–1.05 PU must be maintained in the distribution supply system and transmission systems. Induction motors will run at a greater voltage. If the system voltage is lower, the nominal current will be lower. The recovery voltage after fault clearing will be sluggish at the low voltages. As a result, maintaining safe voltage stages in the power supply is a significant objective.

4. **Reduced power system losses**: The economic benefit of lower conductor losses is adequate to guarantee the fitting of capacitors in older plants or in field pumping processes. The conductor losses are directly related to induced current and are inversely proportional to the square of the power factor, as the current is reduced in proportion to the enhancement in power factor. When shunt capacitors are used for power factor correction, the line current is lowered. Both I^2R and I^2X losses are reduced as a consequence. Depending on the number of hours of full load operation, conductor size, feeder circuit length, and transformer impedance, I^2R losses in industrial power systems range from 3% to 8% of the nominal load current. The highest loads on current circuits vary with respect to time of the day.

5. **Lower carbon footprint**: By lowering the demand charge on the power system by power factor correction, the utility reduces the pressure on the electrical grid, lowering its carbon footprint. Hundreds of tons of carbon emissions can be saved over time by lowering demand on the energy grid, thanks to improvements in power system electrical efficiency via power factor adjustment.

6. **Reactive power support**: Due to a lack of reactive power in distribution systems, the voltage at the load end tends to drop. In such instances, shunt capacitors are used to provide local VAR support. During peak load conditions, long transmission lines have limited available reactive power at the end of the line, necessitating the use of shunt capacitors. The reactive power consumed from the supply is significantly reduced, as are the kVA and current flows. At the load, the power factor has improved.

7. **Generator capacity release**: The capacity of a synchronous generator is limited in both kW and kVA. The kVA limit of the generator may refer to operating with a unity power factor. The generator must produce a kW with a lower power factor when the load is a low power factor device. The output of the generator cannot cross the rated kVA. Adjusting the power factor at the utility side might unleash kVA capabilities [29].

3.27 SOME ENERGY STORAGE DEVICES

Thermostatically regulated home appliances may be leveraged to power system operators and consumers. Reiman et al. [30] used residential AC and electric heaters to increase the power factor in a distribution system, where solar installations due to the feeder power factor drop daily, resulting in higher losses. Transmission and bulk generating systems become more efficient as the distributed feeder power factor enhances. The daily minimum feeder power factor is maximized with a daily optimum dispatch plan. Electric heaters cool down at this period to prepare for a low feeder power factor incident, then restart and rectify the situation. A virtual batteries (VB) model utilizes the thermodynamic energy storage of thermostatic regulated equipment for varying power system demand. A VB charges by switching equipment on and off. The power and energy restrictions, as well as the self-discharging rate, are measured by the properties of the appliances. The energy restrictions are symmetrical about zero and directly relate to the number and voltage of units in use at any specified time. The range of the power constraints is measured by the quantity and heating capacity of the appliances at a moment, and the amount of power necessary to keep the VB energy at its neutral position defines the location of that span relative to zero. The capacity of a specific appliance to participate at a given moment is determined by the thermostat set point in comparison to the ambient or outside temperature. When it is chilly outdoors, for example, an air conditioner is not on or available to switch on. A heater is a device that warms and stores water for immediate consumption and is regulated by a thermostat. To heat water, a resistive element is usually utilized. The temperature set points and dead band of electric heaters are utilized to establish VB characteristics. The environmental temperature should never be higher than the device's thermostat set point. The water heater may not be able to participate in the VB if the water draw is too high. Electric heaters generally employ resistive heating elements with a power factor of one. Another energy-saving strategy is explained later.

To optimize voltage and energy savings in an alternating current induction motor operating under fixed torque and variable load conditions, T and L circuits were selected. The stator losses, which included copper and iron losses, were constant losses associated only with the stator voltage. Meanwhile, the rotor's copper losses were considered variable losses that varied with the load torque. The total electrical-loss formula and the optimum voltage regulation formula were evaluated for the motor. The calculations showed that the stator voltage and rotor load torque had an effect on the motor's overall electrical loss, and that the optimal voltage control changed with the load torque [31]. The overall electrical loss revealed almost no calculation error when the motor's operational voltage was more than 230 V. As the motor load torque increased, the ideal voltage regulation and its error climbed, but the optimum voltage regulation error did not exceed 6 V within the motor's operating voltage range. The overall electrical loss was lower with the optimum voltage regulation mode and a variable load torque compared with a 380 or 220 V constant drive mode. As the load rate increased, the overall electrical loss increased. When the motor's load did not exceed the medium load and the voltage control was excellent, the energy savings were considerable. While the motor was operating without any load, the overall electrical loss was negligible. In order to take advantage of solar energy's low cost, a number of process difficulties must be addressed. Triple intercell bar technology and current-source topology are two further options. To combat damaged or open contacts, the bar employs sidebars, as well as metallurgical short circuits to create short circuit resistance. Furthermore, because of the decreased dispersion of current densities in the technology, a 10% increase in sunlight energy is possible. The factory's energy usage, which produces 300,000 tons of copper cathodes per year, might be regulated as a result of the enhanced dispersion [32]. A targeted parameter model based on industry data is utilized to project the findings. Thanks to this technique, all broken connections are bypassed, and short circuits are avoided or eliminated. The lifetime of the electrodes is also increased, and energy efficiency is improved. A conservative estimate is of US $2.4 million in power savings and US $0.8 million in electrode costs each year. An additional $12.9 million is saved by installing photovoltaic modulated electricity.

3.28 WORKS IN CONNECTION

Though power factor correction is an old technique, several of the following authors have recommended and prescribed a variety of approaches. An adaptive compensation strategy is another way to control switching devices under fluctuating power requirements and uncertain power production, even while dealing with practical voltage settings. To illustrate such a technique, a major power aspect concept is first proposed and established under power transfer conditions, as well as imbalanced and damaged polarities. Second, a common method for generating amplification coefficients for compensation purposes is created, enabling standard commands to be immediately weighed in order to follow a desired grid electricity coefficient value. As a consequence, the technology can be simply implemented to energy converter regulators that will minimize distribution losses and enhance power quality [33].

Iturriaga et al. suggested a model for and control of a power factor corrected dual charge controller. This technique is connected to an electrical network through a conversion stage generated by a comprehensive rectifier, and the constant current demand is provided with a constant operating voltage via a higher efficiency. The key advantage of this design is that it provides good conversion ratios for the voltage level. The control scheme is defined as the average system model and primarily comprises two control loops: current monitoring and voltage profile [34].

Sayed and Massoud provide a review of recent articles on how PFC converters are proposed in electrical vehicles for extended distances. Electric vehicle (EV) charging points are categorized into three parts: main recharging terminals, such as Level 1 (gradual charger), Level 2 (quick charger), and Level 3 (super duper charger), fully electric system, and supplemental charging points. It highlights the latest development in on-power quality adjustment (PFC) converters. The global demand for upgrading green-based refueling infrastructure has sprung from the electrification of the transportation sector. Narrow EVs and long-distance EVs and attempts to promote optimal board charging systems have been the subject of worldwide study. Depending on the EV, the battery voltage ranges from 36 to 900 volts [35]. The battery charger is made of a single conversion or a two-level converter. This study is about only one unidirectional non-isolated converter for the on-power packs. The core PFC topologies and modified PFC topologies are divided into two categories.

Minh and Minh provided an experimental method that reveals that when PFC assistance is used, the output voltage response is more stable than when the simple MPPT algorithm is used. Due to the instability of solar energy (PV), the large penetration of PV in the power grid now poses numerous issues. To address this issue, this research examines the PFC management technique for supporting converters that are integrated with the MPPT algorithm using the existing power grid. After that, an experimental circuit model incorporating two control approaches will be built and tested under real-world settings [36].

Ali et al. presented a solution three-phase boosted compensator (PFC) rectification device that was modeled and controlled, as well as a down-for-maintenance parameterization approach for successfully improving the control performance. While analyzing the shortcomings of the traditional control scheme with complicated control nonlinear systems from smaller scales and conducting an investigation of the constraints of control scheme on harmonic elimination from the model predictive concept, a simplification type of control-governing equations based on a small prototype has been suggested with a fragmented control method. As an outcome, the methodology takes advantage of both simpler control methods and fragmented calculator control to achieve a high power factor and decrease overall harmonic distortion in a variety of categories [37].

Ricardo et al. developed a unique and efficient method of action that allows the utility to modify the system load factor in live time and with preventing financial damage. Decentralized energy PV networks (DGCPSs) have proven to be an effective replacement for traditional energy production by minimizing leakage currents, simplifying transmission networks, and offering extra services to energy systems [38]. Considering these positives, there has been a different sort of DGCPS challenge documented in the literature. From the utility's standpoint, the implementation of a DGCPS may appear to exacerbate the system load factor. This impact is not caused by physical factors such as equipment failure or the load's excessive usage of reactive power, yet it may result in reactive power overcharging (RPEC). Because these fees may increase the time it takes for PV installations to pay for themselves, this study examines three options for dealing with the problem. A revised regulatory structure for power system regulation in DGCPS consumption devices is the first methodology, which has shown to be the most beneficial to users. Traditional solutions that use photovoltaic converters or capacitors to mitigate for reactive power are the other two alternatives.

Ali and Khan, using a unit pattern methodology or a proportional–integral controlled system approach, proposed power factor expansion, zero voltage regulation, and overall transmission loss reduction in power transmission network for load variations. Regulation and compensation are becoming increasingly significant components in maintaining power quality. It is preferable to suggest a compensator to link with power control center (PCC) by employing a shunt compensation through an interface. A DC bus capacitor based on IGBT is used in the suggested compensator with voltage source converters (VSC). In terms of DSTATCOM's performance, it is recommended for load variation with a voltage supply at the point of common coupling and an identity DC connector [39].

Ananda provided a method for increasing the power factor in electrical substations. The substantial inductive load leads to a power factor below 1, causing a significant increase in electrical current. Consequently, the installed power cannot be utilized to its full potential efficiency. The interface is designed to use human machine interface display and a cloud services logger to screen the power factor value in real time [40].

3.29 CONCLUSION

When all components of the power aspects are considered, it is observed that the most essential component for both the energy supplier and the user is the power factor. The utility company is no longer liable for power losses, and the consumer is no longer liable for reduced power factor penalty payments. Inserting correctly sized power capacitors into a plant's circuit can enhance its output power. Because most electrical demands are inductive and require trailing currents, the power factor is weak. Connecting certain devices that consume significant power in parallel to the load enhances power quality. The capacitor is an example of such a device. The capacitor pulls a higher current that partially or totally neutralizes the load current's subsequent reactive part. As a consequence, the load's power factor changes and the power factor of the load improves. Finally, the following things will be discussed.

- To overcome today's low power factor, a less expensive and more efficient method or technology is necessary.

- By connecting capacitors in parallel, some of the solutions in this study affect the power factor. However, SVG is the current generation of power factor correction technology, and it provides a number of advantages over the classic switched capacitor bank style system.
- A microcontroller and a programmable logic controller (PLC) are also devices that are used to control the power factor automatically.
- Power factor adjustment procedures can presumably be used in the power system, yielding stability and reliability of the system.
- Using a microcontroller cuts costs and eliminates the need for additional hardware. As the usage of power electronics grows in popularity; the problem of harmonics arises. Harmonic distortion lowers both power quality and power factor, necessitating the use of power factor methods. As a result, the ideal solution for power factor correction is automatic power factor.

REFERENCES

[1] R. Bayindir, S. Sagiroglu, and I. Colak, "An Intelligent Power Factor Corrector for Power System Using Artificial Neural Networks," vol. 79, pp. 152–160, 2009, doi: 10.1016/j.epsr.2008.05.009.
[2] H. K. Channi, "Overview of Power Factor Improvement Techniques," *Int. J. Res. Eng. Appl. Sci.*, vol. 7, no. 5, pp. 27–36, 2017, [Online]. Available: http://euroasiapub.org/.
[3] S. Kuchibhatla, V. V. Mani, S. Likhitha, S. Vinay, and A. Narendar, "Compensation of Reactive Power and Energy Saving Using Capacitor Banks," *Int. Res. J. Eng. Technol.*, vol. 09, no. 01, pp. 630–633, 2022.
[4] H. Shah, N. Trivedi, and M. Rahish, "Design and Control of Reduction of Harmonics in Power Factor Correction (PFC) Technique," *GIT-J. Eng. Technol.*, vol. 13, pp. 38–44, 2021.
[5] R. Chavhan and D. Champa, "Power Factor Correction (PFC) Preregulator Using Boost Converter for 2 kVA Power Supply," *Int. J. Eng. Appl. Sci. Technol.*, vol. 5, no. 3, pp. 439–445, 2020, doi: 10.33564/ijeast.2020.v05i03.070.
[6] M. Haroon Nadeem, Sohaib Tahir, Mazhar H. Baloch, Ghulam S. Kaloi, Waqas A. Wattoo, M. Yousif and Mehr Gul, "Power Factor Improvements and its Effective Strategy to Optimize the kWh," *Int. J. Comput. Sci. Inf. Secur.*, vol. 15, no. 4, pp. 188–192, 2017.
[7] M. V. Aware, A. G. Kothari, and S. S. Bhat, "Power Factor Improvement Using Active Filter for Unbalanced Three-Phase Non-Linear Loads," *Int. J. Energ. Technol. Policy*, vol. 4, no. 1–2, pp. 103–117, 2006, doi: 10.1504/ijetp.2006.008551.
[8] R. da Silva Benedito, R. Zilles, and J. T. Pinho, "Overcoming the Power Factor Apparent Degradation of Loads Fed by Photovoltaic Distributed Generators," *Renew. Energ.*, vol. 164, pp. 1364–1375, Feb. 2021, doi: 10.1016/j.renene.2020.10.146.
[9] S. Kim, G. Yoo, and J. Song, "A bifunctional utility connected photovoltaic system with power factor correction and UPS facility." *Conference Record of the Twenty Fifth IEEE Photovoltaic Specialists Conference-1996.* IEEE, 1996. doi: 10.1109/PVSC.1996.564386.

[10] Amit G. Shende, S. W. Khubalkar, and P. Vaidya, "Hardware Implementation of Automatic Power Factor Correction Unit for Industry," *J. Phys. Conf. Ser.*, vol, 2089, no. 1, pp. 1–6, 2021.

[11] E. Engineering, and A. Pradesh, "A Review on Improvement of Energy Efficiency in Residential Buildings," *Int. J. Rec. Dev. Sci. Technol.*, vol. 6, no. 1, pp.127–139, 2022.

[12] K. Yasin, Y. Mohammad Mohsin, and M. Monirujjaman Khan, "Automated power factor correction and energy monitoring system." *Second International Conference on Electrical, Computer and Communication Technologies (ICECCT)*. IEEE, 2017, doi: 10.1109/ICECCT.2017.8117969.

[13] M. Raju Ahmed, M. Humayan Kabir Khan, and A. Kumar Karmaker, "Microcontroller Based Power Factor Improvement by Using Switched Single Capacitor Microcontroller-Based Power Factor Improvement Using Switched Single Capacitor," *DUET J.*, vol. 21, no. 1, pp. 21–28, 2018.

[14] T. W. Ngwe, S. Winn, and S. M. Myint, "Design and Control of Automatic Power Factor Correction APFC for Power Factor Improvement in Oakshippin Primary Substation," *Int. J. Trend Sci. Res. Dev.*, vol. 2, no. 5, pp. 2368–2372, 2018, doi: 10.31142/ijtsrd18320.

[15] S. Durgadevi and M. G. Umamaheswari, "Analysis and Design of Single Phase Power Factor Correction using DC-DC SEPIC Converter with Bang-Bang and PSO Based Fixed PWM Techniques," *Energy Procedia*, vol. 117, pp. 79–86, 2017, doi: 10.1016/j.egypro.2017.05.109.

[16] K. Balamurugan, M. Gowsika, T. Monika, and N. Naveen, "Power Factor Correction Using Sepic Dc-Dc Converter in Industrial Motor Drives," *Int. J. Res. Eng. Sci.*, vol. 9, no. 5, pp. 21–28, 2021.

[17] P. Kulshreshtha, S. Gairola, and A. Verma, "Power Factor Correction by Interleaved Boost Converter Using PI Controller," *Int. J. Innov. Res. Creat. Techn.*, vol. 1, no. 5, pp. 481–485, 2016.

[18] S. Busquets-Monge et al., "Design Optimization of a Boost Power Factor Correction Converter Using Genetic Algorithms," *Conf. Proc.—IEEE Appl. Power Electron. Conf. Expo.—APEC*, vol. 2, pp. 1177–1182, 2002, doi: 10.1109/apec.2002.989393.

[19] Rishabh Jain, Shashank Sharma, Mini Sreejeth, Madhusudan Singh, "PLC based power factor correction of 3-phase Induction Motor." *IEEE 1st International Conference on Power Electronics, Intelligent Control and Energy Systems (ICPEICES)*. IEEE, 2016, doi: 10.1109/ICPEICES.2016.7853637.

[20] C. Gaurav, P. R. Patil, and A. R. Singh, "Power Factor Correction of Three Phase Induction Motor Using Switched Capacitor Banks with PLC," *Int. J. Res. Electri. Electron. Eng.*, vol. 2, no. 2, pp. 45–58, 2014.

[21] A. Usama, D. Maksimović, and K. K. Afridi, "A simple control architecture for four-switch buck-boost converter based power factor correction rectifier." *IEEE 18th Workshop on Control and Modeling for Power Electronics (COMPEL)*. IEEE, 2017, doi: 10.1109/COMPEL.2017.8013343.

[22] "PLC Based Power Factor Correction and," pp. 3714–3721, 2016, doi: 10.15680/IJIRSET.2016.0503077.

[23] N. Tshivhase, A. N. Hasan, and T. Shongwe, "A Fault Level-Based System to Control Voltage and Enhance Power Factor through an On-Load Tap Changer and Distributed Generators," *IEEE Access*, vol. 9, pp. 34023–34039, 2021, doi: 10.1109/ACCESS.2021.3061622.

[24] N. Abas, S. Dilshad, A. Khalid, M. S. Saleem, and N. Khan, "Power Quality Improvement Using Dynamic Voltage Restorer," *IEEE Access*, vol. 8, pp. 164325–164339, 2020, doi: 10.1109/ACCESS.2020.3022477.

[25] J. Ye and H. B. Gooi, "Phase Angle Control Based Three-phase DVR with Power Factor Correction at Point of Common Coupling," *J. Mod. Power Syst. Clean Energy*, vol. 8, no. 1, pp. 179–186, 2020, doi: 10.35833/MPCE.2018.000428.

[26] G. Zhang et al., "Control Design and Performance Analysis of a Double-Switched LLC Resonant Rectifier for Unity Power Factor and Soft-Switching," *IEEE Access*, vol. 8, pp. 44511–44521, 2020, doi: 10.1109/ACCESS.2020.2978030.

[27] S. Abdelhady, A. Osama, A. Shaban, and M. Elbayoumi, "A Real-Time Optimization of Reactive Power for An Intelligent System Using Genetic Algorithm," *IEEE Access*, vol. 8, pp. 11991–12000, 2020, doi: 10.1109/ACCESS.2020.2965321.

[28] W. Khamsen, A. Aurasopon, and W. Sa-Ngiamvibool, "Power Factor Improvement and Voltage Harmonics Reduction in Pulse Width Modulation AC Chopper Using Bee Colony Optimization," *IETE Tech. Rev. (Institution Electron. Telecommun. Eng. India)*, vol. 30, no. 3, pp. 173–182, 2013, doi: 10.4103/0256-4602.113478.

[29] A. A. Mon and S. W. Naing, "Power Factor Improvement for Industrial Load by Using Shunt Capacitor Bank," *Int. J. Sci. Eng. Technol. Res.*, vol. 3, no. 15, pp. 3191–3195, 2014.

[30] A. P. Reiman, A. Somani, M. J. E. Alam, P. Wang, D. Wu, and K. Kalsi, "Power Factor Correction in Feeders With Distributed Photovoltaics Using Residential Appliances as Virtual Batteries," *IEEE Access*, vol. 7, pp. 99115–99122, 2019, doi: 10.1109/ACCESS.2019.2928568.

[31] Z. Bin, M. Lili, and D. Hao, "Principle of Optimal Voltage Regulation and Energy-Saving for Induction Motor With Unknown Constant-Torque Working Condition," *IEEE Access*, vol. 8, pp. 187307–187316, 2020, doi: 10.1109/ACCESS.2020.3030936.

[32] E. P. Wiechmann, J. E. Díaz, A. S. Morales, and P. E. Aqueveque, "Advanced Technology to Increase the Use of Photovoltaic Energy in Copper Electrowinning," *IEEE Trans. Ind. Appl.*, vol. 56, no. 2, pp. 2117–2121, 2020, doi: 10.1109/TIA.2020.2966984.

[33] J. P. Bonaldo, V. A. De Souza, A. M. D. S. Alonso, L. D. O. Arenas, F. P. Marafao, and H. K. M. Paredes, "Adaptive Power Factor Regulation Under Asymmetrical and Non-Sinusoidal Grid Condition With Distributed Energy Resource," *IEEE Access*, vol. 9, pp. 140487–140503, 2021, doi: 10.1109/ACCESS.2021.3119335.

[34] S. Iturriaga, et al. "A Control Strategy for a Power Factor Compensator Based on Double-Inductor Boost Converter," *IEEE Int. Autumn Meet Power. Electron. Comput.*, vol. 5, 2021.

[35] S. Sawsan, and A. M. Massoud, "Review on State-Of-The-Art Unidirectional Non-Isolated Power Factor Correction Converters for Short-/Long-Distance Electric Vehicles," *IEEE Access*, vol. 10, pp. 11308–11340, 2022.

[36] Q. D. Minh. "Enhancing Voltage Stability of Photovoltaic Energy Converter Using Power Factor Correction Technique," *J. Mil. Sci. Technol.*, vol. 74, pp. 3–9, 2021.

[37] M. S. Ali, L. Wang, and G. Chen. "Design and Control Aspect of Segmented Proportional Integral-Repetitive Controller Parameter Optimization of the Three-Phase Boost Power Factor Correction Rectifier," *Int. J. Circuit Theory Appl.*, vol. 49, no. 3, pp. 554–575, 2021.

[38] R. da Silva Benedito, R. Zilles, and J. T. Pinho. "Overcoming the Power Factor Apparent Degradation of Loads Fed by Photovoltaic Distributed Generators," *Renew Energ.*, vol. 164, pp. 1364–1375, 2021.

[39] S. Ali, M. Jamil, and M. A. Khan. "Power Quality Analysis and Enhancement Using DSTATCOM for Three-Phase Variable Load." *Advances in Energy Technology.* Springer, Singapore, 2022, 583–596.

[40] G. F. Ananda. "Automatic Capacitor Switching Method for Power Factor Improvement With HMI Interface and Cloud Data Logger." *Proceedings of the International e-Conference on Intelligent Systems and Signal Processing.* Springer, Singapore, 2022.

Chapter 4

Parametric approach of magnetorheological external finishing tool for its better functionality

Ajay Singh Rana, Talwinder Singh Bedi and Hema Gurung

4.1 INTRODUCTION AND BACKGROUND

The ability to alter properties in an organized way of magnetization, current, force and deformation has developed such smart materials as magnetorheological (MR) fluids (Rana et al., 2023). It was initially discovered by Rabinow (Rabinow, 1948). MR materials can be classified in three categories, namely fluids, foams and elastomers (Carlson and Jolly, 2000). Magnetorheological fluids are mostly used in various types of industries, such as automotive, aerospace and machine tooling (Jolly et al., 1998). Their advantages include noiseless operation and ease of control. In present-day manufacturing, magnetorheological fluid is helpful for obtaining better surface finishing on various engineering components, which improves their operative functionality (Bedi and Singh, 2016; Bedi and Rana, 2021). During magnetorheological finishing (MRF), magnetic field–dependent properties such as viscosity, shear rate and yield stress of MR fluid can be controlled by changing the magnetizing current in the electromagnet coils (Jha and Jain, 2004). The MR polishing fluid consists of carbonyl iron particles (CIPs), abrasive particles dispersed in the visco-elastic base medium (mixer of mineral oil and grease), which shows alteration in its rheological properties (Khurana et al., 2017). The behaviour of MR polishing fluid without/with the help of magnetic gradient is shown in Figure 4.1. When no magnetic gradient is applied to the MR polishing fluid, the particles within the fluid are randomly distributed, as shown in Figure 4.1a. Under the magnetic gradient, the CIPs attain a magnetic-dipole moment relative to magnetic field strength and form a chain-like structure in accordance with the magnetic lines of fluxes (Paswan et al., 2017). These chain-like structures of CIPs embed the non-magnetic abrasive particles within or between the fluid's structure, as shown in Figure 4.1b.

Hence, these embedded abrasive particles become active in nature, further rolling over the surface to be finished. Also, this activity helps to control sedimentation of particles with the influence of magnetic field (Bica, 2002). Under the effect of shearing, the indentation force induced by the abrasive particles is very low during magnetorheological finishing (Kordonski and Jacobs, 1996; Shorey et al., 2001). With the change in magnetizing current, the viscosity of this smart fluid

DOI: 10.1201/9781032703046-4

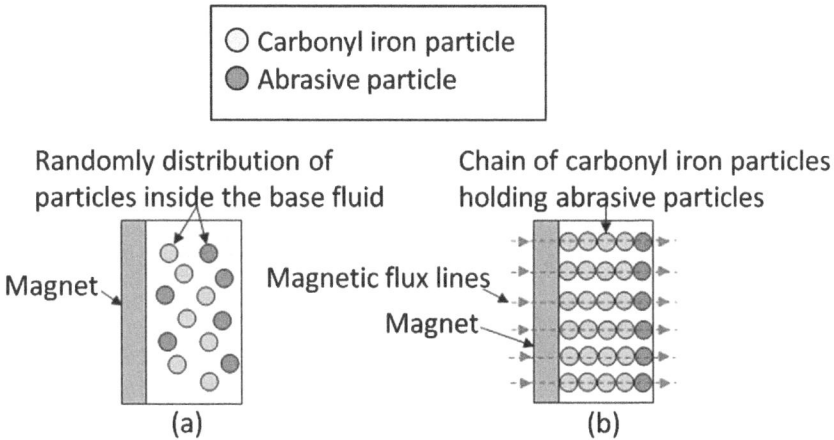

Figure 4.1 Behaviour of MR polishing fluid (a) under no magnetic gradient and (b) with magnetic gradient.

also changes (Sidpara and Jain, 2013). Rana et al. (2020) worked on the magneto-rheological finishing process using a three set of permanent magnets. The results concluded that the maximum decrease in Ra value was 62 nm within 40 minutes during finishing of the aluminium workpiece. Bedi and Kant (2021) developed a magnetorheological finishing setup with different magnetic structures. The authors concluded that 62.74% roughness change was found with curved magnets, whereas 53.81% roughness change was observed with flat magnets during the finishing of a stainless steel workpiece. Rana et al. (2021) performed a fine finishing on a copper cylindrical workpiece using a permanent magnet-based magnetorheological finishing. The results revealed that the final Ra value was achieved as 67 nm from 224 nm after 45 minutes of finishing.

In this chapter, a parametric study with a statistical design of experimentation was conducted for the nanofinishing of the external cylindrical surface of a brass workpiece. With the help of Design Expert 11 Software, response surface methodology was performed to evaluate the effect of different process variables, i.e. workpiece rotational speed, tool linear speed, mesh sizes of SiC and mesh sizes of electrolytic iron particles (EIPs) against the percent change in surface roughness (Ra) value.

4.2 MAGNETOSTATIC SIMULATION

To determine the magnetic flux distribution within the finishing operation, it was important to simulate its results before the experimentation (using Ansoft V13 software). Figure 4.2a represents the simulation results for the brass workpiece when it was in contact with the magnetized magnetorheological polishing fluid. The gap between magnet and workpiece surface was taken as 1 mm. From the

(a)

(b)

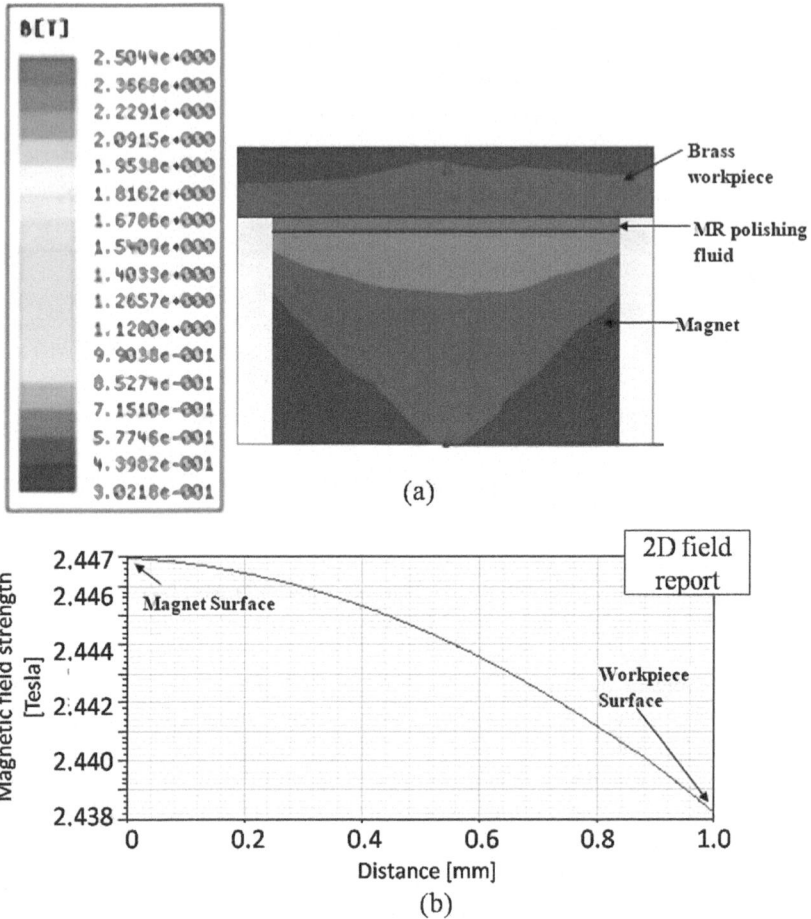

Figure 4.2 (a) Magnetostatic simulation and (b) 2D field report for analysing the allocation of magnetic flux.

simulation results, it was revealed that the higher magnetic flux was found on the magnet surface and decreased to the workpiece surface as shown in Figure 4.2b.

This concluded that magnetorheological polishing would remain attached to the magnet surface instead of the workpiece surface, which is beneficial for obtaining better finishing results.

4.3 EXPERIMENTAL SETUP AND PROCESS VARIABLES

The magnetorheological finishing process, as depicted in Figure 4.3a (Rana et al., 2020), has already demonstrated its ability to finish non-ferromagnetic workpieces, contributing to their improved durability. This setup was installed on the horizontal slides of the lathe machine where the tool post was mounted. The tool post assembly of the lathe machine was replaced by the magnetorheological finishing tool assembly. The three-dimensional (3D) view of a setup along with brass cylindrical workpiece is shown in Figure 4.3b.

During finishing, the magnetorheological polishing is magnetized within the working gap, which further encounters the workpiece surface during the finishing process. For the present experimentations, the different process variables used in this study are workpiece rotational speed (A), tool linear speed (B), mesh size of silicon carbide particles (C) and mesh size of electrolyte iron powders (D). The symbolic representation of A, B, C and D was given to each variable for easy identification. The aforementioned variables were independent controlled variables, whereas the ambient temperature was taken as a dependent controlled variable in

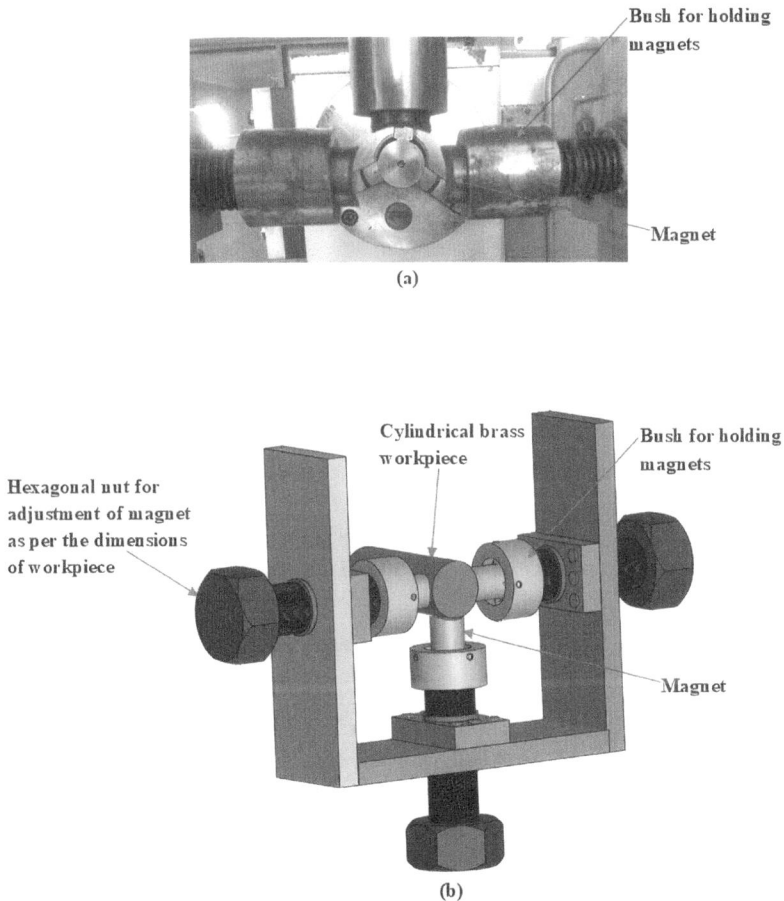

(a)

(b)

Figure 4.3 (a) Actual photograph of tool and (b) 3D drawing of tool along with brass cylindrical workpiece.

the present research. The following are brief descriptions of each of the process variables used in the research.

4.3.1 Workpiece rotational speed (A)

Workpiece rotational speed plays a vital role in the magnetorheological finishing process for achieving a fine surface finishing. When the magnetorheological polishing is magnetized and sticks to the tool surface, then due to the workpiece rotation, the polishing fluid moves as per the rotation of the workpiece. This causes a material removal from the workpiece surface in the form of tiny chips or microchips. At present, the different workpiece rotations, i.e. 400 rpm, 600 rpm and 800 rpm, were used for the experimentation as per the preliminary results.

4.3.2 Tool linear speed (B)

Tool linear speed represents the horizontal movement of the tool with respect to the workpiece surface. The shearing of roughness peaks was only possible when the workpiece rotates and reciprocates. The reciprocation movement of the tool was also known as tool linear motion. This movement was given with the help of the x-axis slide of a lathe machine. At present, the different tool linear speeds, i.e. 10 cm/min, 30 cm/min and 50 cm/min, were used for the experimentation as per the preliminary results.

4.3.3 Mesh sizes of SiC (C)

Abrasive particles used in the present research were silicon carbide of different mesh sizes, i.e. 600, 800 and 1000. The coarser mesh size was 600 whereas finer mesh size was 1000 and are easily available. Scanning electron microscopy images of different mesh sizes of SiC are shown in Figure 4.4a. A coarse range of mesh size (i.e. 600) was suitable for cutting the higher roughness peaks, whereas a finer range of mesh size (i.e. 1000) was suitable for eliminating the waviness roughness peaks. Depending upon the height of roughness peaks on the workpiece surfaces, the abrasive particles perform their function of achieving a smooth surface quality.

4.3.4 Mesh sizes of electrolyte iron powders (D)

In this research, different mesh sizes of electrolyte iron powder, i.e. 200, 300 and 400, were utilizedfor experimentation. The scanning electron microscopy images of different mesh sizes of electrolyte iron powders are shown in Figure 4.4b. When the magnetorheological polishing fluid is magnetized, the iron particles form a rigid structure by holding the abrasive particles on its outer circumference. If there are no iron particles within the MR polishing fluid, then abrasive particles will be randomly circulated inside the base fluid, and no finishing will be performed. The indentation capacity of abrasive particles onto the workpiece surface wholly depends upon the rigid structure of iron particles.

600 mesh size   800 mesh size   1000 mesh size
(a)

200 mesh size   300 mesh size   400 mesh size
(b)

Figure 4.4 Scanning electron microscopy images of different mesh sizes of (a) SiC and (b) EIPs at 500X.

4.4 DESIGN OF EXPERIMENTS

An optimum variable plays an important task in increasing the operative functionality of machined components, which can only be possible by performing a number of experiments according to their sequence. In this chapter, the effect of different process variables such as workpiece rotational speed (A), tool linear speed (B), mesh size of silicon carbide particles (C) and mesh size of EIPs (D) with respect to change in surface roughness Ra value for brass cylindrical workpiece were analysed. The workpiece used in this research found its wide applications in the utensil making and ornament making industries. The nanofinishing of such products was required for their better appearance and performance. The initial surface roughness Ra value of a brass workpiece was found to be between 325 nm and 390 nm, which was measured with Surftest SJ-210 (Mitutoyo) with 0.25 mm as the cut-off length. The difference in Ra value was because of initial grinding of the cylindrical brass workpiece by a surface grinder. In order to counteract the huge difference in Ra value, the percentage change in Ra value was taken as per Equation 4.1.

$$\% \text{ change in Ra value} = \left(\text{Ra}_1 - \text{Ra}_2 \right) / \text{Ra}_1 \qquad (4.1)$$

Where:
Ra$_1$ was the initial Ra value of the cylindrical workpiece
Ra$_2$ was the initial Ra value of the final finished cylindrical workpiece

With the help of Design Expert 11 Software, the response surface methodology (RSM) technique was used, which represents the addition of statistical and numerical data values. This further helps to analyse the engineering models. For creating a model, four variables along with three levels of six axial and central runs were employed for the experimentation. The central composite design in this model was used, and further analysis of variance (ANOVA) with F-test was done to evaluate the regression solution. From this solution, the regression equation was evaluated, which helps to correlate the relation between controlled variables and final surface roughness values. The actual process variables along with their respective levels are depicted in Table 4.1.

For experimentation, the magnetorheological polishing fluid was prepared with the following volumes:

23% of SiC abrasive particles, 17% of carbonyl iron particles and 60% of base fluid, out of which 80% paraffin oil + 20% grease by weight were taken. Other variables such as the working gap were taken as 1 mm, whereas total time for finishing each workpiece was 40 min. Before starting the experimentation, the different sets of experimentation were performed as per given in Table 4.2, which further represents its standard as well as run order.

4.4.1 Regression analysis

Each set of experimentation was conducted, and its corresponding output was calculated in terms of percentage change in Ra value as per Equation 4.1. The response summary for each set of experimentation is shown in Table 4.3.

The consecutive sum of squares was analysed to evaluate the higher polynomial order, shown in Table 4.4.

In this model, the extra values were regarded as significant and the model was considered to be aliased. The consecutive sum of squares determined the increase in complications contributing to the entire model. With this quadratic model, the analysis of variance was evaluated where the value of α was taken as 0.05 for the shown hypothesis. In this p-value, less than or equal to 0.05 is taken as a significant value, whereas if the value is higher than 0.05 it can be eliminated. The full-length ANOVA analysis for the given set of experimentation is shown in Table 4.5.

Table 4.1 Different process variables along with their coded levels

Process variables	Units	Coded levels		
		-1	0	1
Workpiece rotational speeds (A)	rpm	400	600	800
Tool linear speeds (B)	cm/min	10	30	50
SiC abrasives (C)	mesh size	600	800	1000
EIPs (D)	mesh size	200	300	400

Table 4.2 Different sets of experimentation

Std	Run	A:Workpiece rotationalspeed (rpm)	B:Tool linear speed (cm/min)	C:SiC (mesh size)	D:EIPs (mesh size)
20	1	600	50	800	300
25	2	600	30	800	300
1	3	400	10	600	200
28	4	600	30	800	300
5	5	400	10	1000	200
7	6	400	50	1000	200
8	7	800	50	1000	200
9	8	400	10	600	400
4	9	800	50	600	200
19	10	600	10	800	300
18	11	800	30	800	300
21	12	600	30	600	300
16	13	800	50	1000	400
24	14	600	30	800	400
15	15	400	50	1000	400
23	16	600	30	800	200
11	17	400	50	600	400
13	18	400	10	1000	400
2	19	800	10	600	200
30	20	600	30	800	300
10	21	800	10	600	400
26	22	600	30	800	300
12	23	800	50	600	400
27	24	600	30	800	300
17	25	400	30	800	300
3	26	400	50	600	200
6	27	800	10	1000	200
22	28	600	30	1000	300
29	29	600	30	800	300
14	30	800	10	1000	400

The F-value was calculated as 28.04, which clearly showed that the model was significant in nature, i.e. p-value is far less than α value. From Table 4.5, terms such as A, B, C, D, BC, BD, A^2, B^2, C^2 and D^2 were considered significant because their p-values were less than 0.05 (α value). Terms such as AB, AC, AD and CD were considered non-significant because their p-values were greater than 0.05 (α value). In case of lack of fit, the F-value was found to be 4.23, which clearly

Table 4.3 Response summary

Std	Run	A:Workpiece rotational speed (rpm)	B:Tool linear speed (cm/min)	C:SiC (mesh size)	D:EIPs (mesh size)	Ra_1 (nm)	Ra_2 (nm)	Final output (%)
20	1	600	50	800	300	367	121	67
25	2	600	30	800	300	371	186	50
1	3	400	10	600	200	330	145	56
28	4	600	30	800	300	341	177	48
5	5	400	10	1000	200	354	191	46
7	6	400	50	1000	200	370	155	58
8	7	800	50	1000	200	387	185	52
9	8	400	10	600	400	332	169	49
4	9	800	50	600	200	330	191	42
19	10	600	10	800	300	361	97	73
18	11	800	30	800	300	325	230	31
21	12	600	30	600	300	368	236	36
16	13	800	50	1000	400	337	192	43
24	14	600	30	800	400	356	221	38
15	15	400	50	1000	400	361	173	52
23	16	600	30	800	200	326	166	49
11	17	400	50	600	400	390	293	25
13	18	400	10	1000	400	379	243	36
2	19	800	10	600	200	380	190	50
30	20	600	30	800	300	387	201	48
10	21	800	10	600	400	343	185	46
26	22	600	30	800	300	351	179	49
12	23	800	50	600	400	365	280	24
27	24	600	30	800	300	340	177	48
17	25	400	30	800	300	359	208	42
3	26	400	50	600	200	341	208	39
6	27	800	10	1000	200	389	253	35
22	28	600	30	1000	300	377	211	44
29	29	600	30	800	300	350	168	52
14	30	800	10	1000	400	367	231	37

revealed that there was a 6.26% chance that F-value in case of lack of fit occurs because of environmental effects such as noise, vibration and so on. Other than ANOVA, the other calculated values are shown in Table 4.6.

From this, the value of R^2 was found to be 0.9632, which clearly determined the intimacy of experimental data with regression analysis. Hence, the model was

Table 4.4 Consecutive sum of squares

Source	Sum of squares	df	Mean square	F-value	p-value	
Mean vs Total	62,107.50	1	62,107.50			
Linear vs Mean	541.67	4	135.42	1.23	0.3237	
2FI vs Linear	1030.75	6	171.79	1.89	0.1342	
Quadratic vs 2FI	**1601.81**	**4**	**400.45**	**49.53**	**< 0.0001**	**Suggested**
Cubic vs Quadratic	104.33	8	13.04	5.39	0.0194	Aliased
Residual	16.94	7	2.42			
Total	65,403.00	30	2180.10			

Table 4.5 ANOVA for final response

Source	Sum of squares	df	Mean square	F-value	p-value	
Model	3174.22	14	226.73	28.04	< 0.0001	significant
A-Workpiece rotational speed	102.72	1	102.72	12.71	0.0028	
B-Tool linear speed	37.56	1	37.56	4.65	0.0478	
C-SiC	72.00	1	72.00	8.91	0.0093	
D-EIPs	329.39	1	329.39	40.74	< 0.0001	
AB	2.25	1	2.25	0.2783	0.6055	
AC	20.25	1	20.25	2.50	0.1344	
AD	4.00	1	4.00	0.4947	0.4926	
BC	930.25	1	930.25	115.06	< 0.0001	
BD	49.00	1	49.00	6.06	0.0264	
CD	25.00	1	25.00	3.09	0.0991	
A^2	404.26	1	404.26	50.00	< 0.0001	
B^2	1143.55	1	1143.55	141.44	< 0.0001	
C^2	209.45	1	209.45	25.91	0.0001	
D^2	78.13	1	78.13	9.66	0.0072	
Residual	121.28	15	8.09			
Lack of fit	108.44	10	10.84	4.23	0.0626	not significant
Pure error	12.83	5	2.57			
Cor Total	3295.50	29				

represented as good with adequate response values. The predicted R^2 of 0.7644 was in good agreement with the adjusted R^2 of 0.9289; i.e. the difference was less than 0.2. The adequate precision of 23.453 was far greater than 4, so the model was regarded as adequate in nature.

The confidence interval (CI), i.e. 95% CI low and 95% CI high, were the lower and higher limits that elaborated the approximate coefficient of all the factors (Singh et al., 2012). Table 4.7 represents the coefficient values in coded form. In this analysis, the variance inflation factor (VIF) was important for measuring the magnification of variance by using orthogonal design.

The value of VIF was found to be 1.00 in the cases of A, B, C, D, AB, AC, AD, BC, BD and CD, therefore these factors were orthogonal to each other within the model itself. On the other hand, the value of VIF was found to be 2.78 in the case of A^2, B^2, C^2 and D^2, which clearly represents that the factors were not dependent on each other. If the p-values are non-significant, i.e. p-value is greater than 0.05 (α value), then the model is not considered to be significant. In ANOVA (Table 4.5), there were a total of four non-significant terms, i.e. AB, AC, AD and CD. To further improve the model, these non-significant terms can be eliminated, as shown in Table 4.8.

Table 4.6 Other values which were not included in ANOVA

Std. Dev.	2.84	R²	0.9632
Mean	45.50	Adjusted R²	0.9289
C.V. %	6.25	Predicted R²	0.7644
		Adeq. Precision	23.4527

Table 4.7 Other values which were not included in ANOVA

Factor	Coefficient estimate	df	Standard error	95% CI low	95% CI high	VIF
Intercept	49.08	1	0.8833	47.20	50.96	
A-Workpiece rotational speed	−2.39	1	0.6702	−3.82	−0.9604	1.0000
B-Tool linear speed	−1.44	1	0.6702	−2.87	−0.0159	1.0000
C-SiC	2.00	1	0.6702	0.5715	3.43	1.0000
D-EIPs	−4.28	1	0.6702	−5.71	−2.85	1.0000
AB	0.3750	1	0.7109	−1.14	1.89	1.0000
AC	−1.12	1	0.7109	−2.64	0.3902	1.0000
AD	0.5000	1	0.7109	−1.02	2.02	1.0000
BC	7.62	1	0.7109	6.11	9.14	1.0000
BD	−1.75	1	0.7109	−3.27	−0.2348	1.0000
CD	1.25	1	0.7109	−0.2652	2.77	1.0000
A²	−12.49	1	1.77	−16.26	−8.73	2.78
B²	21.01	1	1.77	17.24	24.77	2.78
C²	−8.99	1	1.77	−12.76	−5.23	2.78
D²	−5.49	1	1.77	−9.26	−1.73	2.78

From this table, it has been concluded that the F-value for the model was 34.34 (significant in nature), i.e. the Prob > F value, i.e. < 0.0001, was far less than 0.05 (α value). Hence, the model was significant. The R^2 value after eliminating the non-significant terms was 0.9476, as given in Table 4.9.

The predicted R^2 of 0.8467 was in good agreement with the adjusted R^2 of 0.9200. The adequate precision of 26.2345 was far greater than 4, so the model was regarded as adequate in nature. Table 4.10 represents the coefficient values in coded form after eliminating the non-significant terms. The value of VIF was found to be 1.00 in the cases of A, B, C, D, BC and BD, therefore these factors were orthogonal to each other within the model itself. On the other hand, the value of

Table 4.8 ANOVA for final response after removing the insignificant terms

Source	Sum of squares	df	Mean square	F-value	p-value	
Model	3122.72	10	312.27	34.34	< 0.0001	significant
A-Workpiece rotational speed	102.72	1	102.72	11.30	0.0033	
B-Tool linear speed	37.56	1	37.56	4.13	0.0564	
C-SiC	72.00	1	72.00	7.92	0.0111	
D-EIPs	329.39	1	329.39	36.22	< 0.0001	
BC	930.25	1	930.25	102.30	< 0.0001	
BD	49.00	1	49.00	5.39	0.0315	
A^2	404.26	1	404.26	44.46	< 0.0001	
B^2	1143.55	1	1143.55	125.75	< 0.0001	
C^2	209.45	1	209.45	23.03	0.0001	
D^2	78.13	1	78.13	8.59	0.0086	
Residual	172.78	19	9.09			
Lack of fit	159.94	14	11.42	4.45	0.0543	not significant
Pure error	12.83	5	2.57			
Cor Total	3295.50	29				

Table 4.9 Other values which were not included in ANOVA (after eliminating non-significant terms)

Std. Dev.	3.02	**R^2**	0.9476
Mean	45.50	**Adjusted R^2**	0.9200
C.V. %	6.63	**Predicted R^2**	0.8467
		Adeq Precision	26.2345

Table 4.10 Other values which were not included in ANOVA (after eliminating non-significant terms)

Factor	Coefficient estimate	df	Standard error	95% CI low	95% CI high	VIF
Intercept	49.08	I	0.9367	47.12	51.04	
A-Workpiece rotational speed	−2.39	I	0.7108	−3.88	−0.9012	1.0000
B-Tool linear speed	−1.44	I	0.7108	−2.93	0.0432	1.0000
C-SiC	2.00	I	0.7108	0.5123	3.49	1.0000
D-EIPs	−4.28	I	0.7108	−5.77	−2.79	1.0000
BC	7.63	I	0.7539	6.05	9.20	1.0000
BD	−1.75	I	0.7539	−3.33	−0.1721	1.0000
A²	−12.49	I	1.87	−16.41	−8.57	2.78
B²	21.01	I	1.87	17.09	24.93	2.78
C²	−8.99	I	1.87	−12.91	−5.07	2.78
D²	−5.49	I	1.87	−9.41	−1.57	2.78

VIF was found to be 2.78 in the cases of A^2, B^2, C^2 and D^2, which clearly represents that the factors were not dependent on each other.

The quadratic equations (4.2 and 4.3) represented the correlation between each process variable. Equation 4.2 in its coded form is given as follows:

$$\% \text{ change in Ra} =$$
$$+49.08 - 2.39*A - 1.44*B + 2.00*C - 4.28*D \qquad (4.2)$$
$$+7.63*BC - 1.75*BD - 12.49*A^2 + 21.01*B^2 - 8.99*C^2 - 5.49*D^2$$

Equation 4.3 in its actual form is given as follows:

$$\% \text{ change in Ra} =$$
$$-157.31140 + 0.362792 * \text{Workpiece rotational speed} - 4.48604$$
$$* \text{Tool linear speed} + 0.312462 * \text{SiC} + 0.312946$$
$$* \text{EIPs} + 0.001906 * \text{Tool linear speed} * \text{SiC} - 0.000875 \qquad (4.3)$$
$$* \text{Tool linear speed} * \text{EIPs} - 0.000312 * \text{Workpiece rotational speed}^2$$
$$+0.052522 * \text{Tool linear speed}^2 - 0.000225 * \text{SiC}^2 - 0.000549 * \text{EIPs}^2$$

After analysing each process variable, its contribution against percentage change in Ra value is depicted in Table 4.11.

Table 4.11 Percent contribution of process variable

Process variable	Sum of squares	% contribution
A	102.72	3.28
B	37.56	1.2
C	72.00	2.3
D	329.39	10.54
BC	930.25	29.78
BD	49.00	1.56
A^2	404.26	12.94
B^2	1143.55	36.62
C^2	209.45	6.70
D^2	78.13	2.5

4.5 RESULTS AND DISCUSSION

The actual equation (as per Equation 4.3) evaluated from the regression analysis was directly or indirectly dependent on the experimentation results. The effect of different process variables such as workpiece rotational speed, tool linear speed, mesh sizes of SiC and mesh sizes of EIPs against the percentage change in Ra values was observed. For analysing the results, the effect of each process variable upon the percentage change in Ra value is discussed as follows.

4.5.1 Effect of workpiece rotational speed (A)

Figure 4.5 represents the effect of workpiece rotational speed against percentage change in Ra value. In this plot, the values of workpiece rotational speed (A) are varied, i.e. 400 rpm, 600 rpm and 800 rpm, whereas the other process variables remain constant, i.e. tool linear speed (B) as 30 cm/min, SiC mesh size (C) as 800 and EIPs mesh size (D) as 300.

From this plot, it is clear that with lower speed, i.e. 400 rpm, lesser tangential forces are induced by the abrasive particles, which results in less material removal from the work surface. For higher speed, i.e. 800 rpm, the tangential forces with high impact allow the abrasive particles to move over the surface; rather than removing the material it even produces grooves on the work surface. At 600 rpm speed, the abrasive particles allow a sufficient tangential force on the work by removing the unwanted chips from it.

4.5.2 Effect of tool linear speed (B)

Figure 4.6 represents the effect of tool linear speed against percentage change in Ra value. In this plot, the values of tool linear speed (B) are varied, i.e. 10 cm/min, 30 cm/min and 50 cm/min, whereas the other process variables remain constant,

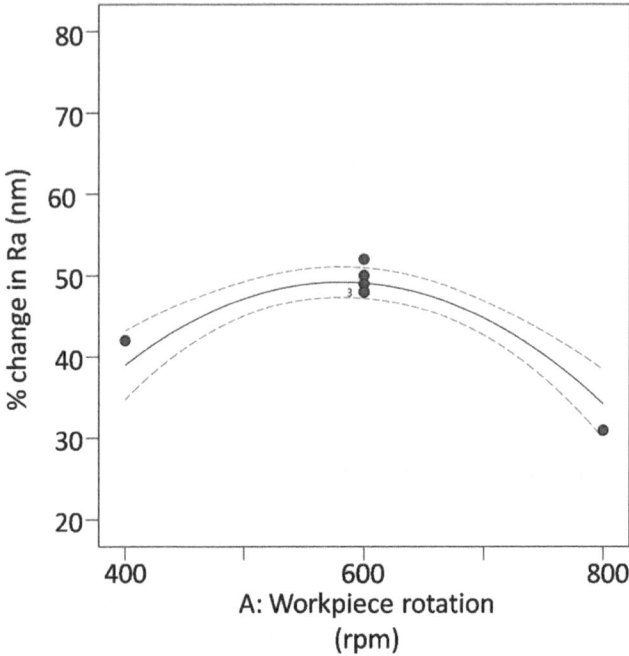

Figure 4.5 Effect of workpiece rotational speed (A) with percentage change in Ra value.

i.e. workpiece rotational speed (A) as 600 rpm, SiC mesh size (C) as 800 and EIPs mesh size (D) as 300. This tool moves in the horizontal direction, which results in producing shearing forces simultaneously with the workpiece rotation.

From this plot, it is clear that with lower speed, i.e. 10 cm/min, the shearing forces induced by the abrasive particles are sufficient to cut the peaks more efficiently from the work material. This is because the travelling speed of the tool in linear direction is very slow, which results in higher material removal. At 30 cm/min speed, the abrasive particles show a lesser material rate because there may be a lesser shearing effect of the abrasive particles finishing cycle. At 50 cm/min, the cutting efficiency of abrasive particle gain increases due to higher shearing action against the tangential forces.

4.5.3 Effect of different mesh sizes of SiC (C)

Figure 4.7 represents the different mesh sizes of SiC against percentage change in Ra value. In this plot, the values of different mesh sizes of SiC (C) are varied, i.e. 600, 800 and 1000, whereas the other process variables remain constant, i.e. workpiece rotational speed (A) as 600 rpm, tool linear speed (C) as 30 cm/min and EIPs mesh size (D) as 300.

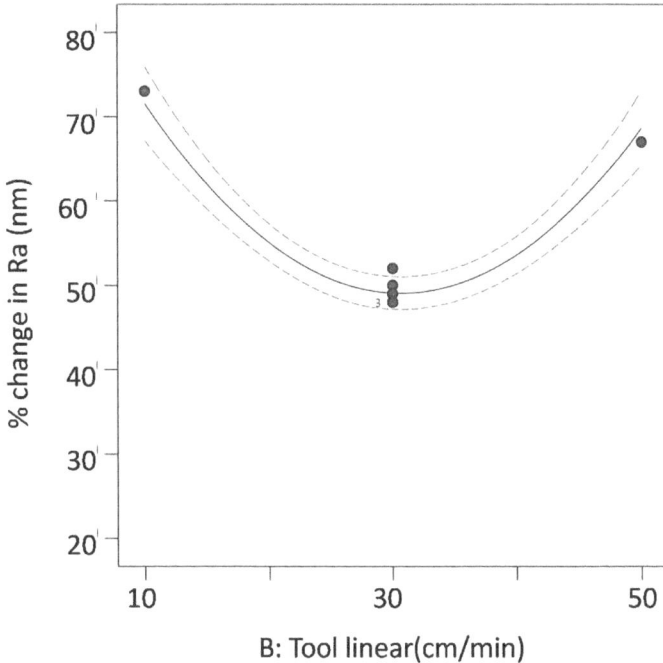

Figure 4.6 Effect of tool linear speed (B) with percentage change in Ra value.

In practicality, it was already proved that the mesh size of abrasive particles would be selected as per the initial surface roughness of any material. However, in this plot, the results are found to be similar in nature, i.e. the 800 mesh size abrasives finish the surface more accurately than do the 600 and 1000 mesh sizes. The sharper corners of abrasive particles cut the peaks more efficiently in the form of tiny chips. These results show that the initial surface roughness is quite important for choosing the mesh size of abrasive particles.

4.5.4 Effect of different mesh sizes of EIPs (D)

Figure 4.8 represents the different mesh sizes of EIPs against percentage change in Ra value. In this plot, the values of different mesh sizes of EIPs (D) are varied, i.e. 200, 300 and 400, whereas the other process variables remain constant, i.e. workpiece rotational speed (A) as 600 rpm, tool linear speed (B) as 30 cm/min and SiC mesh size (C) as 600.

The main role of EIPs is to hold the abrasive particles under the effect of magnetic fields. Due to the magnetic field, the EIPs remain attached to the magnet surface, whereas the abrasive particles are entrapped by the EIPs. The size of EIPs

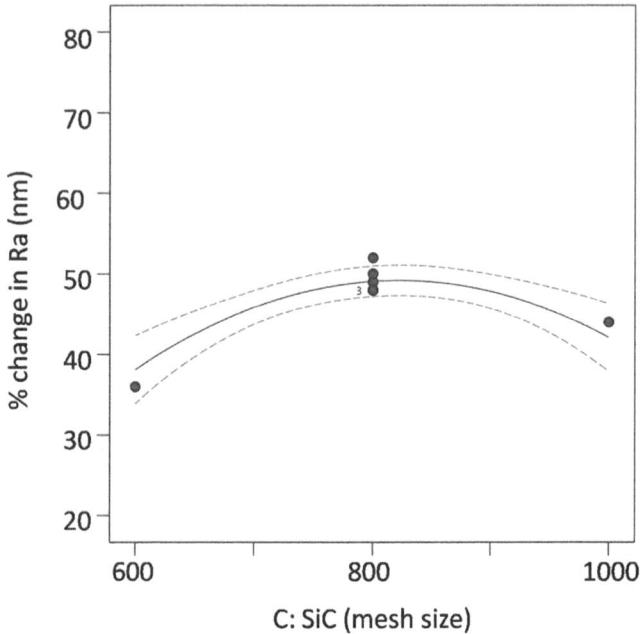

Figure 4.7 Effect of different mesh sizes of SiC (C) with percentage change in Ra value.

will be selected as per the size of abrasive particles. The EIPs of 300 mesh size are found efficient in holding the abrasive particles, and it further approaches the work surface under sufficient indentation forces. Less material removal is achieved with 400 mesh size EIPs, because the size of EIPs is large compared to abrasives size.

4.5.5 Effect of workpiece rotational speed (A) with different mesh sizes of SiC (C)

The 3D contour plots for the effect of workpiece rotational speed with different mesh sizes of SiC abrasives against the percentage change in Ra value is shown in Figure 4.9. In this plot, the values of workpiece rotational speed (A), i.e. from 400 rpm to 800 rpm, and SiC mesh size (C), i.e. from 600 to 1000, are varied, whereas the other process variables remain constant, i.e. tool linear speed (B) as 30 cm/min and EIPs mesh size (D) as 300. From this plot, it is clear that with SiC of 800 mesh size and workpiece rotational speed of 600, the maximum material removal is achieved. This is because of the maximum tangential forces induced by the SiC abrasives, which further help to remove the peaks more efficiently from the work surface.

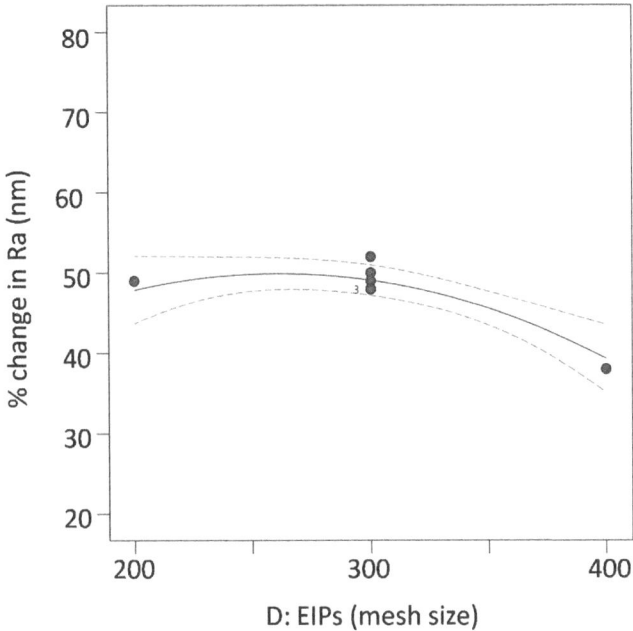

Figure 4.8 Effect of different mesh sizes of EIPs (D) with percentage change in Ra value.

4.5.6 Effect of workpiece rotational speed (A) with different mesh sizes of EIPs (D)

The 3D contour plots for the effect of workpiece rotational speed with different mesh sizes of EIPs against the percentage change in Ra value is shown in Figure 4.10. In this plot, the values of workpiece rotational speed (A), i.e. from 400 rpm to 800 rpm, and EIPs mesh size (D), i.e. from 200 to 400, are varied, whereas the other process variables remain constant, i.e. tool linear speed (B) as 30 cm/min and SiC mesh size (C) as 800. From this plot, it is clear that with EIPs of 300 mesh size and workpiece rotational speed of 600, the maximum material removal is achieved. This is because EIPs forma strong chain under the magnetic field, which further helps the abrasive particles to cut the peaks under the sufficient tangential forces.

4.5.7 Effect of tool linear speed (B) with different mesh sizes of SiC (C)

The 3D contour plots for the effect of tool linear speed with different mesh sizes of SiC abrasives against the percentage change in Ra value is shown in Figure 4.11. In this plot, the values of tool linear speed (B), i.e. from 10 cm/min to 50 cm/

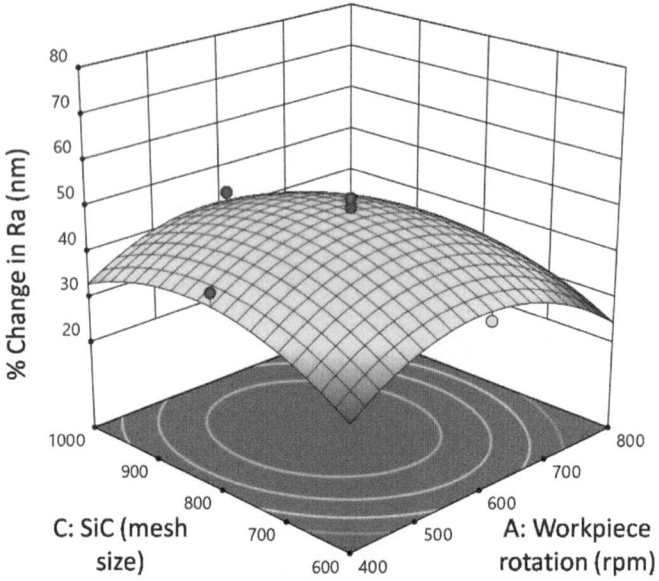

Figure 4.9 Effect of workpiece rotational speed with different mesh sizes of SiC abrasives against the percentage change in Ra value.

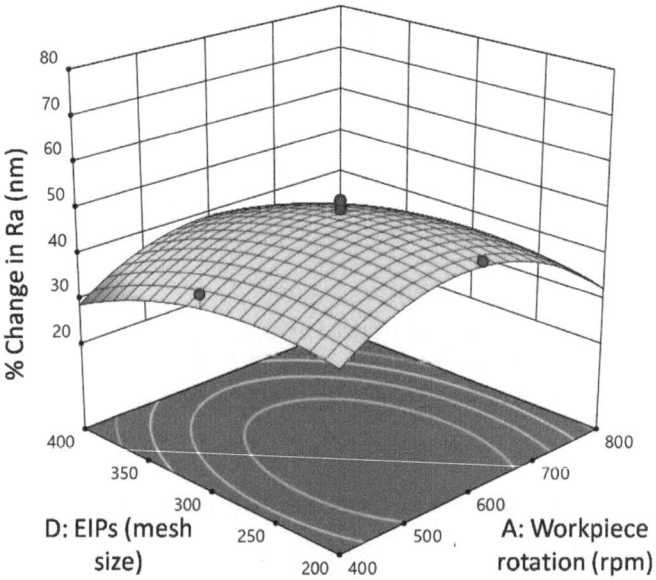

Figure 4.10 Effect of workpiece rotational speed with different mesh sizes of EIPs against the percentage change in Ra value.

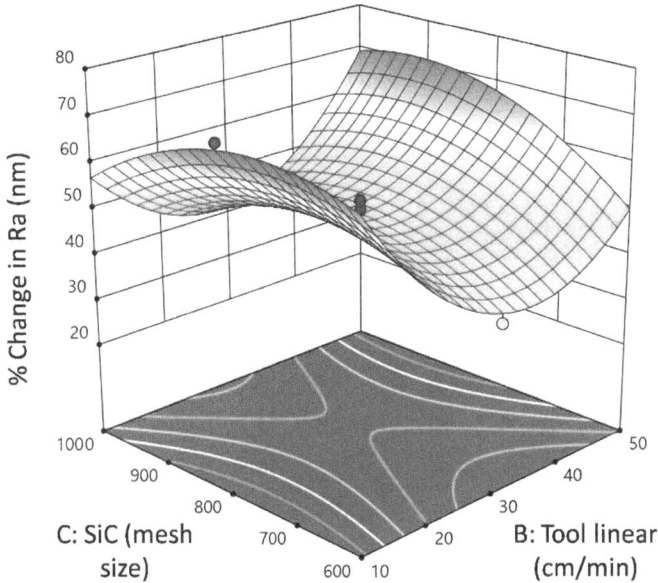

Figure 4.11 Effect of tool linear speed with different mesh sizes of SiC abrasives against the percentage change in Ra value.

min, and SiC mesh size (C), i.e. from 600 to 1000, are varied, whereas the other process variables remain constant, i.e. workpiece rotational speed (A) as 600 rpm and EIPs mesh size (C) as 300. From this plot, it is clear that with SiC of 800 mesh size and tool linear speed of 10 cm/min, maximum material removal is achieved. This is because of the shearing action of the abrasives when the tool moves in the horizontal direction. Hence, the combined effect of workpiece rotation along with tool linear movement provides better finishing results.

4.5.8 Effect of tool linear speed (B) with different mesh sizes of EIPs (D)

The 3D contour plots for the effect of tool linear speed with different mesh sizes of EIPs against the percentage change in Ra value is shown in Figure 4.12. In this plot, the values of tool linear speed (B), i.e. from 10 cm/min to 50 cm/min, and EIPs mesh size (D), i.e. from 200 to 400, are varied, whereas the other process variables remain constant, i.e. workpiece rotational speed (A) as 600 cm/min and SiC mesh size (C) as 800. From this plot, it is clear that with EIPs of 300 mesh size and tool linear speed of 10 cm/min, maximum material removal is achieved. This is because the EIPs provide a sufficient gripping strength to SiC abrasives, which further remove the peaks at low linear tool speed.

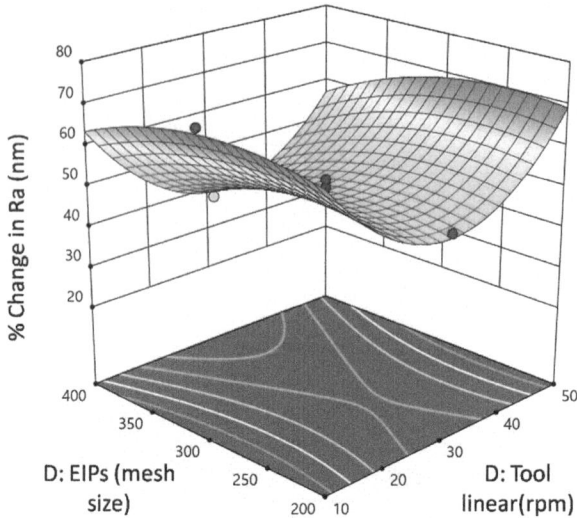

Figure 4.12 Effect of tool linear speed with different mesh sizes of EIPs against the percentage change in Ra value.

4.6 CONFIRMATORY EXPERIMENTATION

For confirming the experimental data, the total four experimental runs (from Table 4.3) were taken randomly, and again the experiments were performed with a finishing time of 40 minutes each and a working gap of 1 mm. After performing the experiments, their results were matched by the putting the optimum variables value of each run in the regression equation (i.e. Equation 4.3). The results of experimentation data and regression equation data (theoretical) were close to each other (maximum 4% error) as depicted in Table 4.12. The optimum process variables, i.e. workpiece rotational speed (A) as 600 rpm, tool linear speed as 10 cm/min, SiC mesh size as 800 and EIPs mesh size as 300, were found where the change in Ra value was 70%. The roughness profiles for initial ground and finished surface are depicted in Figure 4.13a and Figure 4.13b.

4.7 CONCLUSIONS

In this chapter, magnetorheological finishing was found efficient to finish the brass cylindrical workpiece. Some of the brief conclusions from this study are given as follows:

1. From this study, the effect of workpiece rotational speed, tool linear speed, mesh sizes of SiC and mesh sizes of EIPs against percentage change in Ra value was analysed using response surface methodology.

Table 4.12 Confirmatory experimentations analysis

Sr No.	Process variables				% change in Ra value (Experimentation)	% change in Ra value (Theoretical)	% Error
	A	B	C	D			
1	600	50	800	300	65%	69%	3%
2	600	10	800	300	70%	72%	2%
3	800	10	600	400	43%	45%	2%
4	600	30	800	300	53%	49%	4%

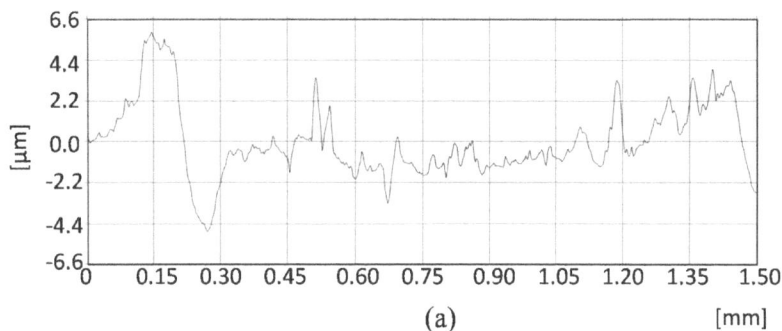

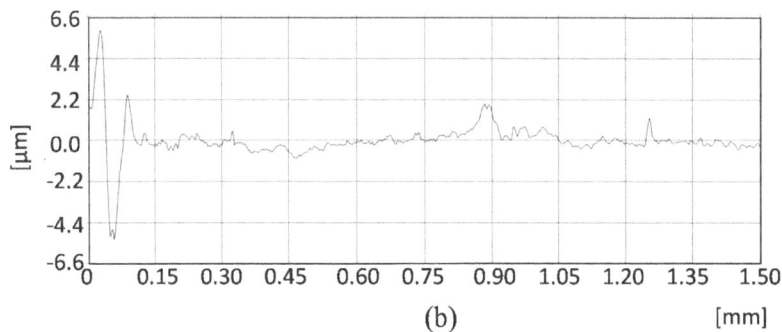

Figure 4.13 Roughness profiles for (a) initial ground and (b) finished surface (experimentation done with optimum process variables, i.e. A= 600 rpm, B= 10 cm/min, C= 800 mesh size and D= 300 mesh size).

2. The optimum process variables, i.e. workpiece rotational speed as 600 rpm, tool linear speed as 10 cm/min, SiC mesh size as 800 and EIPs mesh size as 300, were found to accurately finish the cylindrical components made of brass.

3. The lowest percentage change in surface roughness Ra value was found to be 73%, i.e. 97 nm from 361 nm after 40 minutes of finishing time and 1 mm of working gap.
4. The roughness profiles of the initial and final finished surface clearly showed the difference in surface quality after the present magnetorheological finishing.

REFERENCES

Bedi, T.S. and Kant, R. (2021) Comparative performance of magnetorheological external finishing tools using different magnetic structures. *Materials Today Proceeding*, 41(4), 908–914.

Bedi, T.S. and Rana, A.S. (2021) Surface finishing requirements on various internal cylindrical components: A review. *Journal of Micromanufacturing*. Vol 4(2), 216–228. https://doi.org/10.1177/25165984211035504.

Bedi, T.S. and Singh, A.K. (2016) Magnetorheological methods for nanofinishing-a review. *Particulate Science and Technology*, 34, 412–422.

Bica, I. (2002) Damper with magnetorheological suspension. *Journal of Magnetism and Magnetic Materials*, 241, 196–200.

Carlson, J.D. and Jolly, M.R. (2000) Magnetorheological fluids, foam and elastomers devices. *Mechatronics A*, 10, 555–569.

Jha, S. and Jain, V.K. (2004) Design and development of the magnetorheological abrasive flow finishing (MRAFF) process. *International Journal of Machine Tools and Manufacture*, 44, 1019–1029.

Jolly, M.R., Bender, J.W. and Carlson, J.D. (1998) Properties and application of commercial magnetorheological fluids. *Journal of Intelligent Material Systems and Structures*, 3327, 262–275.

Khurana, A., Singh, A.K. and Bedi, T.S. (2017) Spot nanofinishing using base nose magnetorheological solid rotating core tool. *International Journal of Advanced Manufacturing Technology*, 92(1–4), 1173–1183.

Kordonski, W.I. and Jacobs, S.D. (1996) Magnetorheological finishing. *International Journal of Modern Physics*, 10, 2837–2848.

Paswan, S., Bedi, T.S. and Singh, A.K. (2017) Modeling and simulation of surface roughness in magnetorheological fluid based finishing process. *Wear*, 376–377, 1207–1221.

Rabinow, J. (1948). The magnetic fluid clutch. *AIEE Transactions*, 67, 1308.

Rana, A.S., Bedi, T.S. and Grover, V. (2020) A new permanent magnet type magnetorheological finishing tool for external cylindrical surfaces having different outer diameter. *Advances in Production and Industrial Engineering*, 209–217.

Rana, A.S., Bedi, T.S. and Grover, V. (2021) Fine-finishing of stepped cylindrical workpiece using magnetorheological finishing process. *Materials Today Proceeding*, 41(4), 886–892.

Rana, A.S., Rafiq, I. and Bedi, T.S. (2023) Parametric optimization of curved-rectangular shaped magnetorheological finishing tool for external cylindrical surfaces. *E3S Web of Conferences*, 430, 01263.

Shorey, A.B., Jacobs, S.D., Kordonski, W.I. and Gans, R.F. (2001) Experiments and observations regarding mechanism of glass removal in magnetorheological finishing. *Applied Opicst*, 40, 20–33.

Sidpara, A. and Jain, V.K. (2013) Analysis of forces on the freeform surface in magneto-rheological fluid based finishing process. *International Journal of Machine Tools and Manufacture*, 69, 1–10.

Singh, A.K., Jha, S. and Pandey, P.M. (2012) Parametric analysis of an improved ball end magnetorheological finishing process. *Proceeding of the Institution of Mechanical Engineers Part B: Journal of Engineering Manufacture*, 226(9), 1550–1563.

Chapter 5

Hybrid energy assisted friction stir welding using secondary heating sources

Ankit Mani Tripathi, R. C. Singh, Rajiv Chaudhary and Ravi Kant

5.1 INTRODUCTION

The friction stir welding process is clean, energy-efficient, eco-friendly and versatile. It is a solid-state joining process that was first invented and patented by the Welding Institute in the UK in 1991 and used for the first time to weld alloy, which was difficult to join through the conventional fusion welding process due to poor solidification microstructure liquefaction cracking and porosity in the fusion zone [1, 2]. In addition, the researchers redirected their attention to welding dissimilar materials, aiming to investigate the mechanical and metallurgical properties of the joints and to repeatedly optimize the process parameters. In this regard, Kwon et al. welded Al and Mg plates, maintaining a constant traverse speed while varying the tool's rotational speed. They observed that increasing the rotational speed resulted in a defect-free weld. They have not observed any significant impact of tool rotational speed on strength and ductility of the welded part, which were independent of tool rotational speed [3]. Their paper did not provide a clear explanation for why this occurred. Moreover, Shanmuga et al. have optimized welding parameters such as tool rotational speed, different tool pin profile, plunging force and traverse speed and outlined the effects of these parameters on the dissimilar Al alloy welded joints. They discovered that using tapered hexagonal tool pin profiles offered a smooth flow of the materials to weld centerline, resulting in maximum tensile strength and elongation of the welded joint. In contrast, straight cylindrical tools offered the lowest tensile strength and elongation. Furthermore, they found that the tensile strength increases first, reaches its maximum value, and then decreases when the tool rotational speed and plunging force are increased [4]. Moreover, as previously discussed, friction stir welding (FSW) has many advantages over fusion welding, including good results in welding Al and its alloys. It still faces a lot of challenges, however, in welding intermediate and high melting materials. Welding harder materials places the tool under constant stress and high temperatures, leading to rapid tool deterioration, reduced tool lifespan and increased tool costs [5]. Following this, welding that occurred in the case of dissimilar FSW was the nominal difference in their melting points. Researchers and industrialists have always sought to extend the scope of conventional friction stir welding for welding

DOI: 10.1201/9781032703046-5

harder or higher melting point materials. Considering the causes of tool wear, various perspectives are available in the literature for welding hard or high-melting-point materials. As discussed earlier, if a material is supplied heat by some external means, then its yield strength value decreases, and the materials easily deform plastically with the rotation of the tool. However, if an additional source of heating is introduced in a controlled manner to the workpiece so that the peak temperature remains below its melting temperature, the heat input required from the FSW tool is reduced. This scientific principle is followed by several researchers, and the development of the process is called "hybrid" or "assisted" FSW [6]. Moreover, laser, plasma, gas arc, induction and so forth come under the category of "thermally assisted hybrid" FSW, whereas ultrasonic welding comes under mechanically assisted hybrid FSW. Consequently, as the researchers' primary concern was to extend tool life, they found that using a secondary heat source reduces stresses on the tool and improves material flow. In conclusion, this chapter will provide insights into recent developments in the domain of various hybrid or secondary heating assisted welding processes in terms of their distinct process parameters, experimental methodology, and mechanical and metallurgical characteristics.

5.2 LASER-ASSISTED FSW (LAFSW)

In laser-assisted friction stir welding (LaFSW), the base metal laser spot is used ahead of the tool. Neodymium-doped Yttrium Aluminum Garnet (Nd: YAG) fiber-optic laser, diode laser and CO_2 laser are the most commonly used secondary heating sources. Experimental setup of the laser-assisted preheating system is shown in Figure 5.1. As previously discussed, by using secondary heat sources, the material loses its yield strength. A similar result was obtained by Sun et al. They welded S45C steel with laser-assisted friction stir welding and positioned the laser spot ahead of 10 mm from the tool. In that case, LaFSW increased the welding speed up to 800 mm/min, which was only possible by conventional FSW up to 400 mm/min. Again, they changed the position of the laser spot and found that when base metal was preheated by placing the laser spot on the advancing side, the friction heat generation between tool and material is significantly reduced, and the highest total heat input was found by positioning the laser spot on the retreating side. They observed that preheating on the advancing side had the lowest total heat input, and focusing the laser beam on the retreating side gave the maximum total heat input [7]. Furthermore, Fei et al. investigated the effect of laser power on steel and Al alloy. In measuring the welding strength and thickness of the intermetallic compounds, they found that 950 W laser power is a critical point, where weld strength decreases due to an increase in the size of intermetallic compounds (IMCs). The inevitably larger size of the intermetallic compounds reduces the weld strength at the given weld power, which is shown in Figure 5.2 and Figure 5.3 [8]. Laser-assisted friction stir welding introduces additional non-contact local heating immediately ahead of the weld zone, reducing the mechanical energy exerted

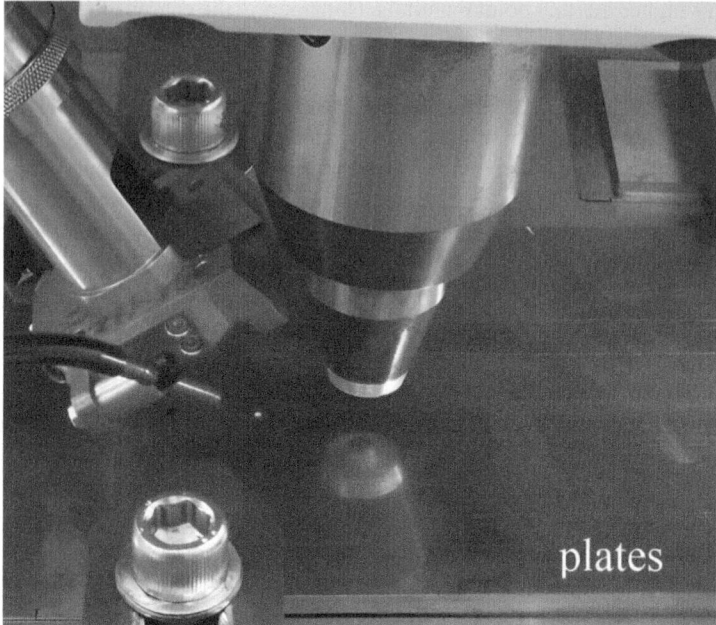

Figure 5.1 Experimental setup of LaFSW used to weld S45C steel, with the laser spot positioned 10 mm ahead of the tool [43].

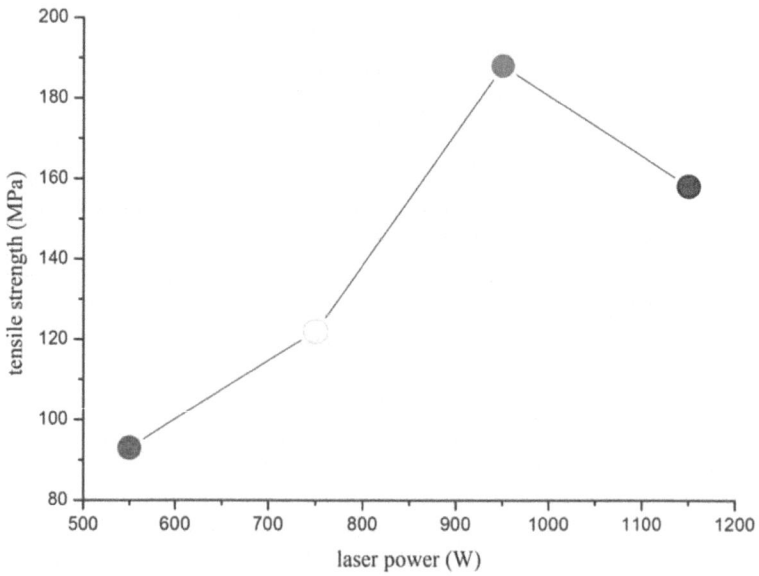

Figure 5.2 Variations of laser power vs tensile strength [8].

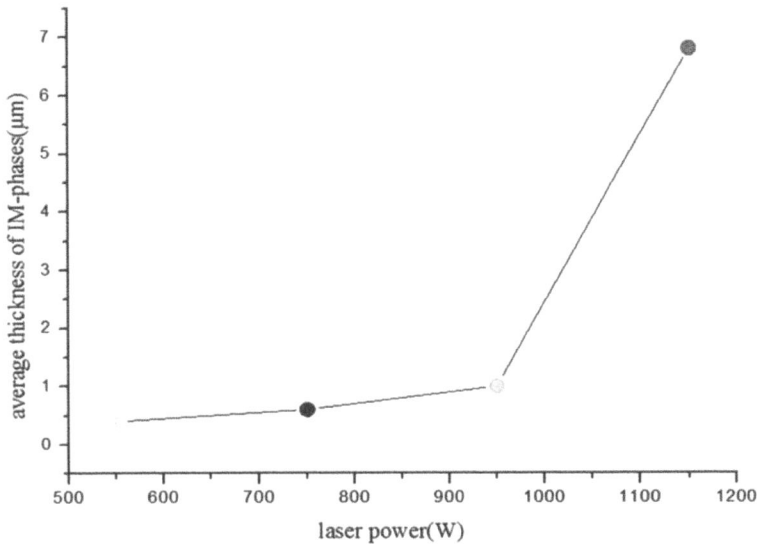

Figure 5.3 Variations of laser power vs thickness of intermetallic phase [8].

on the tool. In addition, Sundqvist et al.'s transient thermal model demonstrated through their mathematical modelling at dissimilar type welding in butt joint configuration that a laser beam reduces the forces on the tool probe and shoulder [9]. Laser power, heat source diameter, and the distance between the heat source and the tool are the three common parameters that significantly influence the process. Ahmad et al., using Abaqus software, optimized the process parameters of LaFSW by considering structural steel as workpiece material. They showed that LaFSW increased the welding speed up to 1500 mm per min, which was higher than the conventional FSW welding process. In addition, when the laser spot was positioned 20 mm ahead of the tool probe, stresses on the tool were reduced by 55% compared to conventional FSW. It was found that reducing the distance between laser spot and tool develops higher peak temperature and reduces the temperature difference between the workpieces to be welded. But it also has some restrictions; locating the heating source nearer to the tool or increasing the laser power will melt the parent metal [5]. Hence, proper optimization of the distance between the tool and the preheating source is necessary to get the maximum joint strength. Moreover, Fei et al. have investigated the metallurgical characteristics of Q235 steel with LaFSW and traditional FSW and found that the strength of the weld nugget was favored LaFSW, which enhances the flow of materials. Thus, a defect-free weld can be obtained, which was one major limitation of using traditional FSW for harder materials. They indented laser spots for 5 s before the plunging stage causes the material to become relatively soft; as a result, the resistance offered by the materials on the tool was reduced and the material flowed easily. The process

Table 5.1 Materials, FSW tools and process parameters to use in LAFSW

References	Sample materials to be welded	FSW tool	Tool rotational speed/RPM	Traverse speed/mm min⁻¹	Distance of laser spot from the rotating tool, mm	Type of laser and operating power/kW
[8]	Q235 steel and 6061-T6 Al alloy (3 mm)	W and Mo alloy	950	23.5	20	Fiber laser, 0.55, 0.75, 0.95, 1.15
[7]	S45C steel	WC-based alloy	600	100–800	5, 10, 15	YAG laser generator, 2 kW

parameters of different reference papers that have been considered in this review of LaFSW are shown in Table 5.1.

5.3 ELECTRICALLY ASSISTED FRICTION STIR WELDING (EAFSW)

In EaFSW, the electric current is directly supplied either to the workpiece or to the tool, which causes the material to be heated due to resistance heating or joule heating. Additionally, the material becomes soft due to the electroplastic effect without any preheating. In the electroplastic phenomenon, the material becomes soft due to moving electrons, and no significant increase in temperature was found during this process [11–13]. The processes of heat generation and material softening are given next in the form of mathematical equations [13].

$$Q_{FSW} = Q_{friction} + Q_{deformation} \qquad (5.1)$$
$$Q_{EAFSW} = Q_{FSW} + Q_{Electric} \qquad (5.2)$$
$$Q_{Electric} = Q_{Joule} + \text{electroplastic softening} \qquad (5.3)$$

In the equations, Q represents heat and the subscripts represent the type of the process.

There are two ways by which an electrical current can flow inside the workpiece. The electric current goes either through the tool inside the workpiece where the tool is an integral part of the circuit, as shown in Figure 5.4a, or the electric current is directly supplied to the workpiece. The tool does not contribute to the passage of electric current through the workpiece and is also not a part of the electrical circuit [13], as shown in Figure 5.4b.

Moreover, Ni et al. [11] provided a different experimental setup to bypass the tool for the flow of current. This was achieved by using a single electrode. They attached two Cu brushes in pressed conditions with the help of springs on the top surface on either side relative to the weld centerline, in which both Cu brushes

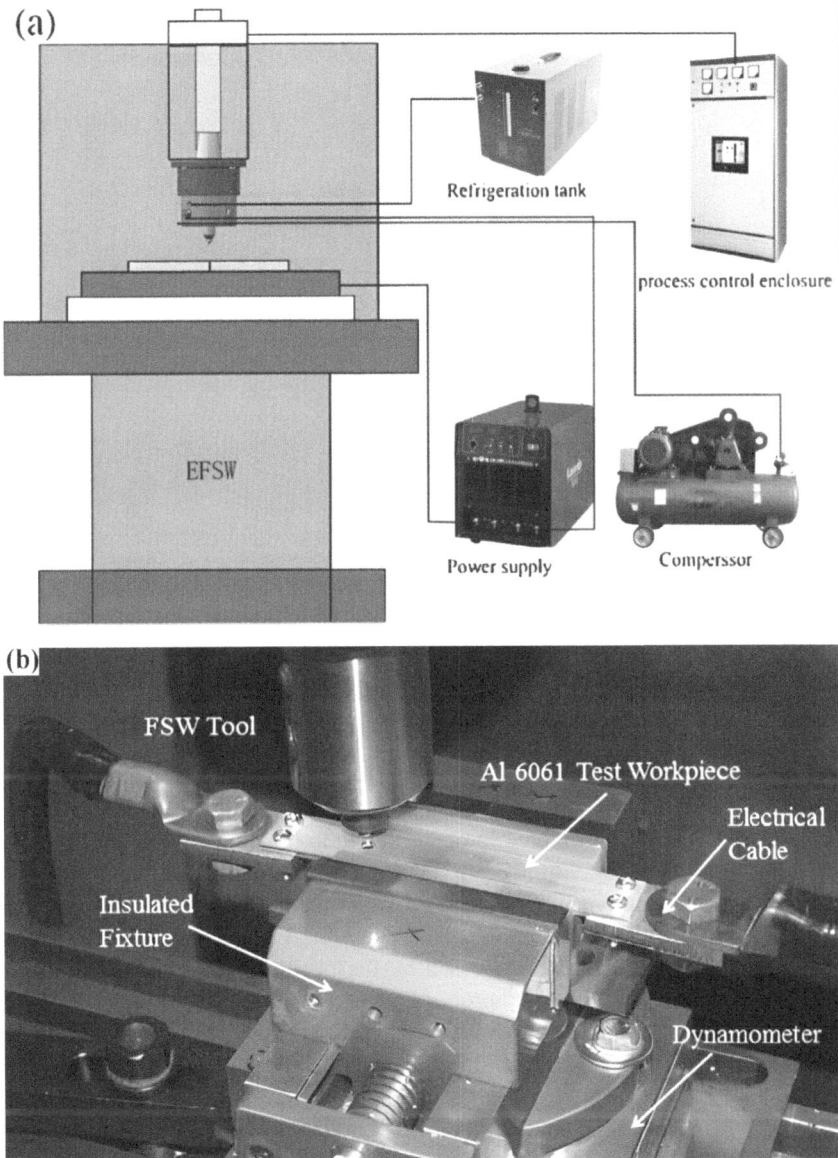

(a)

Refrigeration tank

process control enclosure

EFSW

Power supply

Comperssor

(b)

FSW Tool

Al 6061 Test Workpiece

Electrical Cable

Insulated Fixture

Dynamometer

Figure 5.4 (a) Includes electrically assisted FSW system, insulation system, cooling system and gas protection system. The electrically assisted system was attached to the FSW-TS-S08 bench-type friction stir welding equipment. [12]. (b) Current passed directly to the workpiece [15]. (c) Instead of using the tool as one electrode, two additional copper brushes were spring preloaded to address the concerns of spark generation as they slide on the top surface of the workpiece, serving as the anode and cathode respectively. The copper brushes were mounted to the spindle holder and travel together with the FSW tool in close proximity [11].

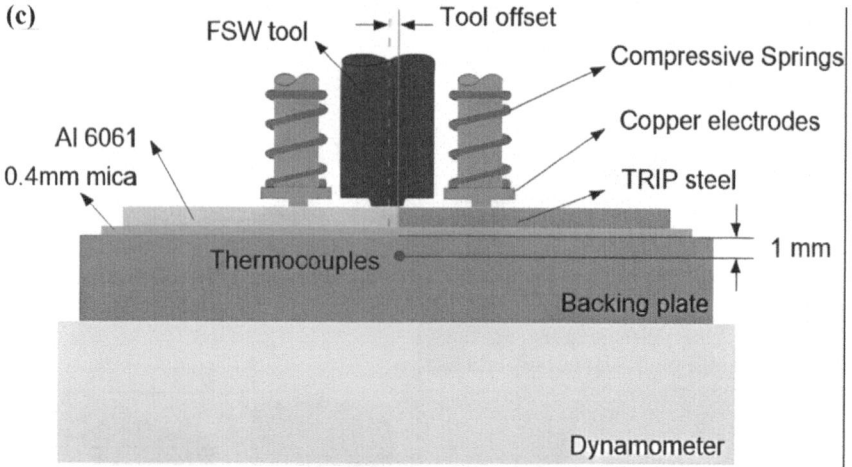

Figure 5.4 (Continued)

were acting as anode and cathode. Both Cu brushes were mounted on the spindle holder, and these were also moving along with the tool as shown in Figure 5.4c. Furthermore, the researchers investigated the effect of elastoplastic and resistance heating on the material welded by EAFSW and found that the plunging force reduced significantly after the flow of current. It gave noticeable results when the tool rotational speed was kept low and the tool was offset towards aluminum. They also investigated the metallurgical characterization and observed that due to the passing of electrical current, the thin intermetallic layer is being formed near the plunge section of the welded joint, as shown in Figure 5.5. This is only possible because of the effect of accelerated atom diffusion and reduced activation energy for a chemical reaction. In their investigation, they also noticed the micro interlock feature in the Al-Fe interface welded joint, which greatly improved the crack initiation and propagation caused by the brittle IMCs. In addition, Chen et al. investigated the mechanical properties of AA2219 and found that just at the time of introduction of electric current, advancing side (AS) and retreating side (RS) had a very high temperature difference. Compared to conventional friction stir welding, due to the current introduction in the weld zone, they observed much improvement in overall hardness, especially in AS. Again, they measured the tensile strength of the welded joint by varying the current supply. When the current range was between 100–400 A, the authors found a 2.74%–7.38% increase in the tensile strength of the welded joint. A significant improvement in the tensile strength of the welded joint 17.11% was observed when the current range was increased to 500–600 A. Furthermore, in the analysis of fractured surfaces, they found that failure in friction stir welded material was occurring in the weld zone (WZ)/thermomechanical affected zone (TMAZ) boundary from the advancing side. The same EAFSW welded material was showing V-shaped ductile fracture

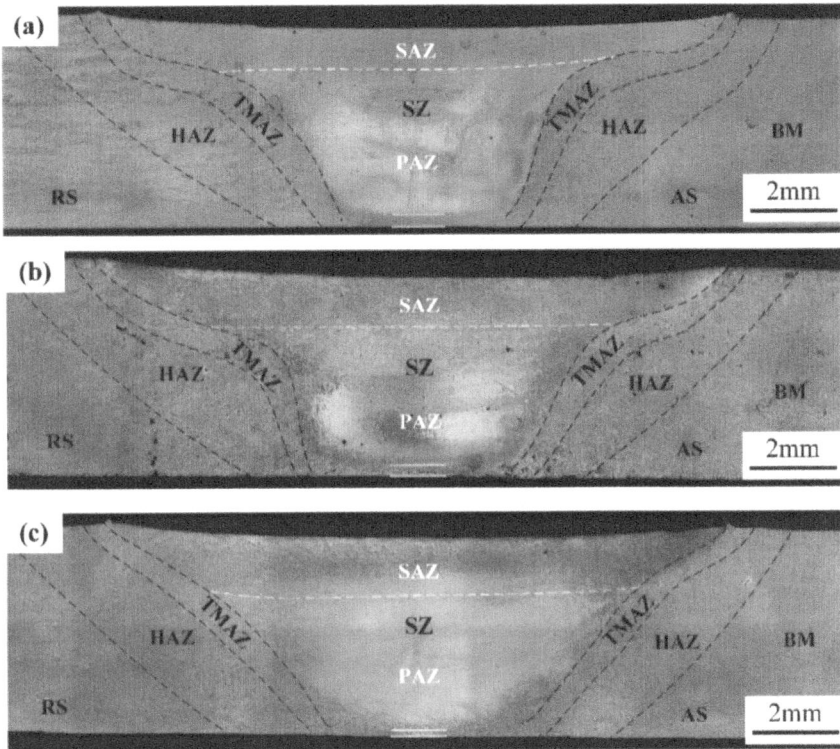

Figure 5.5 Optical images of EAFSW welded joints with a current of (a) 0 A, (b) 100 A, (c) 200 A [14].

behavior and failure starts in the nugget zone from the retreating side, where hardness value was at a minimum. The cross-section of the dimple size of the welded material in the range 500–400 A was found to be relatively large and deep, as shown in Figure 5.6. It was also observed that EAFSW had refined the Al_2Cu precipitate and also distributed this precipitate uniformly, which is the favorable condition in the case of the weld joint strength [12]. Han et al. investigating the mechanical and metallurgical properties of the AZ31B Mg alloy at different current ranges, from 0 to 200 A, welded by EaFSW, found the same results in their investigation as has been previously discussed. They showed that by increasing the temperature, the size of the stir zone increased gradually. In addition to that, the microstructure in the stir zone (SZ) was significantly uniform ahead of the thickness line. The difference in grain size at the advancing side between SZ and TMAZ was also improved due to an increase in temperature, as shown in Figure 5.7. According to their findings, fracture generally initiates in all the welded joints, starting from the area with the lowest hardness value. Further, they showed that as the current increased, the tensile strength of the material had also increased

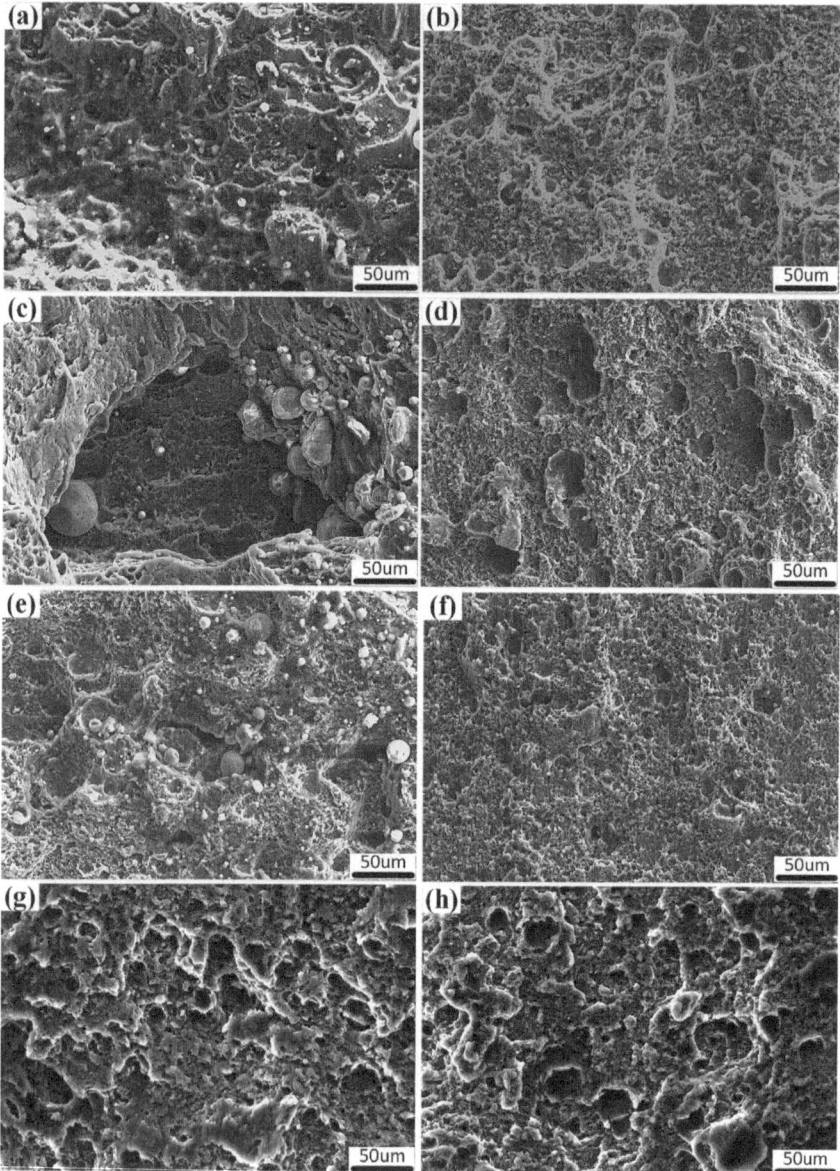

Figure 5.6 Hardness value of the welded joint with different current [14].

and reached the maximum when the current was at 200 A. Due to the increase in current, many effects on the hardness of the welded joint was not found, as shown in Figure 5.8 [14]. Moreover, Jiang et al. examined the microstructural texture and mechanical properties of Ti6Al4V alloy welded with EaFSW and compared

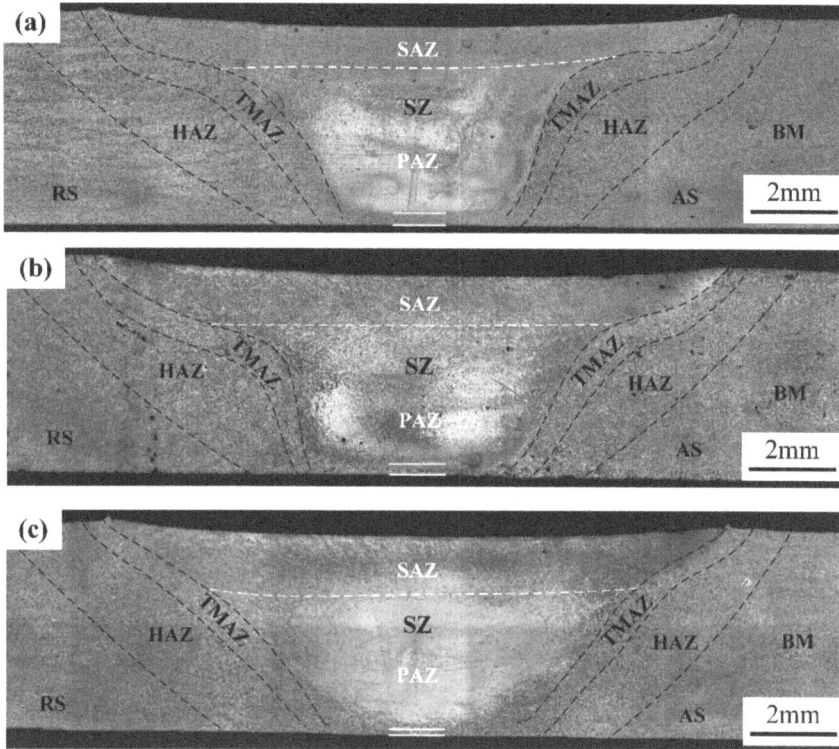

Figure 5.7 Comparison of Al-Fe interface at the weld centerline using EAFSW with traditional FSW [11].

their properties with conventional FSW. It was found in their investigation that in both welding processes, the microstructure of the heat-affected zone (HAZ) and the base metal exhibited a coarse equiaxed structure, while the stir zone displayed a relatively fine structure. In SZ, the grain size in the advancing side was slightly bigger relative to the retreating side. Although the material welded by EaFSW, base metal (BM) and HAZ have strong prismatic texture, the texture of SZ was randomly distributed. Also, in terms of degrees of dynamic recovery in HAZ and dynamic recrystallization in SZ, they found in conventional FSW appeared slightly stronger than in EaFSW [15]. Sengupta et al. showed weld joint strength and weld efficiency by varying the process parameters of EAFSW. They also optimized the process parameters by Taguchi methods to get better weld efficiency, ultimate tensile strength and hardness value. As a result, they found that the weld efficiency and hardness value of the EaFSW welded material was much better than the traditionally FSW welded material [16]. The process parameters of different reference papers included in the review of EaFSW are shown in Table 5.2.

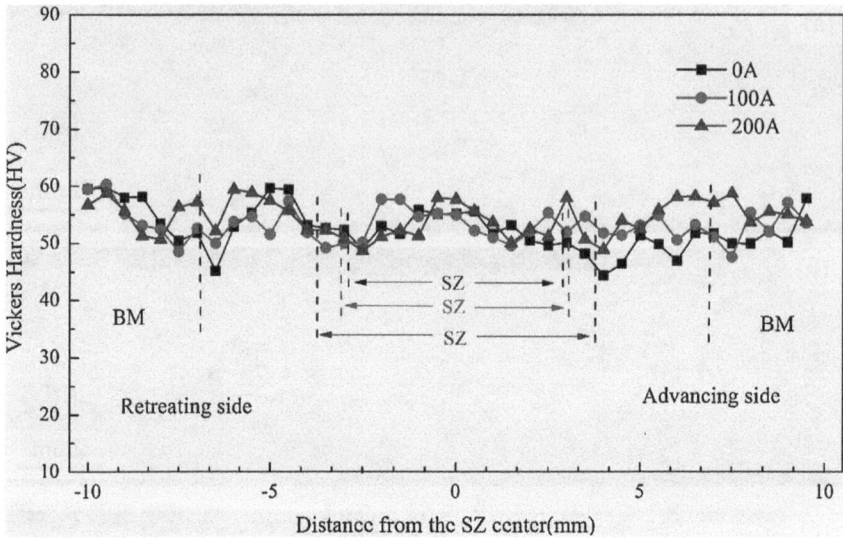

Figure 5.8 Fractured surfaces of EAFSW Al 2219 alloy (a) 0 A, (b) 200 A, (c) 400 A, (d) 600 A [12].

Table 5.2 Materials, FSW tools and process parameters to use in EAFSW

References	Sample materials to be welded	FSW tool	Tool offset (offset: mm)	Tool rotational speed/RPM	Traverse speed/mm min^{-1}	Operating electrical current/A
[15]	Ti$_6$Al$_4$V (3 mm)	. . .	. . .	200	40	100, 200
[11]	Al 6061 to TRIP 780 steel (1.4 mm)	WC-10% CO	1.03, 1.63	1200, 1800	60	560
[12]	AA2219 Al alloy (6 mm)	. . .	. . .	800	160	0–600
[14]	AZ31B Mg alloy (5 mm)	. . .	. . .	1300	50	0, 100, 200
[16]	2062 Grade B plates	WC	. . .	500–1000	10–25	50–150

5.4 INDUCTION-ASSISTED FRICTION STIR WELDING (IAFSW)

A high-power electromagnetic field is produced that induces huge eddy currents in conductive metal, and metal becomes heated because of resistance heating. Thus, it offers less resistance to the tool surface during stirring of the material than conventional FSW, and finally the workpiece welds easily. Induction-assisted friction stir welding is a non-contact type of welding process.

Furthermore, concerning the weld centerline, if there is no loss of induction current, during that time, it heats the metal with its maximum efficiency. For this, the induction coil is placed in such a way that the same current direction is ensured on each side of the workpiece [13]. Moreover, in the context of the experimental process, Sun et al. completed this welding in four stages. In the first stage, both the tool and the heating coil (HC), which were to be welded, were 5 mm above the metal; the heating coil was ahead of the tool and supplying heat where the metal was to be welded. In the second stage, the HC heated the metal for a certain duration, and then it was removed from that place after which the rotating tool was brought there. In the third stage, a particular load was applied to plunge the tool into the sample, initiating a stirring action within the sample. After stirring for some time, the tool was taken out of the sample. Finally, the welding process was completed [17], as shown in Figure 5.9a. Furthermore, Sharma et al. investigated

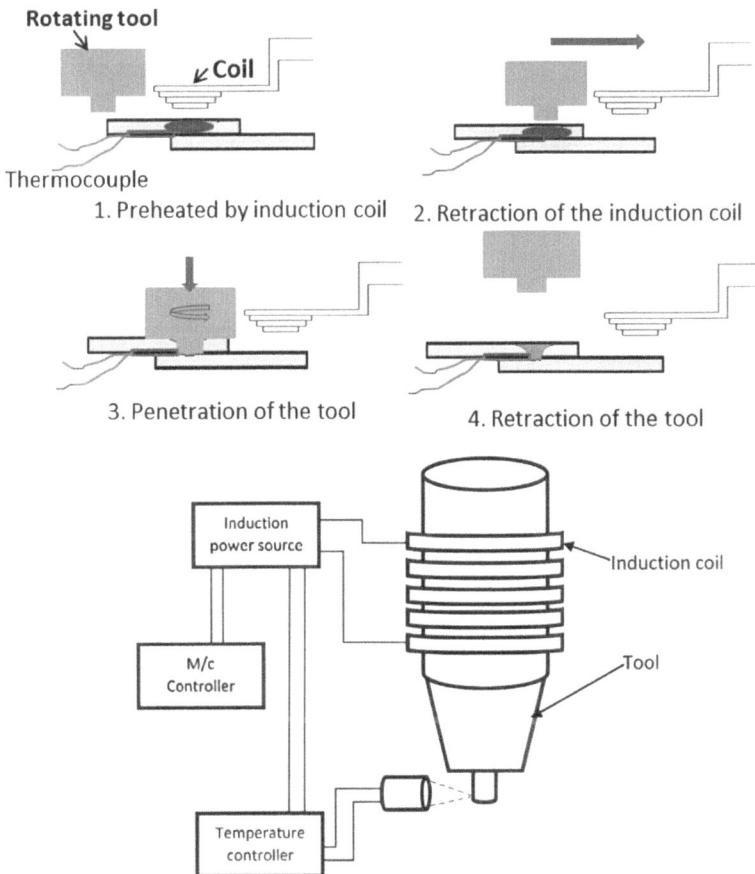

Figure 5.9 (a) The welding work is completed in four stages by IaFSW: first, the surface to be welded is preheated by induction coil; second, the induction coil is retracted; third, the tool is penetrated; and finally, the tool is retracted. (b) Schematic diagram. (c, d, e) Experimental setups for IaFSW. (f) Interfacial and plug failure mode in IaFSW spot weld [17, 18].

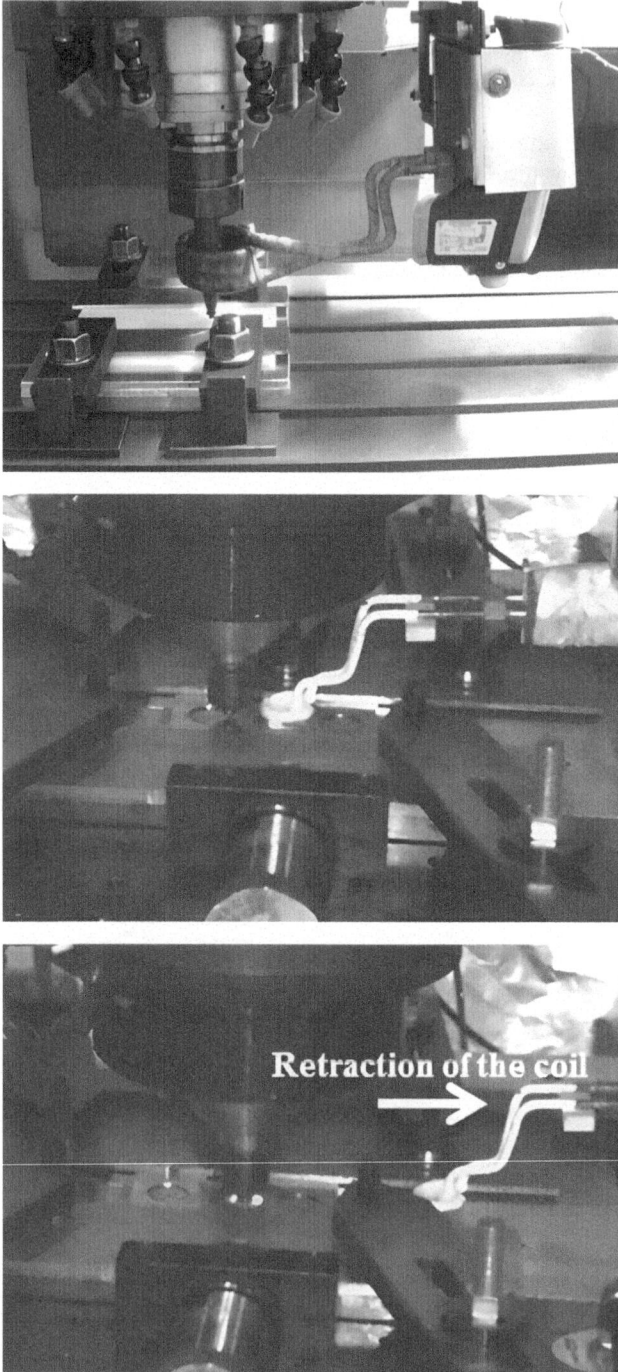

Figure 5.9 (Continued)

Interfacial failure

Plug failure mode

Figure 5.9 (Continued)

the mechanical and metallurgical properties of high-density polyethylene plates by varying the tool rotational speed and tool pin temperature. They found a narrow transition zone between the weld and base metal that exhibited no defects, and the strength was identical to that of the base metal. At almost all process parameters, however, there is a drop in the hardness of the weld zone, and the conversion from brittle to ductile in the joint was observed when the tool pin temperature increased [18], as shown in Figure 5.10. Moreover, Sun et al. investigated the mechanical and microstructural properties of S12C low carbon steel that was welded by high-frequency induction-assisted spot friction stir welding. Keeping the same process parameters, they found that the average grain size of the FSW welded material was smaller. In the same induction-assisted friction stir welding, metals were heated up to 10 s that showed a slight increase in average grain size in the stir zone. Additionally, in induction-assisted friction stir welded joints in which prior heat was given, a larger joint interface was formed, resulting in a significant increase in bonding strength of the welded joint and a considerable increase in the shear tensile load of the welded joint. A fracture in the joint was observed due to plug failure mode rather than interfacial failure mode [17], as shown in Figure 5.9f. IaFSW has some

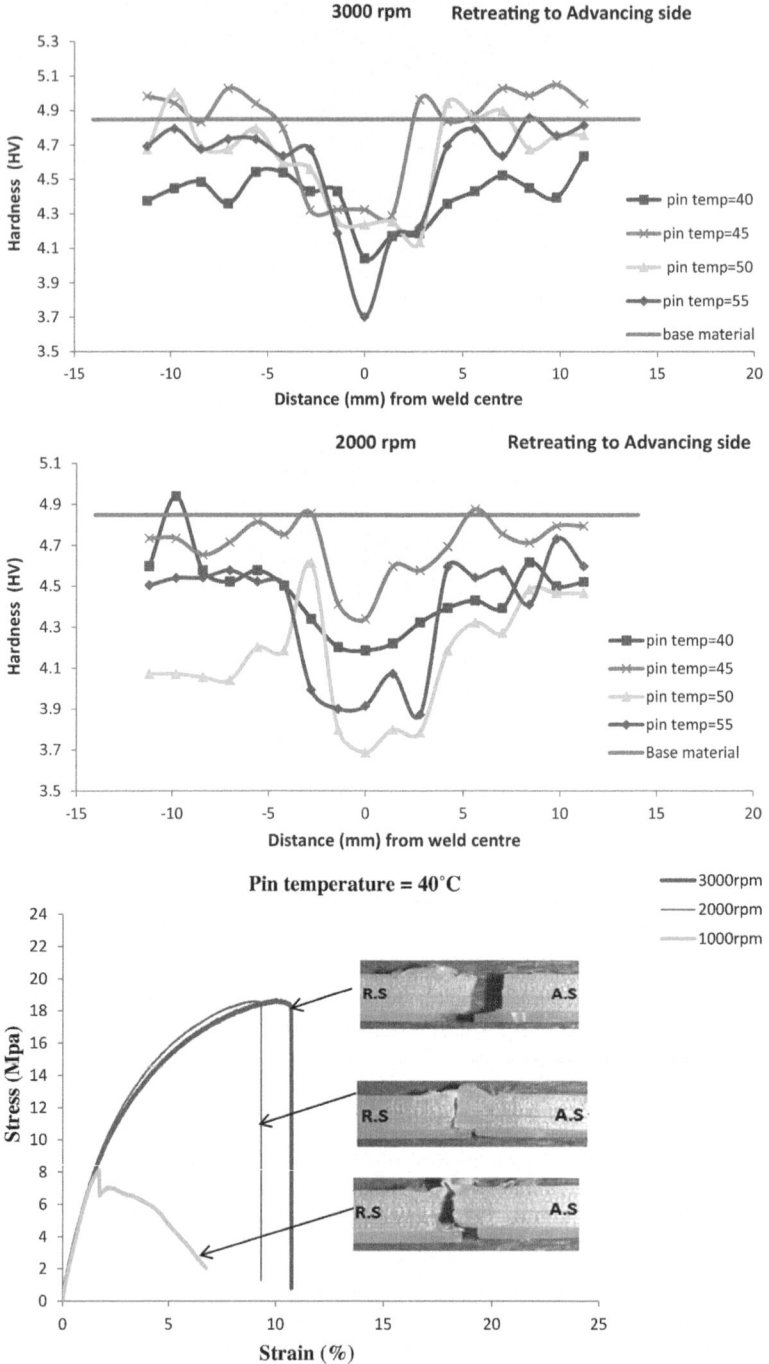

Figure 5.10 Hardness value and stress-strain curve at different tool temperatures [18].

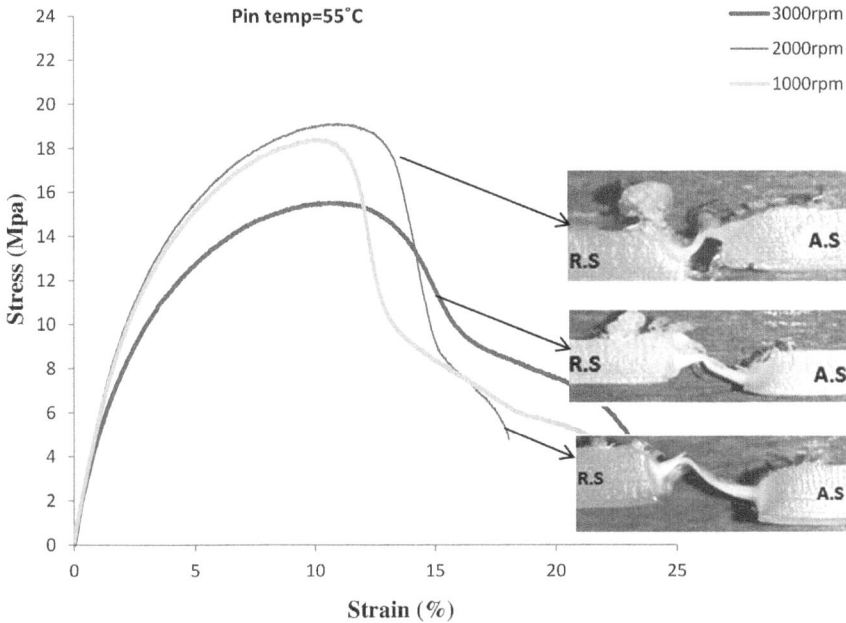

Figure 5.10 (Continued)

Table 5.3 Materials, FSW tools and process parameters to use in IaFSW.

References	Sample materials to be welded	FSW tool	Tool rotational speed/RPM	Traverse speed/mm min^{-1}	Operating power/kW
[17]	S12C low carbon steel (1.6 mm)	WC	800	. . .	. . .
[18]	Thermoplastic (high-density polyethylene plates) (5mm)	H13 tool steel	1000, 2000, 3000	50, 100	. . .

limitations, however, as only electrically conductive material can be welded by this process. The process parameters of different reference papers included in the review of IaFSW are shown in Table 5.3.

5.5 ULTRASONIC VIBRATION ENHANCED FRICTION STIR WELDING (UVEFSW)

In ultrasonic welding, a high-frequency ultrasonic vibratory energy superimposed on static load exceptionally softens the material and much less heat is generated, around 35% to 50% of the melting point temperature, causing the joint to be made through localized plastic deformation at the interface. Furthermore, different components used in UVeFSW are shown by schematic diagrams in Figure 5.11a and b. In UVeFSW, many researchers, using ultrasonic vibration assistance to different materials, studied the changes in their mechanical and metallurgical characteristics, and compared them to the conventional FSW with the same process parameters. In that order, Baradarani et al. observed the effect of UVeFSW on AZ91−C magnesium alloy and compared the change in mechanical and metallurgical properties of the material with conventional FSW on the same parameters, and they observed that both tensile strength and hardness were significantly greater in UVeFSW welded material than in conventional FSW welding welded material, as shown in Figure 5.12a and b. Also, in their microstructural analysis of the workpiece and in this analysis, they found that β-$Mg_{17}Al_{12}$ coarse dendrites are segregated to very fine and partly spherical particles that homogeneously distribute in an α-Mg matrix. Therefore, they concluded that ultrasonic vibrations play a very important role in UVeFSW for this kind of remarkable modification in the microstructure [19]. In addition, Zhong et al. investigated the effect of welding load, temperature and material flow, with or without ultrasonic vibrations, in friction stir welding and found that due to ultrasonic vibration traverse force, tool torque and axial force of welding reduce significantly. At a higher speed, the effect of ultrasonic vibrations on the tool torque was more noticeable than traverse force and axial force. In the microstructural investigation, they observed that ultrasonic vibration improved the material flow as well, and material refilling increased from the RS to the AS. The inner defect was either reduced or eliminated and the strength of the weld joint was also significantly enhanced [20]. In addition, Kumar et al. applied ultrasonic vibrations in the stir zone with a specially designed welding tool, as shown in Figure 5.11b. They showed its effect on plunging force, tool

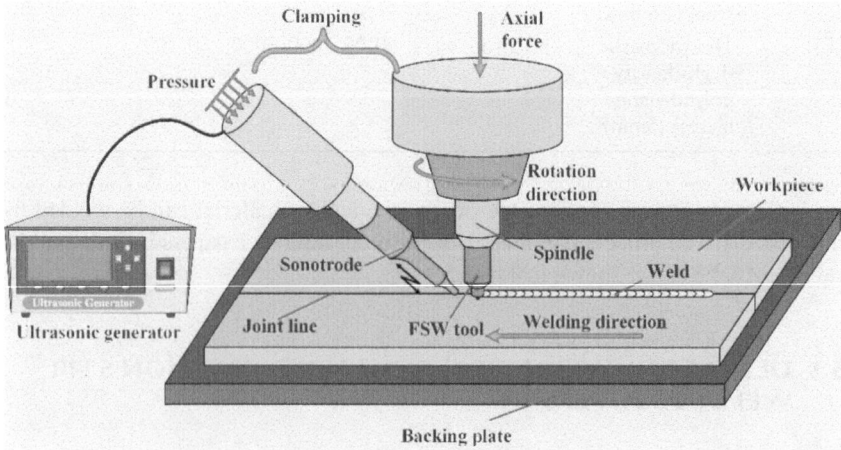

Figure 5.11 Schematic view of the UVeFSW setup [20].

torque and tool input power by varying the welding and rotational speeds, and they found that due to ultrasonic vibrations, the values of these three parameters were reduced by approximately 38%. Consequently, in conventional FSW, IMC phases were found in the stir zone that reduced the weld strength of the welded sample.

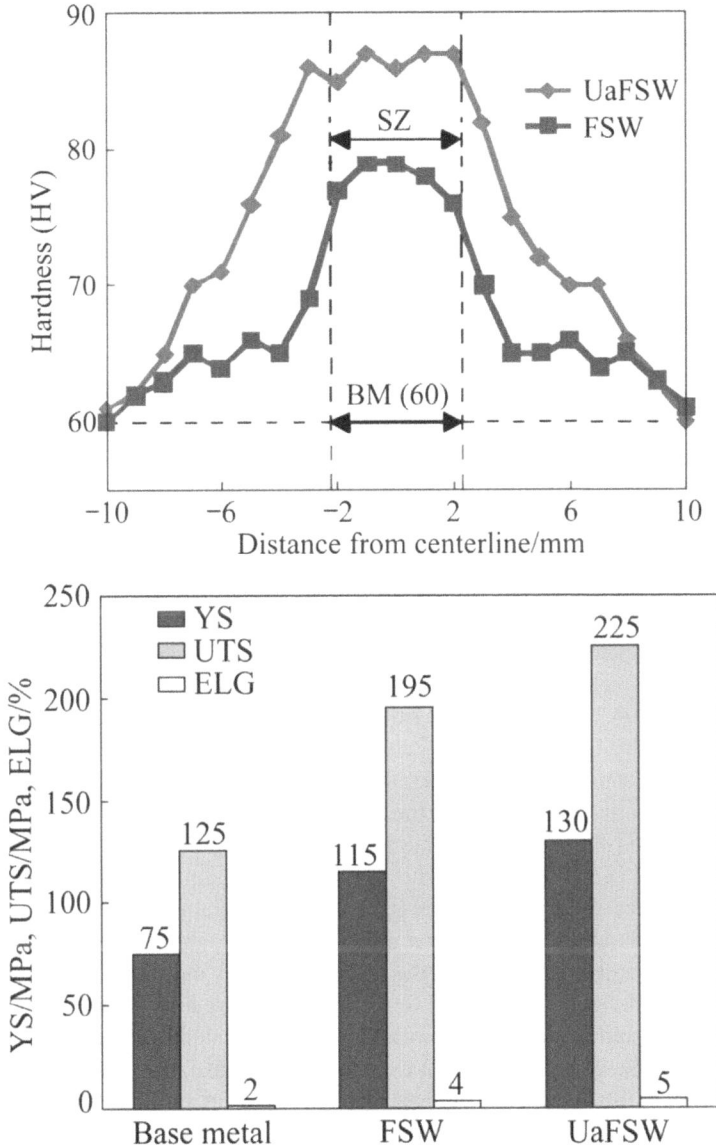

Figure 5.12 (a) Hardness profiles of UVeFSW and FSW joints. (b) Comparison of yield stress (YS) and ultimate tensile strength (UTS) [19].

By using ultrasonic vibration assistance, these unfavorable IMC phases are uniformly fragmented and mixed due to acoustic action, resulting in unprecedented improvements in weld strength, interfacial bonding, and the weld surface quality of the welded sample [21]. Following this, Lu and Wu examined in detail for tunnel defects when aluminum alloy 2024Al-T4 was welded by FSW and UVeFSW. They also explored the elimination of these defects using the marker insert technique based on the "tool sudden stop action" technique and metallographic observation. It was found that due to the time delay of material flow, this particular defect was happening in the welded sample. Furthermore, for a detailed analysis, they divided the weld nugget zone into three different sub-zones, shoulder affected zone (SAZ), pin affected zone (PAZ), and weld bottom zone (WBZ), and observed in their investigation that in PAZ, there was a delay in the flow of material due to which this defect was generated. Two factors were responsible for this defect. The first factor was the insufficient material flow from RS to AS, and the second was due to the shortage of material transfer in the downward direction in SAZ. Ultrasonic vibrations reduced the time delay in material flow by broadening the PAZ; in broader PAZ, the demand for filled material from SAZ was also reduced due to which no tunnel defect was found in the welded joint even at higher welding speed and with less plunging force [22]. Moreover, Tian et al. [23] and Shi et al. [24] showed the mechanical and metallurgical properties of the Al/Cu welded joint that was welded by UVeFSW, and in this case, the results of the welding loads and mixing of materials in the stirring zone were almost the same as previously discussed. Besides, they optimized the process parameters and found when the rotation speed was 500 rpm, the value of weld tensile strength improved approximately by 60.7% compared to the joint welded by conventional FSW [23]. Furthermore, Padhy et al. compared the local microstructure evolution and macrotexture of the Al6061-T6 weld nuggets, welded by UVeFSW with the same material welded by conventional FSW. They observed that the sample in the weld nugget zone was deformed easily welded by UVeFSW than by conventional FSW due to the prior supply of ultrasonic assistance. Electron backscattered diffraction (EBSD) analysis showed ultrasonic vibration had improved the recrystallization process because the grain was refined and there was a variation in the grain orientation along the vicinity of the weld centerline. EBSD showed that application of ultrasonic vibration improves the recrystallization process, causes better grain refinement and brings about variations in grain orientation along the weld center axis [25]. Following this, Alinaghian et al. investigated the longitudinal residual stress as well as weld quality and weld joint strength and found that supplying high frequency and 200-watt power in ultrasonic-assisted welding caused a reduction in tensile residual stress up to 45% with an increase in weld tensile strength. Also, due to weld peening effects, voids and tunnel defects were not seen in weldment [26]. Finally, Wu et al. welded 2024-T3 aluminum alloys using conventional FSW as well as UVeFSW and examined the fatigue strength and life of the joints welded by them. They showed that for 50% survival probability, there was an increase in fatigue strength of the joint welded by UVeFSW; the strength was up to 96.13% of base metal strength. In the microstructural analysis, they observed that in the advancing side, due to material flow near the part above the transverse cross-section of the weld, a weak bond was being formed where the stress concentration reaches its maximum value, and finally, the fatigue crack was initiated at that point. Also, they showed that by using UVeFSW, which narrowed down the weak bonding region and also reduced the stress concentration there in the crack propagation zone, they found fatigue striations in UVeFSW joint fracture where they were parallel to each other and spaced apart. Contrastingly, in conventional FSW, fatigue striations were found to have inappropriate parallelism and enormous spacing between them, as shown in Figure 5.13. By using ultrasonic vibrations, the secondary-phase particles were fragmented into a smaller size, increased in quantity and distributed homogeneously. In conclusion, all these factors enhanced the fatigue performance of the welded joint [27]. The process parameters of different reference papers included in the review of UVeFSW are shown in Table 5.4.

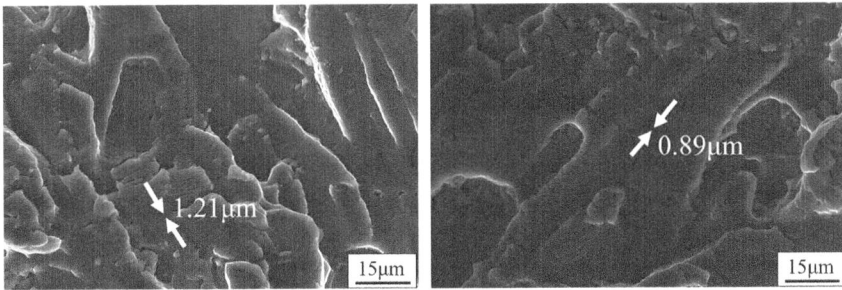

Figure 5.13 Fatigue striation in crack propagation region [27].

Table 5.4 Materials, FSW tools and process parameters to use in UVeFSW

References	Sample materials to be welded	FSW tool	Tool rotational speed/RPM	Traverse speed/mm min⁻¹	Ultrasonic specifications
[19]	AZ91−C Mg alloy (6 mm)	H13 tool steel, heat treated	1400	40	Amplitude 15 μm
[20]	AA6061-T6 to AA2024-T3 plates (6 mm)	. . .	200–800	65–330	Frequency 20 kHz, amplitude 40 μm, output power 500 W
[21]	AA6061-T6 to AZ31B Mg alloy (3 mm)	H13 tool steel	400–1200	50–250	Frequency 20 kHz, power 3000 W, amplitude 25 μm
[22]	Aluminum alloy 2024Al-T4 (3 mm)	. . .	600	80–150	Frequency 20 kHz, output power 300 W, amplitude 40 μm

(Continued)

Table 5.4 (Continued)

References	Sample materials to be welded	FSW tool	Tool rotational speed/RPM	Traverse speed/mm min⁻¹	Ultrasonic specifications
[23]	6061-T6 Al alloy and C11000 copper (2 mm)	H13 tool steel	500–800	60	Frequency 20 kHz, amplitude 25 μm
[25]	Al 6061-T6 plates (6 mm)	Tool steel	800	320	Frequency 20 kHz, output power 300 W, amplitude 40 μm
[26]	AA 6061-T6 (3 mm)	AISI H13	700	50	Frequency 20,347 Hz
[27]	2024-T3 Al alloys (6 mm)	. . .	800	75	Frequency 20 kHz, output power 500 W, amplitude 40 μm

5.6 OTHER HYBRID ASSISTED FRICTION STIR WELDING PROCESSES

In addition to the hybrid friction stir welding discussed previously, some other types of thermally assisted welding processes compiled in this section, such as arc assisted welding, exothermic heat-assisted friction stir welding (EHaFSW) and plasma-assisted friction stir welding. It was observed in this review that the experimental setups were common in all of these welding processes except in EHaFSW, where the preheat source was kept 10–20 mm ahead of the FSW tools, and to weld the dissimilar type workpiece, the higher strength material was kept in the AS concerning the weld centerline. The relatively softer material was kept on the RS. Apart from this, to weld the dissimilar material, the tool was offset towards relatively softer material. This was because their thermophysical properties had huge differences due to heat generation; resistance to mechanical load and flow stress values were different in both plates. Therefore, to reduce the formation of IMCs in the dissimilar type of welding, the tool is generally placed towards the relatively softer material.

We will discuss the major findings of various researchers in this particular field, especially in arc assisted welding, EHaFSW and plasma-assisted welding.

Bang et al. welded Ti and Al alloy samples through gas tungsten arc assisted FSW, where Ti alloy was positioned in the AS and Al alloy in the RS, and the GTAW torch was ahead of 20 mm with the FSW tool. Furthermore, Ti alloy had been preheated by the gas tungsten arc torch so that the melting point temperature difference of the two would be reduced. They compared the mechanical and metallurgical properties of the welded joint by the gas tungsten arc preheat assisted FSW with the traditional FSW welded joint, and found the ultimate tensile strength (UTS) of arc assisted FSW welded joint was 91% of that of base material and elongation of the joint welded by arc assisted FSW was 24% more than the conventional FSW welded joint, whereas fracture of the arc assisted FSW welded joint was ductile in nature [28].

Following this, Bang et al. investigated the mechanical and metallurgical properties of Al6061-T6 and SS400 steel, welded by TIG assisted FSW. They showed that the strength of the welded joints from TIG assisted FSW was about 104% of the strength of Al alloy base metal, and the grain size of welded joints, by TIG assisted FSW in HAZ and TMAZ of aluminum alloy, was fine compared to welded joints with conventional FSW [29]. Subsequently, Bang et al. welded Al5052 Al alloy and DP 590 steel through TIG arc assisted friction stir welding (AAFSW) that had thicknesses of 2.5 mm and 1.4 mm respectively. They observed similar results as were found in other thermally assisted welding processes discussed previously. AAFSW enhanced the plastic flow behavior and also reduced the stresses on the tool. Apart from this, they analyzed Al-Fe IMCs at different TIG currents and found that at 20 A the average maximum tensile strength value was 184 MPa and the size of the IMCs at that current was 2.39 μm. The reason for this behavior was heat generation per unit length of the time was affecting the growth of maximum temperature and IMCs layer thickness [30]. Furthermore, Siva.et al. used exothermic heat-assisted friction stir welding (EHaFSW) to weld the Ni-Al bronze alloy by preheating the exothermic reaction of Al/CuO thermites and comparing the changes in mechanical and metallurgical behavior of the welded joint with conventional FSW joint. It was found that due to EHaFSW, top to center α-phases in the stir zone were seen at an enormous amount; also, transverse tensile strength was increased by 27% and 14% in the joint welded by EHaFSW and traditional FSW respectively compared to the tensile strength of the base metal. They considered the temperature rise to be an important factor behind it. Apart from this, they observed that due to the increase in temperature, the tunnel defects in the welded parts were eliminated, whereas, in the stir zone, the hardness value was the highest in the bottom part of the material welded by EHaFSW, while the material welded by traditional FSW was higher in top and center. For this, the grains of joints welded from EHaFSW were finer than those of traditional FSW. The presence of β-phases in a larger amount was also a major reason for this [31].

Finally, Yaduwanshi et al., by taking the plasma as a preheat source, welded Cu and Al alloy samples through plasma-assisted friction stir welding, in which only harder material was preheated by the plasma heat source. They observed that if Al and Cu were welded by conventional FSW, there would be a greater difference in their temperature and yield strength value, and the difference between these two values was reduced due to preheating by the plasma torch that caused an easier material flow and reduced stresses on the tools. Finally, plasma heating showed satisfactory improvement in the bond strength due to preheating at the interface of the welded joint [6].

The process parameters of different reference papers that have been discussed in this review of arc assisted, exothermic heat-assisted friction stir welding and plasma-assisted FSW heating assisted FSW are shown in Table 5.5.

Table 5.5 Materials, FSW tools and process parameters to use in arc assisted, thermite heating assisted and plasma-assisted FSW

References	Sample materials to be welded	Tool material	Tool rotational speed (RPM)	Tool traverse speed (mm/min)	Specifications/pre-heat source type
[28]	Al6061-T6 Al alloy, titanium alloy (3.5 mm)	WC-12% CO	300–450	60–84	Gas tungsten arc assisted welding (GTAAW)
[30]	Al5052 (2.5 mm), DP590 (1.4 mm)	WC-12% CO	400	60	(GTAAW), arc length 2 mm, distance to tool 20 mm, interface offset 5 mm, torch inclination 60°, Ar shielding gas (99.99%)
[29]	Al6061-T6, SS400 (3 mm)	WC-12% CO (WF20)	400	48	TIG current 60 A
[31]	Ni-Al Bronze Alloy (6 mm)	WC-based alloy tool	1600	100	Exothermic/thermite heat-assisted FSW using Al/Cuo thermites
[6]	Pure copper, AA1100 (6 mm)	. . .	440–815	63–200	Plasma assisted, preheating current 0–65 A

5.7 MODELLING

Generally, the FSW modelling process is categorized into either the area of application that includes flow models or residual stress models or the continuum mechanics approach which includes computational solid mechanics (Lagrangian) based models and computational fluid dynamics (Eulerian) based models. The main objective of any residual model is to calculate the temperatures by applying the given boundary conditions, energy equations, heat sources or other given conditions in Lagrangian (fixed coordinate system) or Eulerian (moving coordinate system) frame to determine how the heat is generated by the rotating tools. Two types of thermomechanical coupled models are mentioned in the literature. The first modeling approach is a fully coupled thermomechanical model that makes realistic predictions about the material flow as well as the formation of the shear layer, heat generation and temperature field during welding. The second approach is semi-coupled thermomechanical models, in which material flow is generally not considered during welding and surface heat flux usually represents complete heat generation, as well as avoiding source terms in energy equations. In addition, it is normally modeled in a Lagrangian frame and the thermal field is calculated before the mechanical field. In this order, residual stresses are easily predicted by this modeling approach, while the fully coupled thermomechanical modeling approach is more suitable to predict the temperature fields and mechanical properties during welding. As previously discussed, fully coupled flow models in FSW are based on either computational fluid dynamics (CFD) or computational solid mechanics (CSM), in which the CSM-based model is capable of predicting the residual stresses on its own. In contrast, the CFD-based model cannot predict residual stress by itself unless it is coupled to any CSM-based model. The CSM-based model is analyzed by arbitrary Lagrangian-Eulerian (ALE) formulation in which dynamic equilibrium equations are solved in an explicit manner [32].

Long and Khanna welded high strength and high melting point materials like steel and titanium alloys through electrically enhanced friction stir welding process (EHFSW) rather than via conventional friction stir welding, a drawback of which was high tool wear and low welding speed. By using finite element modeling, they analyzed the weld characteristics of the welded joint. Modeling hybrid friction stir welding is a very labyrinthine task, so they modeled EHFSW finite element modelling by modifying the thermal–electrical–mechanical finite element (FE) code which was modeled by previous researchers for analyzing the electrical resistance spot welding process.

In addition, they used Fourier's second law as a governing equation for heat flow analysis in FSW for finite element modeling of the process, which is shown mathematically as follows.

$$\left(\rho_C\right)\frac{\partial T}{\partial t} = k\left[\frac{\partial^2 T}{\partial x^2} + \frac{\partial^2 T}{\partial y^2} + \frac{\partial^2 T}{\partial z^2}\right] + q_0 \tag{5.4}$$

Where ρ, c, k, t, T, q_0 and x, y, z are the density of the materials, specific heat, thermal conductivity, time, temperature, inner heat source and coordinate systems, respectively. Material properties c and k both depend on the temperature. At the same time, the inner heat source contained three heat sources:

$$q_0 = q_{0F} + q_{0P} + q_{0R} \tag{5.5}$$

Where, q_{0F} is the friction heat, generated due to friction between tool faces and the workpiece material. q_{0P} is plastic work heat that typically arises from material flow due to stirring by tool pins at the bottom of the workpiece during FSW and q_{0R} is electrical resistance or joule heating. q_{0F} is the main heating source in the upper part of the workpiece. Furthermore, they assume that if all the shearing energy on the tool-workpiece interface is converted into friction heat, then the average friction heat per unit area per unit time can be written in the mathematical form is given in Equation 5.6.

$$q_{0F} = \frac{4\pi^2 \mu PNR^3}{3} \tag{5.6}$$

μ, P, N, and R are friction coefficient, pressure subjected to the tool, tool rotation speed and tool shoulder surface radius, respectively. The researchers also found that the heat generated from friction is proportional to μ, where μ depends on temperature, which is approximately found to be 1 at the time of welding start. The value of μ decreases as the temperature increases. Also, heat generation q_{0P} for time (t) and space (x, y, z) generated in the material due to plastic flow is shown mathematically through the following equation:

$$q_{0P}\left(\mathbf{t},\mathbf{x},\mathbf{y},\mathbf{z}\right)=\eta s_{ij}\dot{\varepsilon}_{ij}^{p}$$

(5.7)

Where $s_{ij}, \dot{\varepsilon}_{ij}^{p}$, and η are the deviatoric stress, incremental plastic strain rate tensors and energy input coefficient, respectively.

Furthermore, they found in their analysis that plastic work heat depends on strain rate. Also, the lower part of the workpiece was found to be relatively colder than the upper part. Because of this temperature difference, the lower cooler surface of the workpiece is responsible for the tool wear and low welding speed. In addition, by preheating using a hybrid heat source, the temperature difference at the bottom surface is minimized and the material is softened enough, which significantly reduces tool wear during stirring and also increases the welding speed. In this case, they used EHFSW, and this joule or electric resistance heat produced by electric current is mathematically shown in the following equation:

$$q_{0R} = I^2 Rt$$

(5.8)

Where I, R and t are the electric currents, the resistance of the tool-workpiece interface and the time for which current is passed, respectively.

Furthermore, the researchers analyzed it in two friction stir welding steps numerically in EHFSW, the first being the tool pin plunge stage and the second the welding speed stage. Through the finite element model given earlier, they measured temperature distribution in these two stages and observed that the significant increase in temperature on both the surface and the lower part of the material was due to both electric current heat and friction heat in the EHFSW process during the plunge stage. Based on this, they expected EHFSW to have less tool wear in welded joints than conventional friction stir welding, but at the same time during the welding stage both conventional FSW and EHFSW had the same temperature distribution, but the EHFSW process was accomplishing the temperature profiles in half the time compared to conventional welding. Finally, they concluded that the welding speed in the EHFSW welding process should be at least double that of the conventional welding process [33].

In addition, the previously discussed modeling approach can be used to model similar materials, but that approach cannot be used to model for dissimilar materials because the thermomechanical properties such as density, specific heat and thermal conductivity will be different in both samples. Yaduwanshi et al. modelled and experimentally validated this approach to weld Al and Cu sheets, two different materials, by friction stir welding. They found that welding Al and Cu through a conventional FSW is a complex process as it contributes to asymmetry in material flow and heat generation due to the different physical and mechanical properties of both materials. Furthermore, they preheated the Cu sample from 500 to 600 K by placing the plasma heating source at the optimal offset distance relative to the weld centerline on the Cu side. In addition, they found a significant reduction in asymmetry in material flow as well as an adequate amount of material flowing around the tool. For modeling, they developed a 3D heat transfer P-FSW model as well as calculated the transient temperature field through this model in which they considered the interaction between tool-workpiece and backing plate within the geometry. In addition, they used the 3D nonlinear heat conduction equation as a governing equation to measure the temperature field, in which the tool was moving along the weld centerline. The governing equations in which the coordinate system is moving are shown next in the form of the mathematical equation:

$$\frac{\partial}{\partial x}\left(k_x \frac{\partial T}{\partial x}\right) + \frac{\partial}{\partial y}\left(k_y \frac{\partial T}{\partial y}\right) + \frac{\partial}{\partial z}\left(k_z \frac{\partial T}{\partial z}\right) + \dot{Q} = \rho c_p \left(\frac{\partial T}{\partial t} - V_T \frac{\partial T}{\partial y}\right) \qquad (5.9)$$

where k specifies the thermal conductivity and $\rho, c_p, V_T, \dot{Q}$ specify the density, specific heat of the material, transverse speed of the tool and rate of heat generation, respectively.

As has been mentioned, the heat generation in the conventional FSW process is due to tool-workpiece friction and the plastic deformation of the stirred material. In addition, the heat generated by the tool-workpiece was treated as surface heat flux and the heat generated by plastic deformation was treated as volumetric heat. They consider this surface heat in the solution domain as the boundary conditions in the FSW process.

In the governing equations, the applied initial boundary condition and the convection and radiation heat loss from the surface are shown mathematically as follows:

$$k\frac{\partial T}{\partial n} = h\left(T - T_0\right) + \varepsilon \sigma \left(T_0^4 - T^4\right) - q_s - q_{pre} \qquad (5.10)$$

Where, q_{pre} is the heat energy provided by the external heat source, which is considered the surface heat flux. They found that plasma heating and friction heating were responsible for generating heat in the P-FSW, where plasma preheats the samples to be welded (q_{pre}). In addition, the actual total heat produced in the FSW process (Q_{FSW}) is due to friction between the tool-workpiece interface (Q_f) and the plastic deformation of the samples (Q_p).

Mathematically, it is represented by the following equations:

$$Q_{FSW} = Q_f + Q_p \tag{5.11}$$

$$Q_{P-FSW} = Q_{FSW} + Q_{pre} \tag{5.12}$$

In addition, they used the flat shoulder and straight cylindrical tools for P-FSW welding and assumed total heat generated from friction is a linear unification of sliding and sticking conditions. This is mathematically shown by the following equations:

$$Q_{FSW} = \delta * Q_{Sticking} + (1-\delta) * Q_{Sliding} = \frac{2}{3}\pi\omega$$
$$\left[\partial\tau_{yield} + (1-\delta)\mu p\right] * \left\{\left(R^3_{sholder} - R^3_{probe}\right) + R^3_{probe} + 3R^2_{probe}H_{probe}\right\} \tag{5.13}$$

Where ω is the angular speed of the tool, μ is the coefficient of friction, H specifies tool probe height, and R specifies the radius of the tool pin and tool shoulder surface, respectively. Included are contact stress variables that establish a relationship between the velocity of the sample contact surface and the velocity of the tool surface. In practice, the value is less than 1, because some friction work in the sample is converted into heat energy. In the same full sliding condition, its value is 0 and in the full sticking condition, it is 1. Contact state variable (δ) is shown mathematically by the following relation:

$$\delta = \frac{V_{matrix}}{V_{tool}} \tag{5.14}$$

Also, heat input by preheating source, in this case plasma heating, can be calculated by the following equation:

$$Q_{pre} = \eta VI \tag{5.15}$$

Where η, V and I are the thermal welding current and voltage, respectively.

Despite this, both materials were forming intermetallic compounds after plasticization and mixing. Moreover, these IMC crystal structures were different from either component (in this case Al and Cu). At the same time, IMC crystal structures also affects joint efficiency in the weld nugget zone. Therefore, it is important to include the effect of IMCs in thermal analysis for the design of material properties. Many researchers in their study found that the formation of IMCs in the nugget zone and their flow pattern were similar to a functionally graded material in the weld zone. In addition, the researchers introduced the concept of time-varying functionally graded material (FGM) to define the weld zone size. In their analysis, they found that when the tool was a plunge to a fixed position along with the thickness in the sample; the weld zone size was increasing but when the tool was moving along the weld centerline there was no significant change in the weld zone size. Furthermore, they analyzed the workpiece by dividing it into three distinct regions – the advancing side (Cu), the retreating side (Al) and the middle region (i.e. FGM). They treated all three regions as three different materials and considered the individual properties of the material in the no-weld zone. Furthermore, the material properties in the weld zone varied according to the mixture rule of the FGM, which depends on various parameters like flow pattern of material. They calculated FGM properties such as specific heat, thermal conductivity, and density from the mathematic relations given as follows:

$$k = k_1 \left[\frac{\left(1 + 3\left(k_1 - k_2\right)v_2\right)}{\left(3k_1 v_2 + \left(k_1 + 2k_2\right)\left(1 - v_2\right)\right)} \right] \tag{5.16}$$

$$c = \frac{\left(c_1 \rho_1 v_1 + c_2 \rho_2 v_2\right)}{\left(\rho_1 v_1 + \rho_2 v_2\right)} \tag{5.17}$$

$$\rho = \rho_1 v_1 + \rho_2 v_2 \tag{5.18}$$

where k, c, and ρ represent thermal conductivity, specific heat and density in the context of FGM. In Equation 5.18, subscript 1 and subscript 2 denote the thermal conductivity, specific heat and mass density of Cu and Al, respectively. v_1 and v_2 are the volume fraction of copper and aluminum, respectively.

They validated this numerical model based on the results of the time-temperature history and the computed isotherm of nugget zones obtained by the P-FSW experiment, and they found that the heat transfer model used to specify the heat input in the P-FSW process gave relatively precise results in predicting the peak temperature during the process; also, the overall error in peak temperature was around 5%–12% with a maximum reliability of 0.84 [6].

5.8 DEFECTS

The welded material from FSW consists of tunneling, kissing bonds, void, joint line remnant, incomplete root penetrations and hooking defects. Generally, insufficient heat generation around the pin, an inappropriate amount of material movement around the pin and an inappropriate amount of material consolidation behind the pin are the main reasons for these defects. Heat generation and stirring of the material depends somewhere on the process parameters. If they are not properly selected or optimized, this generates defects in the welded joint. In this order Dehghani et al. welded Al and steel by FSW, keeping the tool rotation speed constant and varying the welding speed, tool tilt angle, tool pin geometry and plunge depth. They observed the influence on the formation of IMCs, tunnel defects and weld tensile strength. They found that at low traverse speeds, thicker IMCs were present in the weld zone, resulting in reduced weld tensile strength. Additionally, they observed the formation of tunnels at lower welding speeds. When they increased the traverse speed, the amount of IMCs was found to be significantly lower and found only in the upper part of the weld. Because of fewer IMCs, a remarkable increase in weld tensile strength was found, but too high of a traverse speed and a low plunge depth weaken the bond strength, so the traverse speed can be increased to an optimal value only. Furthermore, they observed by varying the diameter of the cylindrical pin from 4 mm to 3 mm, in both cases, tunnel defects were being formed, which means they did not see any significant improvement in the tunnel defects by varying the size of the pin. Again, by tilting the tool to 5 degrees as well as varying the plunge depth, due to the high heat generation and forging force, the tunnel defects were not found in the welded material but the weld strength was very low due to larger IMC thickness. Finally, using a special M3 pin tool, they showed that the welded joint on the 56 mm/min was forming a bell shape nugget. Also, because of the extreme thinness of IMCs at that particular speed, the weld strength was at a maximum due to the tool being a threaded pin. This increased the forging forces as well as helped to send the plasticized material in the downward direction. Moreover, they found weld strength very susceptible to variation of the plunge depth; when the plunge depth decreased, the weld strength decreased significantly [34].

Kim et al. looked at the optimal conditions by varying the plunging force and found that the range of optimal conditions was getting wider as the plunging force increased. Generally, they observed three types of defects. First, when the tool rotational speed was very high, excess heat generation caused a very large amount of flash formation. Second, due to high traverse speed, a cavity or groove-like defect was found, due to insufficient heat generation. Third, cavity defects were generally due to abnormal stirring, which looked distinctly different on the top surface of the AS, which was arising due to less heat generation and decreasing as the plunging force increased [35].

Furthermore, Chen et al. investigated the Al 5456 alloys in which they mainly observed the defects and weld properties generated at different tool tilt angles. As previously discussed, the tilt angle turns the plastically deformed material flow pattern in the stir zone and controls the weld properties accordingly. In addition, they observed that the oxide layer available on the butt surface before welding is dispersed on the grain boundary during FSW, and the failure of the welded joint was induced by this oxide layer [36].

Song et al. welded AA2024-T3 and AA7075-T6 aluminum alloy sheets through FSW by placing them in two lap joint combinations. In the first combination, AA2024 was placed on top and AA7075 placed on the bottom (2024/7075). In the second combination, the position of the plates were reversed (7075/2024). Again, they observed the effect of process parameters (weld speed, lap joint combinations) on the defect features produced in a welded joint (hook, voids), weld joint strength and mechanical properties, and found that hook in both lap weld positions at low welding speed were deflecting upward significantly in the stir zone. In 2024/7075, due to the lap shear test, the weld joint fracture was occurring from the tip of the hook from the AS at the SZ/TMAZ interface where voids were also appearing. At the same time, when the welding speed was high in 7075/2024 joints, the hook geometry was extending horizontally at a long distance in the bottom stir zone on the RS. In addition, the welded joint was showing fracture in three modes – the first shear fracture was happening at the lap joint interface, the second one was a tensile fracture and the last was a mixture of both. They found, in both welded lap joint combinations, the lap shear strength was increasing as the traverse speed increased. Also, at lower traverse speeds in the 7075/2024 joint, there was a higher failure load relative to the 2024/7075 joints, while at higher speeds the failure load result was opposite [37].

Morisada et al. visualized the material flow through X-ray radiographs and investigated the defect formation mechanism in the joint welded by FSW. They observed that the tilt of the material flow around the tool and the sedentary of the material flow in the RS of the tool were correlated with the defect formation. In addition, they also calculated the material flow velocity during FSW based on 3D visualization and found that in the AS where the defects were forming, the material flow velocity was noticeably lower [38].

Chauhan et al. used the coupled Eulerian and Lagrangian method to model the FSW process as well as the volume of fluid principle they used to predict the defects formed during the process. In addition, they validated experimentally observed spindle torque and plunging force with the model. At three different probe heights, they welded the plates by simulation and got optimal results at 2.5 mm heights. Following this, they successfully predicted the defects on different process parameters from the model and also obtained a defect-free welded joint on the 2-degree tilt angle [39].

Huang et al. used a novel multi-functional tool probe with circumferential notches to eliminate hook and cold lap defects from dissimilar lap joints, welded by FSW. They showed, due to the use of noble pins, that turbulent flow regions were relatively larger and that at the bottom of the tool pin plasticized material was converging. They observed that the tensile shear load of the welded joint 6082-T6/2A12-T4 reached 85% of the base alloy 6082-T6 strength. In addition, they obtained a sound weld, as well as a large bonding area and a limited diffusion bonding layer of 2 mm [40].

Moreover, the kissing bond defects appeared at the interface of the stir zone. The main reason for this defect is the inappropriate removal of the oxide layer from the weld interface, the insufficient amount of material movement and very high welding speed.

Generally, phased array ultrasonic NDT is used to detect kissing bond defects. Also, solid-state filling/fusion filling plus FSW, fusion welding and friction plug welding are the major methods to repair commonly welded joints by FSW in engineering applications. Incomplete root penetration defects appear normally below the weld centerline in the stir zone. The main reasons this defect forms is the difference in plate thickness, poorly designed tools and reduced pin length.

5.9 APPLICATIONS

Miles et al. found in the starting phase of FSW that this welding process has more potential than conventional welding to weld sheets made of aluminum, Cu, Cu alloy, Ti and Mg alloy. They suggested the future researcher use it in the transportation industry because Al is three times lighter than steel, and its use in the transportation sector will result in high speed, high payload, low emissions and low fuel consumption [41].

Wahid et al. analyzed aluminum alloys (AAS), mainly the 5xxx and 6xxx series, because of their high strength-to-weight ratio and high corrosion resistance properties. They are primarily used for marine applications and shipbuilding, usually for constructing hulls, superstructures, deck panels, bulkheads, stack enclosure, keen, gunwales, monohull vessels and superstructures. The researchers found that it was difficult to weld these materials through conventional fusion welding because of the enormous difference in their chemical and mechanical properties. In addition, corrosion of Al alloys was also a serious concern. They also found in their analysis that welding Al alloy by FSW compared to fusion welding showed unprecedented increases in weld joint strength and ductility. Additionally, considering the advantages of FSW in the field of shipbuilding, they suggested hybrid friction stir welding has immense potential [42].

Wang et al. studied aluminum alloy, which is used extensively in the aerospace industry, and found that FSW is widely accepted for fabricating aluminum alloys, including the joining of large-volume fuel tanks and rocket tanks. This is because the FSW welded joint exhibits notable properties such as low distortion, fewer defects compared to fusion welding and fine equiaxed grain due to dynamic recrystallization in the stir zone [43].

In conclusion, the application of FSW has been reported in previously published literature, summarized in the next section.

Engine and chassis cradles, wheel rims, track bodies, cryogenic tanks for space shuttles, helicopter landing platform goods wagon, container bodies and refrigerator panels are some of the applications in which this welding can be used extensively.

5.10 CONCLUSION

In the literature review of hybrid assisted friction stir welding, it was observed that welding dissimilar materials through assisted welding compared to conventional FSW welding, whether it was thermally assisted or mechanically assisted. Both improved the mechanical and metallurgical properties of the welded joint.

Furthermore, the material flowed smoothly and applied less resistance to the tool compared to conventional FSW, significantly enhancing tool life. Apart from this, in most of the cases of FSW, the plate was either in the butt joint configuration or in the lap joint configuration. There is huge scope for future researchers to weld circular pipe or any other hollow shape joint configurations and to investigate their mechanical and metallurgical properties and optimize their process parameters.

In conclusion, much progress has been made in the field of friction stir welding process so far. Still, welding high melting point materials while also achieving lower cost and good quality is a challenging task. There is still a lot of work to be done in this area, such as reducing tool costs by the development of new tools, minimizing stresses caused by the welding and optimizing process parameters, particularly to weld the advanced high strength materials.

REFERENCES

[1] R. S. Mishra and Z. Y. Ma, "Friction stir welding and processing," Mater. Sci. Eng. R, 50, 2005, 1–78.

[2] W. M. Thomas, K. I. Johnson, and C. S. Wiesner, "Friction stir welding—recent developments in tool and process," Materials, 5(7), 2003, 485–490.

[3] Y. J. Kwon, I. Shigematsu, and N. Saito, "Dissimilar friction stir welding between magnesium and aluminum alloys," Mater. Lett., 62(23), 2008, 3827–3829.

[4] N. Shanmuga Sundaram and N. Murugan, "Tensile behavior of dissimilar friction stir welded joints of aluminum alloys," Mater. Des., 31(9), 2010, 4184–4193.

[5] B. Ahmad, A. Galloway, and A. Toumpis, "Numerical optimization of laser-assisted friction stir welding of structural steel," Sci. Technol. Weld. Join., 24(6), 2019, 548–558.

[6] D. K. Yaduwanshi, S. Bag, and S. Pal, "Numerical modeling and experimental investigation on plasma-assisted hybrid friction stir welding of dissimilar materials," Mater. Des., 92, 2016, 166–183.

[7] Y. F. Sun, Y. Konishi, M. Kamai, and H. Fujii, "Microstructure and mechanical properties of S45C steel prepared by laser-assisted friction stir welding," Mater. Des., 47, 2013, 842–849.

[8] X. Fei, J. Li, W. Yao, and L. Jin, "Study of temperature on microstructure and mechanical properties on Fe/Al joint in laser-assisted friction stir welding," AIP Adv., 8(7), 2018.

[9] J. Sundqvist, K. H. Kim, H. S. Bang, H. S. Bang, and A. F. H. Kaplan, "Numerical simulation of laser preheating of friction stir welding of dissimilar metals," Sci. Technol. Weld. Join., 23(4), 2018, 351–356.

[10] S. L. Campanelli, G. Casalino, C. Casavola, and V. Moramarco, "Analysis and comparison of friction stir welding and laser assisted friction stir welding of aluminum alloy," Materials (Basel), 6, 2013, 5923–5941.

[11] X. Liu, S. Lan, and J. Ni, "Electrically assisted friction stir welding for joining Al 6061 to TRIP 780 steel," J. Mater. Process. Technol., 219, 2015, 112–123.

[12] S. Chen, H. Zhang, X. Jiang, T. Yuan, Y. Han, and X. Li, "Mechanical properties of electric-assisted friction stir welded 2219 aluminum alloy," J. Manuf. Process., 44, 2019, 197–206.

[13] G. K. Padhy, C. S. Wu, and S. Gao, "Auxiliary energy-assisted friction stir welding—status review," Sci. Technol. Weld. Join., 20 (8), 2015, 631–649.

[14] Y. Han, X. Jiang, S. Chen, T. Yuan, H. Zhang, Y. Bai, Y. Xiang, and X. Li, "Microstructure and mechanical properties of electrically assisted friction stir welded AZ31B alloy joints," J. Manuf. Process., 43, 2019, 26–34.

[15] X. Jiang, Y. Han, S. Chen, Y. Bai, T. Yuan, and X. Wang, "Microstructure and texture investigation on electrically assisted friction stir welded titanium alloy," Mater. Sci. Technol. (United Kingdom), 2020, 1–11.

[16] K. Sengupta, D. K. Singh, A. K. Mondal, D. Bose, and B. Ghosh, "Analysis of mechanical property of electrically assisted friction stir welding to enhance the efficiency of joints," Mater. Today Proc., 2020, no. XXXX.

[17] Y. F. Sun, J. M. Shen, Y. Morisada, and H. Fujii, "Spot friction stir welding of low carbon steel plates preheated by high-frequency induction," Mater. Des., 54, 2014, 450–457.

[18] B. Vijendra and A. Sharma, "Induction heated tool-assisted friction-stir welding (i-FSW): A novel hybrid process for joining of thermoplastics," J. Manuf. Process., 20, 2015, 234–244.

[19] F. Baradarani, A. Mostafapour, and M. Shalvandi, "Effect of ultrasonic-assisted friction stir welding on microstructure and mechanical properties of AZ91−C magnesium alloy," Trans. Nonferrous Met. Soc. China (English Ed.), 29(12), 2019, 2514–2522.

[20] Y. B. Zhong, C. S. Wu, and G. K. Padhy, "Effect of ultrasonic vibration on welding load, temperature, and material flow in friction stir welding," J. Mater. Process. Technol., 239, 2017, 273–283.

[21] S. Kumar, C. Wu, and S. Gao, "Process parametric dependency of axial downward force and macro- and microstructural morphologies in ultrasonically assisted friction stir welding of Al/Mg alloys," Metall. Mater. Trans. A Phys. Metall. Mater. Sci., 51(6), 2020, 2863–2881.

[22] X. C. Liu and C. S. Wu, "Elimination of tunnel defect in ultrasonic vibration enhanced friction stir welding," Mater. Des., 90, 2016, 350–358.

[23] W. Tian, H. Su, and C. Wu, "Effect of ultrasonic vibration on thermal and material flow behavior, microstructure and mechanical properties of friction stir welded Al/Cu joints," Int. J. Adv. Manuf. Technol., 107(1–2), 2020, 59–71.

[24] L. Shi, C. S. Wu, and Z. Sun, "An integrated model for analyzing the effects of ultrasonic vibration on tool torque and thermal processes in friction stir welding," Sci. Technol. Weld. Join., 23(5), 2018, 365–379.

[25] G. K. Padhy, C. S. Wu, S. Gao, and L. Shi, "Local microstructure evolution in Al 6061-T6 friction stir weld nugget enhanced by ultrasonic vibration," Mater. Des., 92, 2016, 710–723.

[26] I. Alinaghian, S. Amini, and M. Honarpisheh, "Residual stress, tensile strength, and macrostructure investigations on ultrasonic-assisted friction stir welding of AA 6061-T6," J. Strain Anal. Eng. Des., 53(7), 2018, 494–503.

[27] M. Wu, C. S. Wu, and S. Gao, "Effect of ultrasonic vibration on fatigue performance of AA 2024-T3 friction stir weld joints," J. Manuf. Process., 29, 2017, 85–95.

[28] H. S. Bang, H. J. Song, and S. M. Joo, "Joint properties of dissimilar Al6061-T6 aluminum alloy/Ti-6%Al-4%V titanium alloy by gas tungsten arc welding assisted hybrid friction stir welding," Mater. Des., 51, 2013, 544–551.

[29] H. S. Bang, "Effect of tungsten-inert-gas preheating on mechanical and microstructural properties of friction stir welded dissimilar al alloy and mild steel," 48(1), 2016, 152–159.

[30] H. S. Bang, S. M. Hong, A. Das, and H. S. Bang, "A prediction of Fe-Al IMC layer thickness in TIG-assisted hybrid friction stir welded Al/steel dissimilar joints by numerical analysis," Int. J. Adv. Manuf. Technol., 106(1–2), 2020, 765–778.

[31] S. Siva, S. Sampathkumar, and J. Sudha, "Microstructure and mechanical properties of exothermic-reaction-assisted friction-stir-welded nickel-aluminum bronze alloy," J. Mater. Eng. Perform., 28(4), 2019, 2256–2270.

[32] J. H. Hattel, M. R. Sonne, and C. C. Tutum, "Modelling residual stresses in friction stir welding of Al alloys—a review of possibilities and future trends," Int. J. Adv. Manuf. Technol., 76(9–12), 2015, 1793–1805.

[33] X. Long and S. K. Khanna, "Modelling of electrically enhanced friction stir welding process using finite element method," Sci. Technol. Weld. Join., 10(4), 2005, 482–487.

[34] M. Dehghani, A. Amadeh, and S. A. A. Akbari Mousavi, "Investigations on the effects of friction stir welding parameters on intermetallic and defect formation in joining aluminum alloy to mild steel", Mater. Des., 49, 433–441, 2013.

[35] Y. G. Kim, H. Fujii, T. Tsumura, T. Komazaki, and K. Nakata, "Three defect types in friction stir welding of aluminum die casting alloy," Mater. Sci. Eng. A 415(1–2), 2006, 250–254.

[36] H. Bin Chen, K. Yan, T. Lin, S. Ben Chen, C. Y. Jiang, and Y. Zhao, "The investigation of typical welding defects for 5456 aluminum alloy friction stir welds," Mater. Sci. Eng. A 433(1–2), 2006, 64–69.

[37] Y. Song, X. Yang, L. Cui, X. Hou, Z. Shen, and Y. Xu, "Defect features and mechanical properties of friction stir lap welded dissimilar AA2024-AA7075 aluminum alloy sheets," Mater. Des., 55, 2014, 9–18.

[38] Y. Morisada, T. Imaizumi, and H. Fujii, "Clarification of material flow and defect formation during friction stir welding," Sci. Technol. Weld. Join., 20(2), 2015, 130–137.

[39] P. Chauhan, R. Jain, S. K. Pal, and S. B. Singh, "Modeling of defects in friction stir welding using coupled Eulerian and Lagrangian method," J. Manuf. Process., 34, 2018, 158–166.

[40] Y. Huang, L. Wan, X. Meng, Y. Xie, Z. Lv, and L. Zhou, "Probe shape design for eliminating the defects of friction stir lap welded dissimilar materials," J. Manuf. Process., 35, 2018, 420–427.

[41] M. P. Miles, J. Pew, T. W. Nelson, and M. Li, "Formability of friction stir welded dual-phase steel sheets," Frict. Stir Weld. Process. III—Proc. a Symp. Spons. by Shap. Form. Comm. Miner. Met. Mater. Soc. TMS, 18, 2005, 91–96.

[42] M. A. Wahid, A. N. Siddiquee, and Z. A. Khan, "Aluminum alloys in marine construction: characteristics, application, and problems from a fabrication viewpoint," Mar. Syst. Ocean Technol., 15(1), 2020, 70–80.

[43] G. Wang, Y. Zhao, and Y. Hao, "Friction stir welding of high-strength aerospace aluminum alloy and application in rocket tank manufacturing," J. Mater. Sci. Technol., 34(1), 2018, 73–91.

Chapter 6

An overview of robot assisted additive manufacturing

Sumitkumar Rathor, Avneesh Kumar, Ravi Kant and Ekta Singla

6.1 INTRODUCTION

6.1.1 Additive manufacturing

Additive manufacturing (AM), also referred to as 3D printing, is a process that builds a three-dimensional object by adding layers of material until the desired shape is achieved. The manufacturing industry is shifting towards AM, either as a replacement for subtractive manufacturing processes or as a hybrid technology that combines both [1]. This evolution is mainly because of advancements in digitalization and material innovation. The advantages of AM over conventional manufacturing processes have made it a preferred choice for industries such as aerospace, automotive, energy, and biomedical. These industries often require parts that are difficult, expensive, or even impossible to produce with traditional methods. AM technologies, inspired by the principles of Industry 4.0, offer intrinsic advantages like design flexibility, sustainable production practices, and faster time to market [2].

AM provides significant advantages over traditional manufacturing methods by minimizing the amount of equipment and tooling required. This results in reduced inventory costs, improved material utilization, decreased manufacturing time, and lower production waste. As a result, AM technologies have become increasingly popular across various industries and applications. In the biomedical industry, AM has revolutionized different areas of the healthcare system by enhancing product quality and reducing manufacturing costs [3]. AM technologies are particularly useful for producing patient-specific implants, improving cardiology and orthopaedic procedures, and advancing dental care. AM technologies are already being used in the automotive and aerospace industries for full-scale production processes.

There is enormous potential for improving processes in defence through the use of AM technologies. AM has proven to be particularly advantageous in operational fields like mobility, sustainability, reduced logistics, field repair, and maintenance [4, 5]. Additionally, with the standardization of advanced machinery, incorporating AM into various industries has become more profitable. As a result, an increasing

DOI: 10.1201/9781032703046-6

number of industrial facilities have begun to integrate advanced AM systems into their production processes. The process flow of AM can be divided into several stages, which are briefly discussed next.

1. Design
   The first step in the AM process is to design the object using computer-aided design (CAD) software. This step involves creating a 3D model of the object, which is typically saved in a file format such as STL.
2. Preparation
   After the design is finalized, the subsequent step involves preparing the file for printing. This process typically involves converting the file into a format that is compatible with the 3D printer. Often, the model must be sliced into layers to enable the printer to produce one layer at a time.
3. Printing
   The actual printing process begins once the file is prepared. The printer reads the file and begins printing the object layer by layer. The printer can use a variety of materials, including plastics, metals, and ceramics.
4. Post-Processing
   Once the printing is complete, the object may require post-processing. This can include removing support structures, sanding, or polishing the surface, and painting or coating the object.
5. Inspection and Testing
   In order to verify that the object produced through the AM process meets the necessary standards, the last stage involves conducting an inspection and testing procedure. This may involve evaluating the dimensions of the object, assessing its strength and durability, and performing other quality control checks.

6.1.2 Role of robotics in additive manufacturing

The use of robots to automate additive manufacturing processes has become increasingly popular due to the numerous advantages they offer, including enhanced speed, precision, and consistency [6]. Although robots have been employed in manufacturing for a significant period, their significance has grown with the advent of 3D printing technologies like directed energy deposition, material extrusion, and cold spray. Robots can handle a broader range of materials than traditional manufacturing methods can. Robot helps to create parts using materials such as metals, plastics, and composites, which would be difficult or impossible to produce using traditional methods [7]. Robots can produce parts with complex geometries, internal structures, and other features that would also be difficult or impossible to produce using traditional methods. The use of robots is more cost-effective than traditional manufacturing methods. Robots can be reprogrammed quickly and efficiently to produce a complex part, which means they can be used for multiple production runs with minimal setup costs [8]. Robots are used in a variety of ways in additive manufacturing systems, including the following:

- Material handling
  One of the essential roles of robotics in additive manufacturing is material handling. The 3D printing process requires raw materials, such as plastic and metal powders, which must be carefully handled and stored. Robots can be programmed to handle and transport these materials safely and efficiently, reducing the risk of contamination or damage to the materials.
- Part handling
  Robots are also used for part handling in additive manufacturing. After a part has been printed, it needs to be removed from the printer and inspected. Robots can be programmed to remove the parts from the printer and transport them to a designated location for inspection or post-processing.
- Precision and accuracy
  Robots are known for their precision and accuracy, which is why they are useful for additive manufacturing systems. The 3D printing process requires precise and accurate movements to ensure that each layer is added correctly. Robots can be programmed to make these movements with high accuracy, ensuring that each layer is added precisely where it needs to be.
- Quality control
  Robots are also used for quality control in additive manufacturing. After a part has been printed, it must be inspected to ensure it meets the required standards. Robots can be programmed to inspect the parts using cameras or sensors and identify defects or inconsistencies.
- Post-processing
  After a part has been printed, it often requires post-processing, such as sanding, polishing, and painting. Robots can be programmed to carry out these post-processing tasks, ensuring that they are carried out consistently and to a high standard.

6.1.2.1 Robot Configurations Used for Additive Manufacturing

Several types of robot configurations used in manufacturing applications are discussed next, and some popular robot configurations can be seen in Figure 6.1.

- Cartesian robot
  These robots have three linear axes (X, Y, and Z) and can move in a straight line along each axis. Cartesian robots are commonly used in additive manufacturing because of their precise and accurate movement in three-dimensional space. It can move the extruder or printing head with great accuracy, which is necessary for producing high-quality prints. Cartesian robots are also able to operate at high speeds, which allows for faster production times.

- SCARA robot
  These robots have four axes (X, Y, Z, and rotational) and often are used for tasks such as picking and placing objects, positioning the printing bed, and removing finished parts from the printer. They can also be used to perform secondary operations on printed parts, such as sanding, painting, and polishing. The rotational motion of the SCARA robot arm allows it to cover a larger workspace than a Cartesian robot with a similar range of motion. This can be beneficial in certain additive manufacturing applications where a larger workspace is required. Additionally, SCARA robots are known for their speed and precision, making them suitable for high-speed, high-accuracy applications such as additive manufacturing.
- Delta robot
  Delta robots have a unique parallel link structure that allows them to move very quickly and precisely. They can move in three dimensions, making them well suited for additive manufacturing applications that require intricate, three-dimensional printing. The parallel link structure of delta robots allows them to maintain a constant orientation while moving, which can be important for applications that require precise positioning of the printing head. Another advantage of delta robots is their relatively small size, which makes them ideal for use in applications where space is limited. This can be particularly important in additive manufacturing, where printers are often designed to fit on a desktop or in a small workshop.
- Collaborative robot
  Collaborative robot, or Cobot, is a type of robot that is designed to work alongside human workers in a shared workspace. Cobots are used for tasks that require human-robot collaboration, such as loading and unloading materials, removing finished parts from the printer, and performing quality control checks on printed parts. Cobots are designed to be safe for human workers to interact with and are equipped with sensors and safety features that allow them to detect when a human worker is in their workspace. This makes them ideal for additive manufacturing applications where human workers are involved in the production process.
- Dual-arm robot
  Dual-arm robots are a type of robot that has two arms, each of which can move independently of the other. In additive manufacturing, dual-arm robots are used for tasks that require more complex manipulation of the materials or objects being printed. For example, dual-arm robots can be used to hold and manipulate objects in a way that allows for more precise printing, such as holding an object steady while the printing head moves around it. They can also be used for tasks such as material mixing and handling, which can be important in certain additive manufacturing applications. Dual-arm robots are also capable of performing more intricate tasks than single arm robots can, such as assembly or disassembly of complex parts. This can be important in additive manufacturing applications where printed parts need to be assembled or to create the final product.

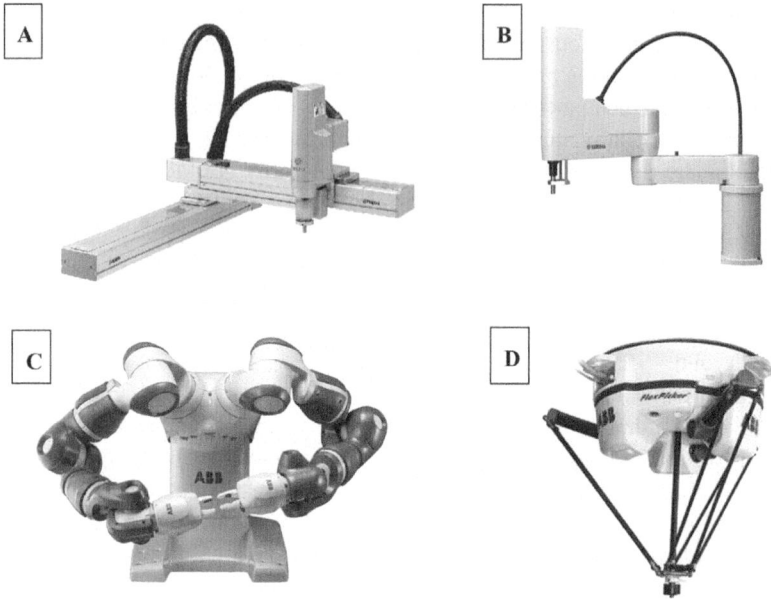

Figure 6.1 Common types of robots: (A) Cartesian robot, (B) SCARA robot (Courtesy of Yamaha Motor Co., Ltd.), (C) dual-arm robot, and (D) delta robot (Courtesy of ABB).

6.2 ROBOTIC ADDITIVE MANUFACTURING: ADVANCE PROCESSES

Recently, directed energy deposition (DED), material extrusion AM, and cold spray AM processes have been developed using robotic configurations.

6.2.1 Directed energy deposition

The melting of metal powder or wire using a laser or electron beam and depositing the molten material onto a substrate is involved in this process [9]. This process is known for repairing or adding material to existing parts and creating large, complex parts with high material utilization rates. Robots can automate the material deposition process and precisely move the laser or electron beam as the power source, ensuring consistent part quality. The process parameters in DED play a critical role in determining the quality of the deposited material and the overall efficiency of the process. Each process parameter must be considered for DED's path-planning strategy in this response. The discussion on the process parameters for laser-based DED using powder materials gives an idea of the complexity of path generation. The robotic DED systems can be seen in Figure 6.2. The following paragraphs highlight the importance and details of various process parameters.

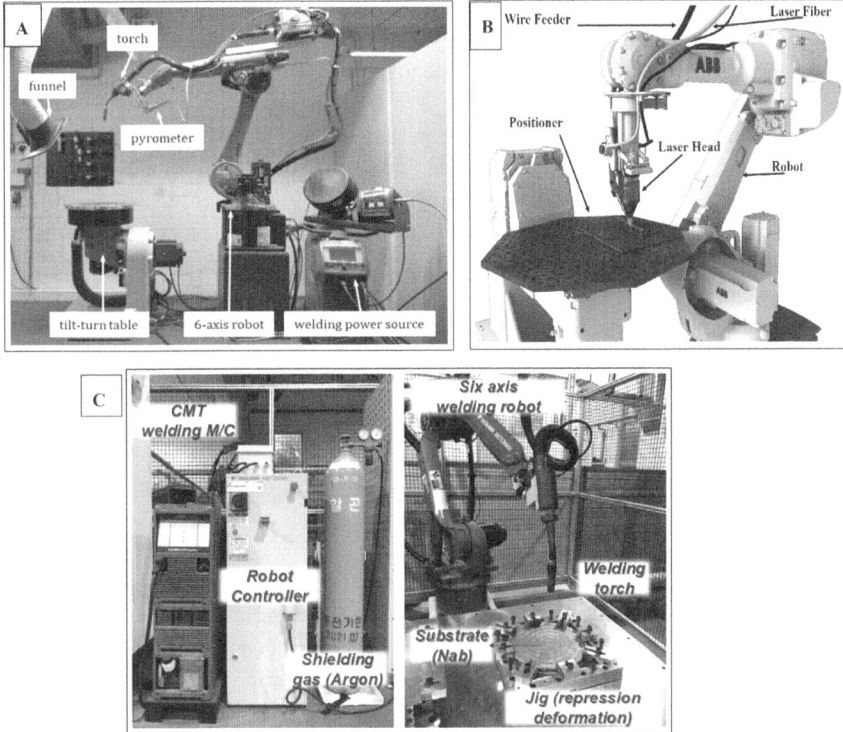

Figure 6.2 Robotic systems used for DED process: (A) arc-DED [10], (B) laser wire [11], and (C) wire arc [12].

- Laser power

  The laser power used in DED is an important process parameter that affects the deposition rate, quality of the deposited material, and energy input to the substrate. High laser power leads to a faster deposition rate but can also cause thermal distortion and stress in the material. Low laser power can lead to poor bonding between the deposited layers. Therefore, a careful balance between laser power and the deposition rate is crucial in achieving optimal results.

- Travel speed

  The travel speed of the deposition nozzle in DED determines the rate of material deposition and affects the quality of the deposited material. A higher travel speed leads to faster deposition, poor bonding, and reduced accuracy. A lower travel speed results in higher accuracy and better bonding, leading to longer processing times.

- Powder flow rate

  The powder flow rate is an important parameter determining the amount of material deposited. It affects the deposition rate, surface roughness, and quality of the deposited material. A higher powder flow rate results in faster deposition but can also lead to poor surface finish and porosity. A lower powder flow rate yields a better surface finish and reduced porosity. However, it can also result in slower deposition.

- Layer thickness
Layer thickness is an important parameter that determines the deposited material's accuracy, surface finish, and overall quality. A thinner layer leads to higher accuracy and better surface finish but can result in the slower deposition. A thicker layer results in faster deposition, leading to reduced accuracy and poorer surface finish.

- Laser spot size
Laser spot size is an important parameter affecting the deposited material's accuracy and quality. A smaller laser spot size leads to higher accuracy and better surface finish that cause slower deposition rates. A larger laser spot size results in faster deposition rates but can reduce accuracy and lead to a poorer surface finish.

6.2.2 Material extrusion

Material extrusion AM involves melting a plastic filament and depositing the molten material layer by layer to create a three-dimensional object. This process is ideal for creating low-cost, large, and complex parts [13]. Robots can automate the material extrusion process and precisely control the extruder, ensuring consistent part quality. Path planning is an essential aspect of this process that involves determining the optimal path for the extruder nozzle to follow to create a high-quality part. Several process parameters must be considered when developing the path-planning strategy in this context. Some robotic AM systems for material extrusion can be seen in Figure 6.3. The important process parameters for material extrusion AM are discussed further.

- Extrusion temperature
The temperature of the extrusion material is a critical parameter that affects the flow characteristics and deposition behaviour. A higher temperature generally leads to better adhesion between layers and increases the risk of warping and distortion. A lower temperature can result in poor bonding between layers, reducing the risk of defects such as stringing or oozing. Therefore, the path-planning strategy should be optimized based on the specific extrusion temperature used in the process.

- Layer height
The layer height is the thickness of each deposited layer, affecting the final part's overall resolution and surface finish. A smaller layer height can result in a smoother surface finish and higher resolution, increase the printing time, and require more support structures. A larger layer height can reduce the printing time and support material usage resulting in a rougher surface finish and lower resolution. The path-planning strategy should be optimized based on the desired layer height and the limitations of the printing equipment.

- Extrusion speed

  The extrusion speed is the rate at which the material is deposited, which affects the overall print time and the quality of the part. A slower extrusion speed can improve surface quality and accuracy and increase printing time. A faster extrusion speed can reduce the printing time and lower surface quality and accuracy. The path-planning strategy should be optimized based on the desired extrusion speed and the limitations of the printing equipment.

- Nozzle diameter

  The diameter of the extrusion nozzle affects the resolution and surface finish of the final part. A smaller nozzle diameter can result in a higher resolution and smoother surface finish and increase printing time. A larger nozzle diameter can reduce the printing time but can result in a rougher surface finish and lower resolution. The path-planning strategy should be optimized based on the specific nozzle diameter used in the process.

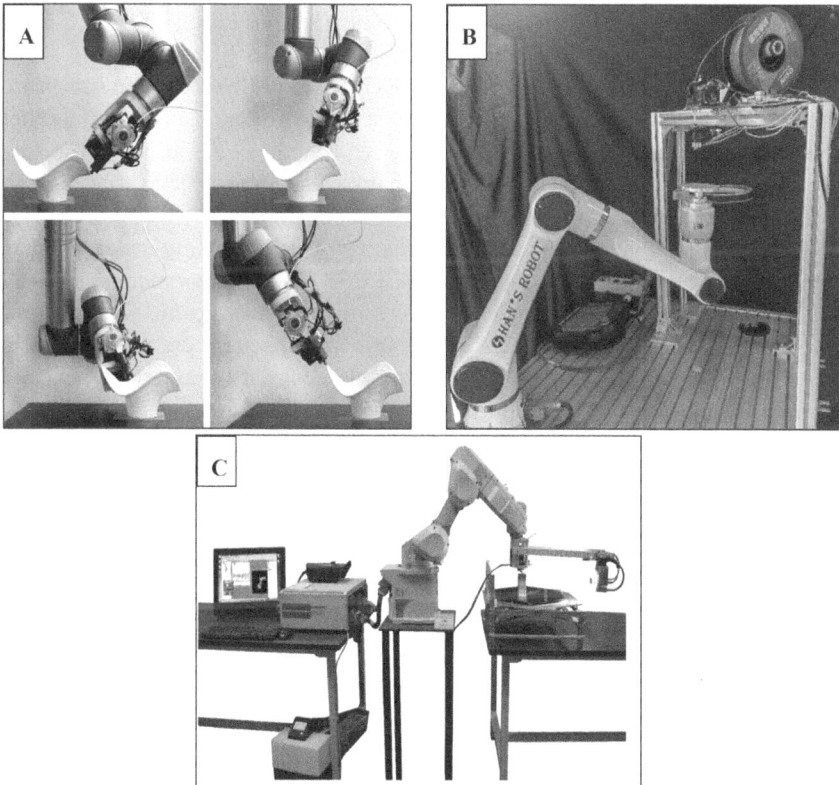

Figure 6.3 Robotic systems used for material extrusion process: (A) UR5 robot [14], (B) Elfin5 series manipulator [15], and (C) Mitsubishi robot [16].

6.2.3 Cold spray additive manufacturing

Cold spray (CS) is a solid-state material deposition process developed in the 1980s primarily for coating purposes. CS uses high-pressure gas (nitrogen, air, or helium) to propel feedstock powders at supersonic velocities (330–1600 m/s), resulting in deposition when the particles collide with a substrate (typically metal). The kinetic energy of the particles is the dominant factor in the deposition during CS, rather than thermal energy, which makes it a unique and highly efficient process. The feedstock powders remain solid throughout the deposition process. The bonding mechanism is governed due to mechanical interlocking and/or metallurgical bonding through localized plastic deformation at the particle-substrate and inter-particle interfaces. CS technology offers advantages over conventional high-temperature deposition techniques, including avoiding oxidation and phase transformation [17–19]. Cold spray additive manufacturing (CSAM) is an innovative process that utilizes CS technology to build parts completely by cold-sprayed material or add features to existing parts. Compared to other AM techniques, CSAM typically employs larger nozzle exit diameters ranging from 0.5 to 5 mm, resulting in smoother Gaussian-shaped deposition profiles [20]. However, CSAM may not be ideal for creating parts with features smaller than 0.5 mm in scale. However, the layer thickness can still be precisely controlled for large geometric features. Despite the lower spatial resolution, CSAM offers significantly higher deposition rates than other AM techniques. The technology has already been successfully utilized to manufacture complex parts with sophisticated surfaces, such as turbine blades and aerofoil-shaped leading edges [21].

CSAM is a relatively new technology that involves depositing metal powders onto a substrate using high-pressure gas. The CSAM has many potential applications, including repairing damaged parts, creating complex geometries, and producing lightweight materials [22, 23]. Robotic systems are increasingly used in CSAM to automate the process and increase efficiency. These systems typically consist of a robot arm with a spray nozzle attached, which moves along a programmed path while depositing metal powder. The robot arm also has sensors and cameras to ensure accurate deposition and quality control.

The successful implementation of CSAM requires precise coordination and control of spray parameters and robot path planning to ensure the production of high-quality deposits. This involves carefully selecting processing gas pressure, gas type, gas temperature, and powder feed rate for the spray parameters and considering the traverse speed of the nozzle, spray angle, standoff distance, scanning step, and trajectory of nozzle for robot path planning. The quality of the deposit heavily relies on these factors, making it imperative to have a comprehensive understanding of them. By regulating these parameters, one can create an effective CSAM manufacturing strategy and produce high-quality products.

The processing gas parameters are critical in CSAM as they directly dictate the velocity of particle impact, ultimately affecting the deposit properties. Helium, nitrogen, and air typically process gases in CSAM, with gas pressure and temperature ranging from 0.5 to 7.0 MPa and 25 to 1100 °C, respectively. Literature reports that higher gas temperature, gas pressure, or lower gas molecular weight results increase in particle impact velocity [24–27]. Although helium is the most effective gas for achieving significantly higher particle velocity than air or nitrogen, its higher cost is a drawback. Additionally, increasing gas temperature is a more effective method for achieving higher particle impact velocity than increasing gas pressure [28, 29].

The rate at which powder is introduced per unit of time into the nozzle is known as the powder feed rate. It has three significant effects on the properties of CSAM deposits. Firstly, it influences the particle velocity and flow stream. The particle velocity decreases as the powder feed rate increases due to gas-particle interaction at high powder loads. This leads to an increase in deposit porosity and a reduction in deposition efficiency, tensile strength, and hardness. The powder feed rate affects the thickness and profile of the single-track deposit, resulting in sharper track profiles and thicker deposits at higher powder feed rates. The high feed rates cause high localized residual stresses between deposits and the substrate, causing delamination of the deposit from the substrate during spraying [30–33]. These three effects suggest that the powder feed rate should be optimized for the CSAM process.

The current state of AM poses challenges in achieving high geometrical accuracy for parts produced using CSAM. These challenges are primarily due to the size limitations of features and the influence of nozzle process parameters on the final deposit geometry. Several methods have been explored, including Gaussian function fitting, artificial neural networks, and partial differential equations, to model the shape of a single-track profile. However, these models lack validation data for full-size 3D structures, indicating a need for further research. To fully realize the potential of CSAM, it is essential to investigate the impact of robot path planning on 3D geometry and materials microstructure. Furthermore, to meet the expectations of design freedom associated with CSAM, research efforts should be directed toward developing robotic path-planning strategies that incorporate multiple robotic arms and kinematics. Current state of robotic systems used for CSAM can be seen in Figure 6.4. The following process parameters must be considered during robot path planning in CSAM.

- Traverse speed
  The speed of the nozzle affects the amount and duration of feedstock powder hitting the target surface per unit of time. As a result, it directly impacts the cross-sectional profile and thickness of a single-track deposit. The slow nozzle movement creates a thicker deposit and a thinner cross-sectional profile. At the same time, a higher powder feed rate can achieve the same results. The flow characteristics of the processing gas passing through a De-Laval nozzle cause a radial variation in particle velocity and deposition efficiency. It results in a Gaussian-shaped cross-sectional profile. The speed at which the nozzle moves also significantly impacts the microstructure and properties of the deposit. When the nozzle traverse speed decreases, the deposit density increases, but the mechanical properties, such as deposit strength

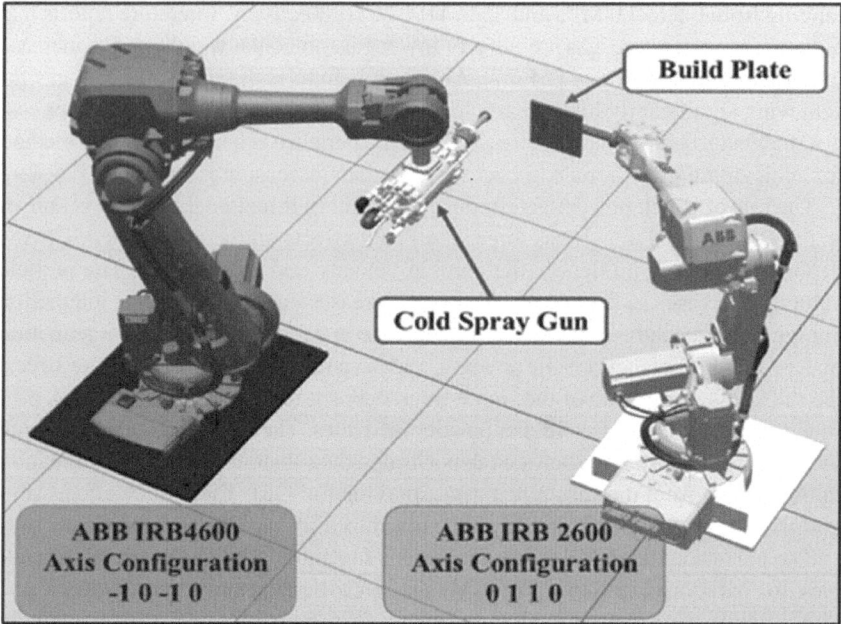

Figure 6.4 Robotic systems used for cold spray additive manufacturing [42].

and elastic modulus, decrease the adhesion strength [34, 35]. The deposit and/or substrate heating by the high-temperature impinging jet affect the nozzle traverse speed. Reducing the traverse speed results in higher deposit and substrate temperatures, which contributes to the deposition process but causes thermal stresses at the interface [36].

• Scanning steps

In CSAM, the deposit is formed by stacking many single-track deposits in an overlapping strategy. The interval between the centrelines of two single-track deposits is defined as the scanning step, which significantly impacts uniformity and surface morphology of deposits. An adequately chosen scanning step leads to a homogeneous deposit and smooth surface. The scanning step is typically half the width of a single-track deposit to ensure flatness. The effect of the scanning step on deposit properties is not well understood, and more investigations on this topic are urgently needed. It is important to note that the properties of deposits are influenced by the stacking of single-track deposits and not just the scanning step. Therefore, the investigators should consider the effects of both factors.

• Spray angle

The spray angle refers to the angle between the central axis of the nozzle and the target surface. The normal velocity component contributes to particle deposition when spraying at an angle. In contrast, the tangential velocity component can detach the deposited particles. Decreasing spray angle reduces the normal velocity component and increases the tangential velocity component. This deteriorates the reducing deposition efficiency, deposit strength, deposit quality, and adhesion strength. It also increases porosity in the deposit, resulting in deposit inhomogeneity [37, 38]. Therefore, spraying at other angles than normal to the substrate surface must be avoided. However, it is possible to achieve vertical walls and sharp 90° corners in CSAM by adjusting the spray angle during the edge pass and optimizing the traverse speed and corner radius using two robots [39].

- Standoff distance
  The distance between the nozzle outlet and the target surface is called the standoff distance. The intensity of the jet core in a supersonic-free jet gradually decreases along the central axis of jet due to momentum exchange with atmosphere. Inside the jet core, particles rapidly accelerate due to positive drag forces and then decelerate as the gas exits the diverging section of the nozzle due to negative drag forces [40]. The impact velocity of particles and deposition efficiency increase as the standoff distance increases until an optimal operating point is reached, after which they decrease. This phenomenon has been experimentally confirmed for materials such as aluminium, copper, and titanium deposits [41]. More investigation is still required for detailed understanding of the effect of standoff distance on CSAM deposit properties, including deposition efficiency, for which there is no consistent conclusion to date. Most existing work has been done at 10–40 mm as the standoff distance.

- Nozzle trajectory
  The purpose is to ensure that the deposited material has a homogeneous density and properties. When spraying on a flat surface, a simple zigzag pattern is used as it can yield a homogeneous deposit and is easy to define. For complex curved surfaces, the trajectory must ensure that the nozzle speed, spray angle, standoff distance, and other processing parameters are kept constant to avoid any inhomogeneity. Nozzle trajectory development and novel trajectory strategies are strongly encouraged to be investigated, as research in this area has been very limited to date.

6.3 ROBOTIC ADDITIVE MANUFACTURING: ADVANCE MATERIALS

Robotic systems are used to print various materials, including metals, polymers, ceramics, and composites. Advanced materials, such as carbon fiber-reinforced polymers and metal matrix composites, are enabled for robotic additive manufacturing due to facilitating the need for complex geometry structures or parts.

6.3.1 Carbon fiber-reinforced polymers

Carbon fiber-reinforced polymers (CFRPs) are composite materials from carbon fibers and a polymer matrix. CFRPs have high strength-to-weight ratios and are used in the aerospace, automotive, and sporting goods industries. Robots can lay down carbon fiber tapes or print with carbon fiber-reinforced filaments to create high-strength, lightweight parts. The study was conducted using a robotic system equipped with a composite extrusion 3D printing head capable of printing continuous carbon fiber strands along with thermoplastic matrix material. The robotic system was also equipped with a vision system for feature recognition, which allowed the system to identify the features and adjust the printing process accordingly, as can be seen in Figure 6.5(A). The system was tested on several specimens, including rectangular plates, T-joints, and L-shaped brackets. The study results showed that the proposed feature-based approach effectively produced high-quality CFRP structures with improved mechanical properties. The system could accurately place the carbon fibers along the features, significantly improving inter-layer adhesion and reducing warping. A three-step feature-based 3D printing strategy considers volume, patch, and wrap features.

6.3.2 Metal matrix composites

Metal matrix composites (MMCs) are composite materials made from a metal matrix and reinforcing materials, such as ceramics, carbon fibers, and metallic whiskers. MMCs have high strength, stiffness, and wear resistance and are used in the aerospace, automotive, and defence industries [43]. Robotic systems are used to print MMCs using AM processes like DED.

The investigation was carried out to study the influence of TiB2 content on the microstructural features and hardness of TiB2/AA7075 composites manufactured by the robotic laser metal deposition (LMD) process [44]. This study was important because TiB2/AA7075 composites are widely used in various industries such as aerospace, automotive, and sports equipment due to their high strength-to-weight ratio, excellent corrosion resistance, and good mechanical properties. The findings of the study were significant because they provided important insights into the effects of TiB2 content on the microstructural features and hardness of TiB2/AA7075 composites. The study showed that increasing the TiB2 content in the composite leads to a more refined microstructure, with TiB2 particles being uniformly distributed in the AA7075 matrix. The TiB2 particles acted as reinforcement, leading to an increase in the hardness of the composite. The prior discussion can enlighten the need for robotic systems to achieve the required properties of MMC. The robotic system used for AM of MMC can be seen Figure 6.5 (B).

6.3.3 Functionally graded materials

Functionally graded materials (FGMs) are advanced engineering materials with unique properties that vary gradually and continuously throughout the material's volume [45]. The properties of FGMs are graded so that they can perform optimally under different

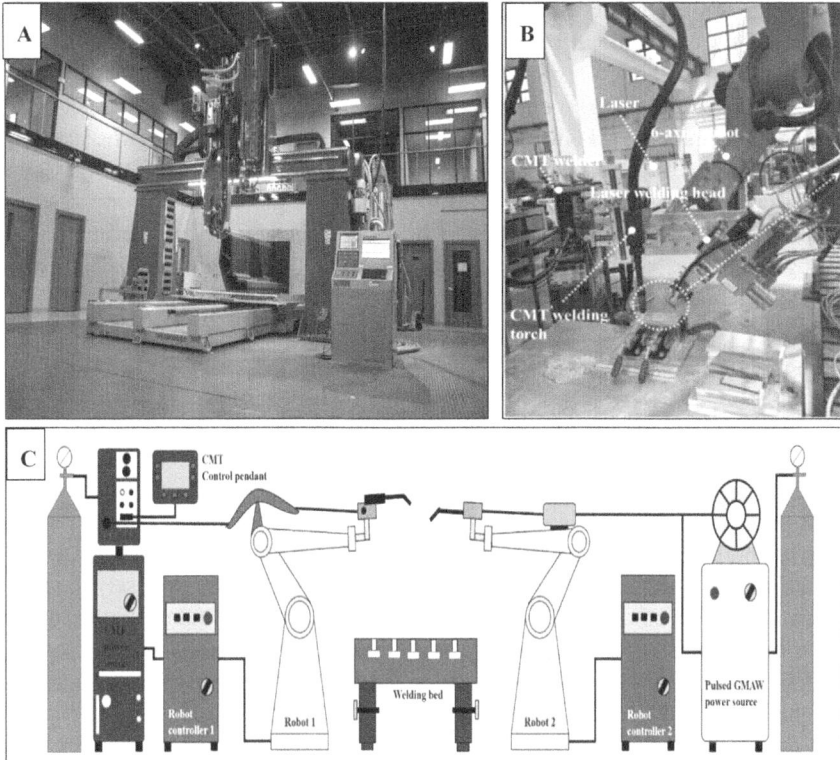

Figure 6.5 An experimental setup for (A) CFRP [Courtesy of Thermwood Corporation], (B) MMC [46], and (C) FGM [47].

environmental and loading conditions. FGMs can be made by combining two or more materials with different chemical, physical, or mechanical properties in varying proportions. The main advantage of FGMs is that they can withstand high stresses and extreme temperatures, making them useful in various applications, including aerospace, automotive, biomedical, and structural engineering. For example, FGMs can be used to make heat-resistant components for gas turbine engines, lightweight and strong parts for aircraft, or dental implants that mimic the properties of natural teeth.

FGMs can be classified into two main types, functionally graded composites (FGCs) and functionally graded coatings (FGCOs). The FGCs are made by combining two or more materials, such as ceramics and metals, in varying proportions. The resulting composite material exhibits a gradual change in properties, such as thermal expansion, modulus of elasticity, and thermal conductivity, along a certain direction. They can be used to make structural components with different properties on the inside and outside or to make coatings that protect against wear, corrosion, or high temperatures. The FGCOs, on the other hand, are thin layers of material that are applied to a substrate, such as a metal, ceramic, or polymer. The properties of the coating vary gradually from one end to the other, resulting in a gradient of properties that can improve the performance of the substrate. These coatings can be used to improve the wear resistance, corrosion resistance, or thermal stability of materials, such as cutting tools, bearings, and engine parts.

The development of FGMs is still ongoing, and researchers are investigating new ways to tailor the properties of these materials for specific applications. One of the challenges in making FGMs is to control the gradients in properties and ensure that the materials are homogeneous and defect-free. Advanced manufacturing techniques, such as robotic additive manufacturing, can help to overcome these challenges and produce complex FGMs with precise control over their properties; the system can be seen in Figure 6.5 (C).

6.4 PATH-PLANNING STRATEGIES IN ADDITIVE MANUFACTURING

Path-planning strategies are important in additive manufacturing because they help to optimize the process of building a part layer by layer [48]. Additive manufacturing processes like 3D printing involve depositing material precisely to create a three-dimensional object. The path that the printing nozzle takes to deposit the material is crucial, as it determines the final shape and properties of the object.

Path-planning strategies aim to optimize the path that the printer takes to deposit the material. They consider the part's geometry, the material used, printing speed, and the required surface finish [49]. By optimizing the path, these strategies can reduce the time and material required to build a part while also improving its accuracy and surface finish. Additionally, path-planning strategies can help to overcome some of the limitations of additive manufacturing processes, such as the tendency for parts to warp or distort during printing. By carefully planning the printer's path, these strategies can help minimize these issues and improve the overall quality of the printed part. Different path planning strategies are summarized in Table 6.1.

Trajectory planning and path planning are related but distinct concepts in robotics and control systems. Path planning refers to determining a feasible path from a starting point to a goal point in a given environment while avoiding obstacles or other constraints. Path planning typically involves generating a geometric representation of the environment and then using algorithms to search for a path that satisfies the given constraints. Trajectory planning, conversely, refers to determining the specific trajectory or motion that a robot or other system should follow to execute a given task. Trajectory planning considers the system's dynamics, such as its velocity, acceleration, and jerk. It generates a smooth, continuous motion that satisfies any constraints on the nozzle motion.

Also, trajectory planning is a crucial aspect of wire-arc additive manufacturing (WAAM). It determines the path the welding torch follows to deposit material and creates the desired shape. The software tool described by Zhang et al. uses geometric and optimization algorithms to generate optimal welding trajectories. The

Table 6.1 Path-planning strategies introduced in additive manufacturing

Strategy	Goals and considerations	Authors
Layer-based path planning	Layer-by-layer printing, considers path geometry	[50, 51]
Infill path planning	Considers infill structures within the part, such as honeycomb and lattice structures	[52–54]
Zigzag path planning	To reduce warping and improve surface finish	[55–57]
Spiral path planning	To reduce printing time and improve part accuracy	[58, 59]
Hilbert curve path planning	Printing along the Hilbert curve to reduce printing time	[60, 61]
Adaptive slicing path planning	It dynamically adjusts the thickness of each layer based on the geometry of the part, which can help to improve the surface finish and reduce printing time	[62, 63]
Support structure path planning	To print overhang or complex geometries	[64–66]
Continuous deposition path planning	Printing the part in a continuous path without stopping or retracting the printing head, which can help to improve printing speed and reduce material waste	[67, 68]

implementation of the trajectory-planning algorithm was discussed, including the mathematical models and optimization techniques used to generate optimal welding trajectories [69]. The authors have provided several examples of the software tool, demonstrating how it can generate trajectories for complex geometries and produce high-quality parts.

The algorithms consider the robotic arm's kinematics and the material's physical properties to generate optimal paths for the robot to follow. In the case of WAAM, the paths generated by the algorithm determine the trajectory of the welding torch as it deposits material onto the workpiece. In the case of pure object manipulation, the paths determine the trajectory of the end-effector of the robotic arm as it moves the object to a desired location. One of the key features of the algorithm presented by Ferreira et al. is its ability to generate multi-directional paths for WAAM processes [70]. This means the robot can deposit material in multiple directions rather than being constrained to one direction. This allows for more complex geometries to be produced more accurately and efficiently. The mathematical models used in the algorithm include the use of B-spline curves to represent the paths. The authors also discuss the implementation of the algorithm on a robotic arm and provide examples of its use in both WAAM and pure object manipulation.

A path-planning algorithm for semi-constrained Cartesian paths incorporating multiple tool centre points (TCPs) was discussed by Schmitz et al. [71]. The robot's configuration space is the set of all possible configurations that it can reach, considering the constraints on its movement. The trajectory-planning algorithm presented in the paper uses configuration space to generate optimal paths for the robot. The algorithm works by first generating a set of waypoints along the semi-constrained Cartesian path. These waypoints are then used to generate a series of configurations that the robot can reach. The algorithm then uses an optimization technique to generate the optimal path for the robot to follow, considering the constraints on its movement and the need to avoid obstacles. Schmitz et al. provide a detailed description of the mathematical models used in the algorithm, including the use of quaternions to represent rotations and the use of the inverse kinematics of the robot to generate configurations. The authors also provide examples of the algorithm, demonstrating its ability to generate trajectories over semi-constrained Cartesian paths for various robotic manipulators.

In Malhan et al., a trajectory-planning approach was introduced for mobile robots in large-scale assembly and additive manufacturing applications in construction [72]. The trajectory-planning method is based on a contour tracking control approach, allowing the robot to follow a desired path while maintaining a desired orientation. This paper describes the trajectory-planning approach use of a geometric path representation, which allows for the creation of smooth and continuous paths that the robot can follow. The path representation is created using a series of connected Bezier curves, which are used to define the contour of the object being assembled or manufactured. The robot then follows the path by controlling its position and orientation using sensor feedback. This work also discusses using a proportional-integral-derivative (PID) controller to regulate the robot's position and orientation along the path. The PID controller allows precise control of the robot's movements, which is necessary for large-scale assembly and additive manufacturing applications.

6.5 FUTURE PERSPECTIVE OF ADVANCE ROBOTIC SYSTEMS FOR ADDITIVE MANUFACTURING

The future of robots in additive manufacturing is exciting. With technological advances, robots are set to play an even bigger role in this industry. In this section, we discuss how robots are shaping the future of additive manufacturing.

6.5.1 Collaborative robots

The future of collaborative robots in the industry, particularly additive manufacturing, is promising. Cobots are designed to work alongside human operators, providing several advantages such as improved efficiency, reduced costs, and increased safety. They can also be easily reprogrammed to perform different tasks, making them ideal for small-batch production or prototyping. However, several challenges need to be addressed to realize the potential of collaborative robots in additive manufacturing.

- Work safety
  Cobots are designed to work safely alongside humans, but ensuring their safety requires advanced sensing and control systems. Cobots must be able to detect and avoid obstacles in their workspace, and they must be able to shut down or slow down their operations when they detect a human operator in their vicinity. Developing reliable and robust safety protocols is essential for the widespread adoption of cobots in additive manufacturing.
- Compatibility
  Cobots must be compatible with existing additive manufacturing equipment. New cobot-compatible equipment must be developed for complex paths for 3D printing. Ensuring compatibility requires close collaboration between cobot manufacturers and equipment manufacturers to develop standards and protocols for seamless integration.
- Artificial intelligence
  Cobots must be able to learn from their interactions with human operators and improve their performance over time. This requires advanced artificial intelligence and machine learning algorithms to analyse data from sensors and other sources to optimize cobot performance. Developing these algorithms requires a significant investment in research and development.
- Communication
  Cobots must communicate seamlessly with other systems in the additive manufacturing process. This requires robust and reliable communication protocols to handle large amounts of data and enable real-time monitoring and control of cobot operations. Developing these protocols requires close collaboration between cobot manufacturers, equipment manufacturers, and software developers.
- Cost
  Cobots are still relatively expensive compared to traditional robots, which can limit their adoption in small and medium-sized businesses. Reducing the cost of cobots requires economies of scale and advances in manufacturing processes that can reduce the cost of components and assembly.

6.5.2 Autonomous robots

The future of the manufacturing industry is expected to be heavily influenced by the rise of autonomous robots. These robots can perform tasks such as printing, assembly, inspection, and maintenance without the need for human intervention. However, several challenges still need to be addressed before autonomous robots can become more widely used in additive manufacturing.

- Accuracy and precision
  One of the key challenges of autonomous robots in additive manufacturing is ensuring they can achieve the necessary levels of accuracy and precision. 3D printing requires precise movements, and errors or deviations can lead to defects in the final product. Autonomous robots need to be able to perform these movements with the same level of precision as human operators while also adapting to changes in the printing environment.
- Material handling
  Another challenge is material handling. Additive manufacturing requires various materials, such as powders, filaments, and resins, which must be stored, transported, and loaded into the printing equipment. Autonomous robots need to handle these materials safely and efficiently while also ensuring that they are used in the correct quantities and locations.
- System integration
  Autonomous robots integrate with the broader additive manufacturing system, including CAD software, printers, and other equipment. This requires high compatibility and coordination between different components, which can be challenging to achieve in practice.
- Quality control
  Autonomous robots need to be able to monitor the printing process and detect any defects or errors in real time. This requires advanced sensors and software, which can analyse the data and make adjustments as necessary. Ensuring consistent quality control is critical in additive manufacturing, as defects can lead to product failures or safety issues.
- Cost and scalability
  Autonomous robots are often expensive to develop and implement, requiring significant investment in hardware and software. Additionally, they may not be suitable for all types of additive manufacturing applications, particularly those that require a high degree of customization or flexibility.

6.5.3 Mobile robots

Mobile robots will play a crucial role in increasing productivity, efficiency, and safety in different industries. With technological advancements, mobile robots will become more intelligent, autonomous, and adaptable to different environments, leading to increased use in various industries. Additionally, the adoption of mobile robotics will lead to new job opportunities in areas such as robotics programming and maintenance.

- Safety
  Safety is a significant concern when it comes to mobile robotics. With robots working alongside human workers, accidents occur, and it is essential to ensure that robots are safe to work with.

- Navigation
  Navigation is a crucial challenge for mobile robots, particularly in environments with changing layouts. Robots need to be able to navigate through different environments, avoiding obstacles and reaching their destination safely.
- Power supply
  Mobile robots require a reliable power supply to operate efficiently. As the use of mobile robots increases, it is essential to find more efficient ways of powering robots, particularly in environments with limited power sources.

Industries are keen to use robots as they have necessary potential to scale and improve the productivity of AM processes. NASA has developed a spider-like robot called the "In-Space Manufacturing (ISM) Robotic System." The robot is designed to 3D print structures in space. The ISM Robotic System has six legs, each with a 3D printer nozzle at the end. This can be seen in Figure 6.6 (A). The robot is capable of printing structures up to 14 feet in diameter and 8 feet tall. The spider-like design allows the robot to move easily in zero gravity environments and access hard-to-reach areas. The use of additive manufacturing in the development of the ISM Robotic System offers several advantages, including increased efficiency and greater flexibility in design. The ability to print structures in space also eliminates the need to transport large and bulky parts from Earth, which can be expensive and time-consuming.

Adidas FUTURECRAFT.STRUNG is a 3D-printed shoe that utilizes a new technology called Strung using a robotic manipulator, as can be seen in Figure 6.6 (B). The Strung is a process where strands of material are individually woven together to create customized textiles for each shoe. The process allows for precise control over the shoe's fit, comfort, and performance. The FUTURECRAFT.

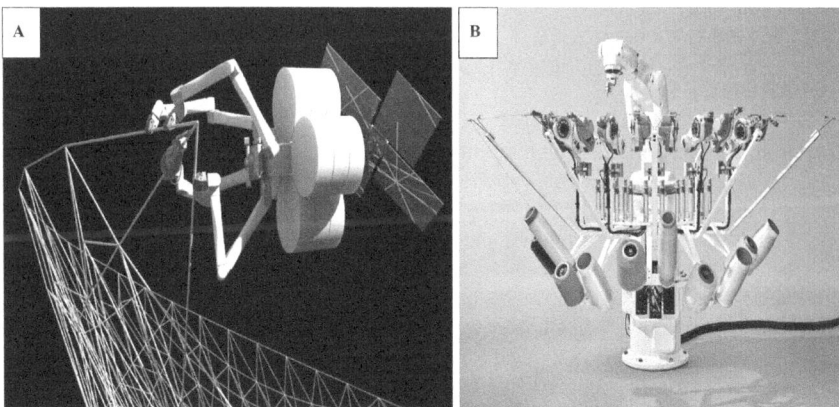

Figure 6.6 Use of robots in (A) space (Courtesy of NASA) and (B) Strung robot in the shoe industry (Courtesy of Adidas).

STRUNG shoe is created using a 3D-printed midsole and outsole, with the Strung textile upper. The midsole and outsole are printed using Adidas' 4D technology, which uses digital light projection to create a lattice structure that provides support and cushioning. The Strung textile is then woven onto the 3D-printed base to create a seamless upper. The shoe is designed to fit the unique shape of each individual foot, providing personalized comfort and support. The Strung textile also allows for greater breathability and flexibility, while the 3D-printed midsole and outsole provide durability and stability.

6.6 CONCLUSIONS

This chapter has highlighted the various additive manufacturing processes such as DED, material extrusion, and cold spray, which have revolutionized the way advanced materials are produced. These techniques have led to the development of innovative materials such as MMC, FGM, and CFRP, which have unique properties and applications in various industries, including aerospace, automotive, and sports. The recent advancements in additive manufacturing have also seen the integration of robots in the manufacturing process, as demonstrated by NASA and Adidas. The use of robots has increased production efficiency and reduced human error, resulting in high-quality products.

Overall, the future of additive manufacturing looks promising, with more advancements expected in the coming years. The use of advanced materials and robotic systems in additive manufacturing will undoubtedly revolutionize various industries, leading to increased efficiency and reduced costs. The integration of machine learning and artificial intelligence in additive manufacturing also holds great potential to improve the overall production process. Therefore, further research and development are necessary to explore the full potential of robots for additive manufacturing in various industries.

REFERENCES

[1] Butt, J.; Shirvani, H. Additive, Subtractive, and Hybrid Manufacturing Processes. *Advances in Manufacturing and Processing of Materials and Structures*, **2018**, *18* (4), 187–218. https://doi.org/10.1201/b22020-9.
[2] Lemu, H. G. On Opportunities and Limitations of Additive Manufacturing Technology for Industry 4.0 Era. *Lecture Notes in Electrical Engineering*, **2019**, *484*, 106–113. https://doi.org/10.1007/978-981-13-2375-1_15/COVER.
[3] Kumar, R.; Kumar, M.; Chohan, J. S. The Role of Additive Manufacturing for Biomedical Applications: A Critical Review. *Journal of Manufacturing Processes*, **2021**, *64*, 828–850. https://doi.org/10.1016/J.JMAPRO.2021.02.022.
[4] Segonds, F. Design by Additive Manufacturing: An Application in Aeronautics and Defence. *Virtual and Physical Prototyping*, **2018**, *13* (4), 237–245. https://doi.org/10.1080/17452759.2018.1498660.

[5] Busachi, A.; Erkoyuncu, J.; Colegrove, P.; Martina, F.; Watts, C.; Drake, R. A. Review of Additive Manufacturing Technology and Cost Estimation Techniques for the Defence Sector. *CIRP Journal of Manufacturing Science and Technology*, **2017**, *19*, 117–128. https://doi.org/10.1016/J.CIRPJ.2017.07.001.

[6] Urhal, P.; Weightman, A.; Diver, C.; Bartolo, P. Robot Assisted Additive Manufacturing: A Review. *Robotics and Computer-Integrated Manufacturing*, **2019**, *59*, 335–345. https://doi.org/10.1016/J.RCIM.2019.05.005.

[7] Esmaeilian, B.; Behdad, S.; Wang, B. The Evolution and Future of Manufacturing: A Review. *Journal of Manufacturing Systems*, **2016**, *39*, 79–100. https://doi.org/10.1016/J.JMSY.2016.03.001.

[8] Kortenkamp, D.; Simmons, R.; Brugali, D. Robotic Systems Architectures and Programming. *Springer Handbooks*, **2016**, 283–306. https://doi.org/10.1007/978-3-319-32552-1_12/COVER.

[9] Ahn, D. G. Directed Energy Deposition (DED) Process: State of the Art. *International Journal of Precision Engineering and Manufacturing-Green Technology*, **2021**, *8* (2), 703–742. https://doi.org/10.1007/S40684-020-00302-7.

[10] Müller, C.; Müller, J.; Kloft, H.; Hensel, J. Design of Structural Steel Components According to Manufacturing Possibilities of the Robot-Guided DED-Arc Process. *Buildings*, **2022**, *12* (12). https://doi.org/10.3390/buildings12122154.

[11] Abuabiah, M.; Mbodj, N. G.; Shaqour, B.; Herzallah, L.; Juaidi, A.; Abdallah, R.; Plapper, P. Advancements in Laser Wire-Feed Metal Additive Manufacturing: A Brief Review. *Materials*, **2023**, *16* (5), 1–23. https://doi.org/10.3390/ma16052030.

[12] Kim, J.; Kim, J.; Pyo, C. Comparison of Mechanical Properties of Ni-Al-Bronze Alloy Fabricated through Wire Arc Additive Manufacturing with Ni-Al-Bronze Alloy Fabricated through Casting. *Metals*, **2020**, *10* (9), 1164. https://doi.org/10.3390/MET10091164.

[13] Zhuo, P.; Li, S.; Ashcroft, I. A.; Jones, A. I. Material Extrusion Additive Manufacturing of Continuous Fibre Reinforced Polymer Matrix Composites: A Review and Outlook. *Composites Part B: Engineering*, **2021**, *224*, 109143. https://doi.org/10.1016/J.COMPOSITESB.2021.109143.

[14] Mitropoulou, I.; Bernhard, M.; Dillenburger, B. Print Paths Key-Framing: Design for Non-Planar Layered Robotic FDM Printing. *Proceedings—SCF 2020: ACM Symposium on Computational Fabrication*, **2020**. https://doi.org/10.1145/3424630.3425408.

[15] Jing, X.; Lv, D.; Xie, F.; Zhang, C.; Chen, S.; Mou, B. A Robotic 3D Printing System for Supporting-Free Manufacturing of Complex Model Based on FDM Technology. *Industrial Robot*, **2023**, *50* (2), 314–325. https://doi.org/10.1108/IR-05-2022-0136.

[16] Shen, H.; Sun, W.; Fu, J. Multi-View Online Vision Detection Based on Robot Fused Deposit Modeling 3D Printing Technology. *Rapid Prototyping Journal*, **2019**, *25* (2), 343–355. https://doi.org/10.1108/RPJ-03-2018-0052.

[17] Alkhimov, A. P.; Kosarev, V. F.; Papyrin, A. N. A Method of "Cold" Gas-Dynamic Deposition. *SPhD*, **1990**, *35*, 1047.

[18] Papyrin, A.; Kosarev, V.; Klinkov, S.; Alkhimov, A.; Fomin, V. M. *Cold Spray Technology*; Elsevier, 2006.

[19] Kumar, A.; Singh, H.; Kant, R.; Rasool, N. Development of Cold Sprayed Titanium/Baghdadite Composite Coating for Bio-Implant Applications. *Journal of Thermal Spray Technology*, **2021**, *30* (8), 2099–2116. https://doi.org/10.1007/S11666-021-01269-W/FIGURES/19.

[20] Nault, I. M.; Ferguson, G. D.; Nardi, A. T. Multi-Axis Tool Path Optimization and Deposition Modeling for Cold Spray Additive Manufacturing. *Additive Manufacturing*, **2021**, *38* (November 2020), 101779. https://doi.org/10.1016/j.addma.2020.101779.

[21] Nault, I. M.; Ferguson, G. D.; Champagne, V.; Nardi, A. Design of Residual-Stress-Compensating Molds for Cold Spray Additive Manufacturing Applications. *Journal of Thermal Spray Technology*, **2020**, *29* (6), 1466–1476. https://doi.org/10.1007/s11666-020-00984-0.

[22] Fan, N.; Cizek, J.; Huang, C.; Xie, X.; Chlup, Z.; Jenkins, R.; Lupoi, R.; Yin, S. A New Strategy for Strengthening Additively Manufactured Cold Spray Deposits through in-Process Densification. *Additive Manufacturing*, **2020**, *36*, 101626. https://doi.org/10.1016/J.ADDMA.2020.101626.

[23] Yin, S.; Cizek, J.; Yan, X.; Lupoi, R. Annealing Strategies for Enhancing Mechanical Properties of Additively Manufactured 316L Stainless Steel Deposited by Cold Spray. *Surface and Coatings Technology*, **2019**, *370*, 353–361. https://doi.org/10.1016/J.SURFCOAT.2019.04.012.

[24] Kumar, A.; Kant, R.; Singh, H. Tribological Behavior of Cold-Sprayed Titanium/Baghdadite Composite Coatings in Dry and Simulated Body Fluid Environments. *Surface and Coatings Technology*, **2021**, *425*, 127727. https://doi.org/10.1016/J.SURFCOAT.2021.127727.

[25] Lee, H.; Shin, H.; Lee, S.; Ko, K. Effect of Gas Pressure on Al Coatings by Cold Gas Dynamic Spray. *Materials Letters*, **2008**, *62* (10–11), 1579–1581. https://doi.org/10.1016/j.matlet.2007.09.026.

[26] Yin, S.; Suo, X.; Liao, H.; Guo, Z.; Wang, X. Significant Influence of Carrier Gas Temperature During the Cold Spray Process. *Surface Engineering*, **2014**, *30* (6), 443–450. https://doi.org/10.1179/1743294414Y.0000000276.

[27] Cao, C.; Han, T.; Xu, Y.; Li, W.; Yang, X.; Hu, K. The Associated Effect of Powder Carrier Gas and Powder Characteristics on the Optimal Design of the Cold Spray Nozzle. *Surface Engineering*, **2020**, *36* (10), 1081–1089. https://doi.org/10.1080/02670844.2020.1744297.

[28] Zahiri, S. H.; Fraser, D.; Gulizia, S.; Jahedi, M. Effect of Processing Conditions on Porosity Formation in Cold Gas Dynamic Spraying of Copper. *Journal of Thermal Spray Technology*, **2006**, *15* (3), 422–430. https://doi.org/10.1361/105996306X124437.

[29] Assadi, H.; Kreye, H.; Gärtner, F.; Klassen, T. Cold Spraying—A Materials Perspective. *Acta Materialia*, **2016**, *116*, 382–407. https://doi.org/10.1016/j.actamat.2016.06.034.

[30] Meyer, M.; Yin, S.; Lupoi, R. Particle in-Flight Velocity and Dispersion Measurements at Increasing Particle Feed Rates in Cold Spray. *Journal of Thermal Spray Technology*, **2017**, *26* (1–2), 60–70. https://doi.org/10.1007/s11666-016-0496-3.

[31] Grujicic, M.; Saylor, J. R.; Beasley, D. E.; DeRosset, W. S.; Helfritch, D. Computational Analysis of the Interfacial Bonding between Feed-Powder Particles and the Substrate in the Cold-Gas Dynamic-Spray Process. *Applied Surface Science*, **2003**, *219* (3–4), 211–227. https://doi.org/10.1016/S0169-4332(03)00643-3.

[32] Ozdemir, O. C.; Widener, C. A.; Carter, M. J.; Johnson, K. W. Predicting the Effects of Powder Feeding Rates on Particle Impact Conditions and Cold Spray Deposited Coatings. *Journal of Thermal Spray Technology*, **2017**, *26* (7), 1598–1615. https://doi.org/10.1007/s11666-017-0611-0.

[33] Taylor, K.; Jodoin, B.; Karov, J. Particle Loading Effect in Cold Spray. *Journal of Thermal Spray Technology*, **2006**, *15* (2), 273–279. https://doi.org/10.1361/105996306X108237.

[34] Meyer, M. C.; Yin, S.; McDonnell, K. A.; Stier, O.; Lupoi, R. Feed Rate Effect on Particulate Acceleration in Cold Spray under Low Stagnation Pressure Conditions. *Surface and Coatings Technology*, **2016**, *304*, 237–245. https://doi.org/10.1016/j.surfcoat.2016.07.017.

[35] Tan, A. W. Y.; Sun, W.; Phang, Y. P.; Dai, M.; Marinescu, I.; Dong, Z.; Liu, E. Effects of Traverse Scanning Speed of Spray Nozzle on the Microstructure and Mechanical Properties of Cold-Sprayed Ti6Al4V Coatings. *Journal of Thermal Spray Technology*, **2017**, *26* (7), 1484–1497. https://doi.org/10.1007/s11666-017-0619-5.

[36] Moridi, A.; Hassani Gangaraj, S. M.; Vezzu, S.; Guagliano, M. Number of Passes and Thickness Effect on Mechanical Characteristics of Cold Spray Coating. *Procedia Engineering*, **2014**, *74*, 449–459. https://doi.org/10.1016/j.proeng.2014.06.296.

[37] Candel, A.; Gadow, R. Trajectory Generation and Coupled Numerical Simulation for Thermal Spraying Applications on Complex Geometries. *Journal of Thermal Spray Technology*, **2009**, *18* (5–6), 981–987. https://doi.org/10.1007/s11666-009-9338-x.

[38] Singh, R.; Rauwald, K. H.; Wessel, E.; Mauer, G.; Schruefer, S.; Barth, A.; Wilson, S.; Vassen, R. Effects of Substrate Roughness and Spray-Angle on Deposition Behavior of Cold-Sprayed Inconel 718. *Surface and Coatings Technology*, **2017**, *319*, 249–259. https://doi.org/10.1016/j.surfcoat.2017.03.072.

[39] Li, W. Y.; Yin, S.; Wang, X. F. Numerical Investigations of the Effect of Oblique Impact on Particle Deformation in Cold Spraying by the SPH Method. *Applied Surface Science*, **2010**, *256* (12), 3725–3734. https://doi.org/10.1016/j.apsusc.2010.01.014.

[40] Vargas-Uscategui, A.; King, P. C.; Yang, S.; Chu, C.; Li, J. Toolpath Planning for Cold Spray Additively Manufactured Titanium Walls and Corners: Effect on Geometry and Porosity. *Journal of Materials Processing Technology*, **2021**, *298* (December), 117272. https://doi.org/10.1016/j.jmatprotec.2021.117272.

[41] Pattison, J.; Celotto, S.; Khan, A.; O'Neill, W. Standoff Distance and Bow Shock Phenomena in the Cold Spray Process. *Surface and Coatings Technology*, **2008**, *202* (8), 1443–1454. https://doi.org/10.1016/j.surfcoat.2007.06.065.

[42] Rech, S.; Trentin, A.; Vezzù, S.; Vedelago, E.; Legoux, J. G.; Irissou, E. Different Cold Spray Deposition Strategies: Single- and Multi-Layers to Repair Aluminium Alloy Components. *Journal of Thermal Spray Technology*, **2014**, *23* (8), 1237–1250. https://doi.org/10.1007/s11666-014-0141-y.

[43] Kállai, Z.; Schüppstuhl, T. Robot-Guided, Feature-Based 3D Printing Strategies for Carbon Fiber Reinforced Plastic. *Tagungsband des 4. Kongresses Montage Handhabung Industrieroboter*, **2019**, 178–186. https://doi.org/10.1007/978-3-662-59317-2_18.

[44] Singh, L.; Singh, B.; Saxena, K. K. Manufacturing Techniques for Metal Matrix Composites (MMC): An Overview. *Advances in Materials and Processing Technologies*, **2020**, *6* (2), 224–240. https://doi.org/10.1080/2374068X.2020.1729603.

[45] Lei, Z.; Bi, J.; Chen, Y.; Chen, X.; Tian, Z.; Qin, X. Effect of TiB2 Content on Microstructural Features and Hardness of TiB2/AA7075 Composites Manufactured by LMD. *Journal of Manufacturing Processes*, **2020**, *53*, 283–292. https://doi.org/10.1016/J.JMAPRO.2020.02.036.

[46] Ahsan, M. R. U.; Tanvir, A. N. M.; Ross, T.; Elsawy, A.; Oh, M. S.; Kim, D. B. Fabrication of Bimetallic Additively Manufactured Structure (BAMS) of Low Carbon Steel and 316L Austenitic Stainless Steel with Wire + Arc Additive Manufacturing. *Rapid Prototyping Journal*, **2020**, *26* (3), 519–530. https://doi.org/10.1108/RPJ-09-2018-0235.

[47] Naebe, M.; Shirvanimoghaddam, K. Functionally Graded Materials: A Review of Fabrication and Properties. *Applied Materials Today*, **2016**, *5*, 223–245. https://doi.org/10.1016/J.APMT.2016.10.001.

[48] Ma, S.; Chen, X.; Jiang, M.; Li, B.; Wang, Z.; Lei, Z.; Chen, Y. Surface Morphology, Microstructure and Mechanical Properties of Al–Mg–Sc Alloy Thin Wall Produced by Laser-Arc Hybrid Additive Manufacturing. *Thin-Walled Structures*, **2023**, *186* (January), 110674. https://doi.org/10.1016/j.tws.2023.110674.

[49] Zhao, D.; Guo, W. Shape and Performance Controlled Advanced Design for Additive Manufacturing: A Review of Slicing and Path Planning. *Journal of Manufacturing Science and Engineering, Transactions of the ASME*, **2020**, *142* (1). https://doi.org/10.1115/1.4045055/1046939.

[50] Jiang, J.; Ma, Y. Path Planning Strategies to Optimize Accuracy, Quality, Build Time and Material Use in Additive Manufacturing: A Review. *Micromachines*, **2020**, *11* (7), 633. https://doi.org/10.3390/MI11070633.

[51] Li, Y.; He, D.; Yuan, S.; Tang, K.; Zhu, J. Vector Field-Based Curved Layer Slicing and Path Planning for Multi-Axis Printing. *Robotics and Computer-Integrated Manufacturing*, **2022**, *77*, 102362. https://doi.org/10.1016/J.RCIM.2022.102362.

[52] Jin, Y.; He, Y.; Fu, G.; Zhang, A.; Du, J. A Non-Retraction Path Planning Approach for Extrusion-Based Additive Manufacturing. *Robotics and Computer-Integrated Manufacturing*, **2017**, *48*, 132–144. https://doi.org/10.1016/J.RCIM.2017.03.008.

[53] Li, Y.; Yuan, S.; Zhang, W.; Zhu, J. A New Continuous Printing Path Planning Method for Gradient Honeycomb Infill Structures. *International Journal of Advanced Manufacturing Technology*, **2023**, 1–16. https://doi.org/10.1007/S00170-023-11065-1/FIGURES/C.

[54] Liu, T.; Yuan, S.; Wang, Y.; Xiong, Y.; Zhu, J.; Lu, L.; Tang, Y. Stress-Driven Infill Mapping for 3D-Printed Continuous Fiber Composite with Tunable Infill Density and Morphology. *Additive Manufacturing*, **2023**, *62*, 103374. https://doi.org/10.1016/J.ADDMA.2022.103374.

[55] Zhao, G.; Zhou, C.; Lin, D. Tool Path Planning for Directional Freezing-Based Three-Dimensional Printing of Nanomaterials. *Journal of Micro and Nano-Manufacturing*, **2018**, *6* (1). https://doi.org/10.1115/1.4038452/368797.

[56] Biegler, M.; Wang, J.; Kaiser, L.; Rethmeier, M. Automated Tool-Path Generation for Rapid Manufacturing of Additive Manufacturing Directed Energy Deposition Geometries. *Steel Research International*, **2020**, *91* (11), 2000017. https://doi.org/10.1002/SRIN.202000017.

[57] Dwivedi, R.; Kovacevic, R. Automated Torch Path Planning Using Polygon Subdivision for Solid Freeform Fabrication Based on Welding. *Journal of Manufacturing Systems*, **2004**, *23* (4), 278–291. https://doi.org/10.1016/S0278-6125(04)80040-2.

[58] Amal, M. S.; Justus Panicker, C. T.; Senthilkumar, V. Simulation of Wire Arc Additive Manufacturing to Find out the Optimal Path Planning Strategy. *Materials Today: Proceedings*, **2022**, *66*, 2405–2410. https://doi.org/10.1016/J.MATPR.2022.06.338.

[59] Zhai, X.; Chen, F. Path Planning of a Type of Porous Structures for Additive Manufacturing. *Computer-Aided Design*, **2019**, *115*, 218–230. https://doi.org/10.1016/J.CAD.2019.06.002.

[60] Zhou, B.; Tian, T. A Path Planning Method of Lattice Structural Components for Additive Manufacturing. *International Journal of Advanced Manufacturing Technology*, **2021**, *116* (5–6), 1467–1490. https://doi.org/10.1007/S00170-021-07092-5/TABLES/4.

[61] Manoharan, M.; Kumaraguru, S. Path Planning for Direct Energy Deposition with Collaborative Robots: A Review. *2018 Conference on Information and Communication Technology, CICT 2018*, **2018**. https://doi.org/10.1109/INFOCOMTECH.2018.8722362.

[62] Papacharalampopoulos, A.; Bikas, H.; Stavropoulos, P. Path Planning for the Infill of 3D Printed Parts Utilizing Hilbert Curves. *Procedia Manufacturing*, **2018**, *21*, 757–764. https://doi.org/10.1016/J.PROMFG.2018.02.181.

[63] Ding, D.; Pan, Z.; Cuiuri, D.; Li, H.; Larkin, N. Adaptive Path Planning for Wire-Feed Additive Manufacturing Using Medial Axis Transformation. *Journal of Cleaner Production*, **2016**, *133*, 942–952. https://doi.org/10.1016/J.JCLEPRO.2016.06.036.

[64] Ding, D.; Pan, Z.; Cuiuri, D.; Li, H.; van Duin, S. Advanced Design for Additive Manufacturing: 3D Slicing and 2D Path Planning. *New Trends in 3D Printing*, **2016**. https://doi.org/10.5772/63042.

[65] Jiang, J.; Stringer, J.; Xu, X. Support Optimization for Flat Features via Path Planning in Additive Manufacturing. **2019**, *6* (3), 171–179. https://doi.org/10.1089/3DP.2017.0124; https://home.liebertpub.com/3dp.

[66] Liu, J.; To, A. C. Deposition Path Planning-Integrated Structural Topology Optimization for 3D Additive Manufacturing Subject to Self-Support Constraint. *Computer-Aided Design*, **2017**, *91*, 27–45. https://doi.org/10.1016/J.CAD.2017.05.003.

[67] Wang, X.; Wang, A.; Li, Y. A Sequential Path-Planning Methodology for Wire and Arc Additive Manufacturing Based on a Water-Pouring Rule. *International Journal of Advanced Manufacturing Technology*, **2019**, *103* (9–12), 3813–3830. https://doi.org/10.1007/S00170-019-03706-1/TABLES/3.

[68] Diourté, A.; Bugarin, F.; Bordreuil, C.; Segonds, S. Continuous Three-Dimensional Path Planning (CTPP) for Complex Thin Parts with Wire Arc Additive Manufacturing. *Additive Manufacturing*, **2021**, *37*, 101622. https://doi.org/10.1016/J.ADDMA.2020.101622.

[69] Zhang, C.; Shen, C.; Hua, X.; Li, F.; Zhang, Y.; Zhu, Y. Influence of Wire-Arc Additive Manufacturing Path Planning Strategy on the Residual Stress Status in One Single Buildup Layer. *International Journal of Advanced Manufacturing Technology*, **2020**, *111* (3–4), 797–806. https://doi.org/10.1007/S00170-020-06178-W/FIGURES/8.

[70] Ferreira, R. P.; Vilarinho, L. O.; Scotti, A. Development and Implementation of a Software for Wire Arc Additive Manufacturing Preprocessing Planning: Trajectory Planning and Machine Code Generation. *Welding in the World*, **2022**, *66* (3), 455–470. https://doi.org/10.1007/s40194-021-01233-w.

[71] Schmitz, M.; Wiartalla, J.; Gelfgren, M.; Mann, S.; Corves, B.; Hüsing, M. A Robot-Centered Path-Planning Algorithm for Multidirectional Additive Manufacturing for WAAM Processes and Pure Object Manipulation. *Applied Sciences (Switzerland)*, **2021**, *11* (13). https://doi.org/10.3390/app11135759.

[72] Malhan, R. K.; Thakar, S.; Kabir, A. M.; Rajendran, P.; Bhatt, P. M.; Gupta, S. K. Generation of Configuration Space Trajectories Over Semi-Constrained Cartesian Paths for Robotic Manipulators. *IEEE Transactions on Automation Science and Engineering*, **2022**. https://doi.org/10.1109/TASE.2022.3144673.

Chapter 7

A review on fundamentals of cold spray additive manufacturing

Gidla Vinay, Ravi Kant and Harpreet Singh

7.1 INTRODUCTION

Cold gas dynamic spray, also known as cold spray (CS), is a versatile technique for the deposition of a few hundred microns to mm layer coatings. It belongs to the family of thermal spray processes, a relatively new technique developed and patented in late 1990s by the Institute of Theoretical and Applied Mechanics, Russian Academies of Sciences (ITAM, RAS) [1]. In CS, coatings are generated by the plastic deformation of feedstock particles onto a substrate thereby generating layer-by-layer build-up of deforming particles bonded with each other. The feedstock is of typically micron-size particles, where they flew at supersonic velocities with the help of a convergent divergent nozzle (Figure 7.1), and the driving force to generate these speeds is a carrier gas of typically compressed helium, nitrogen or air [2]. Since it is a solid-state deposition process there is no melting of feedstock during the entire process, which makes it the most suitable technique to generate coatings of materials that are thermally sensitive, expensive materials, refractory materials and the like, and the generated coating properties are close to their bulk counterpart. Almost all plastically deformable materials can be coated using CS and materials like ceramics, and polymers can be coated using some special approaches. Nozzle design and carrier gas are typical system parameters, while particle size, process pressure, process temperature, and feed rate are some of the process parameters which determine the properties of the coatings developed.

The layer-by-layer deposition feature of the CS process not only restricts it to being a coating deposition method but also establishes it as an additive manufacturing (AM) technique with a primary emphasis on both repairing and manufacturing parts. [2]. This feature marks the entry of cold spray additive manufacturing (CSAM) into one of the additive manufacturing processes. Advantages like rapid build time, no melting, no need of controlled atmosphere and precision (using robots) makes CS a promising AM technique for repairing as well as making complete parts. Recent developments are going in a way to make CS as "manufacturing at need" kind of AM technology where any part can be repaired or made immediately with an onsite installed CSAM, thereby reducing the operation/production

DOI: 10.1201/9781032703046-7

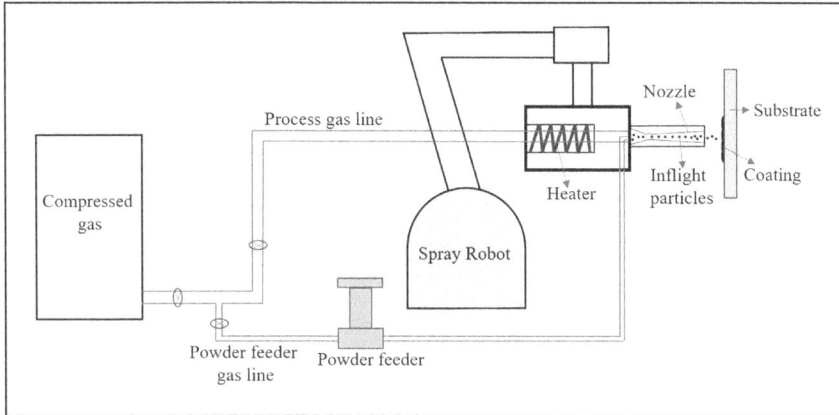

Figure 7.1 Schematic diagram of a CSAM system.

halt time. Significant developments have been made in understanding the bonding mechanism behind CS, and researchers from all over the globe contributed in expanding the window of deposition of materials, thickness achievable and 3D part-making strategies [2–6].

In the initial days due to the usage of low process pressures, coatings were generated with a hand-held coating gun. However, with the increase in process knowledge, the use of higher pressures for achieving higher velocities and the need for better accuracy paved the way for attaching a robot arm to the coating gun. The gun is fixed to a robot arm, and the movement of the arm helps in generating coating along the workpiece (Figure 7.1). Since the non-deposited/rebounded feedstock powder can be recycled or reused because they are collected in powder collectors and no flaming is involved in the process, in a way CS is a sustainable as well as green technology. Throughout this chapter, readers will be introduced to CS process fundamentals like bonding mechanism, process parameters, nozzle design, post-processing techniques and their effect on coating properties.

7.2 BONDING MECHANISM

In the initial spraying of feedstock using cold spray, it was widely understood that particles are not melting, as the operating temperatures are quite below melting temperature; however, the bonding mechanism behind the cold spray was not understood. It is believed that the pure kinetic energy of particles helps in bonding. So, the early trials were more focused on expanding the number of materials deposited and modifying the nozzle geometry to achieve higher kinetic energies [1, 2]. It is difficult to understand the bonding mechanism in cold spray due to the micron-size feedstock particle bombards with the substrate at very high speeds,

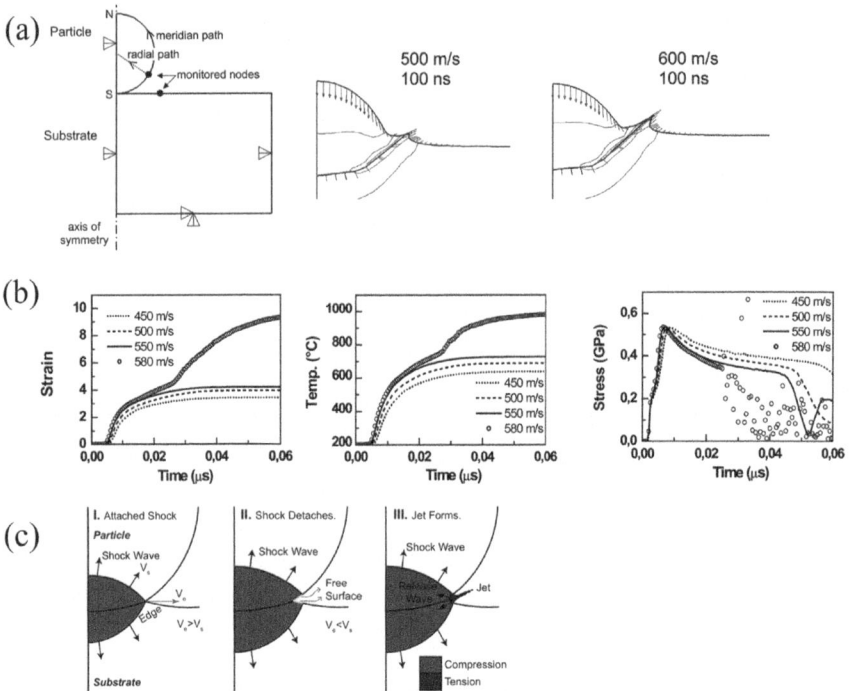

Figure 7.2 (a) Initial condition of simulation and impact of particle at 500m/s, 600m/s; (b) temporal development of stress, temp and PEEQ with different velocities for Cu particle [7]; (c) schematic of jetting in hydrodynamic viscosity theory [8].

where typical impact time would be in nanoseconds, and it is believed that strain rate would be between 10^6 to 10^9 [7]. Later in 2003, Assadi et al. [7] proposed a hypothesis for bonding mechanism with the help of finite element–based impact simulations, naming it adiabatic shear instability (ASI). Later work was done on expanding the knowledge based on this theory, and it was widely accepted among the cold spray community as the bonding mechanism. Recently Hassani et al. [8] have challenged this bonding mechanism by saying that ASI is not necessary to fulfil bonding criteria. It has attracted comments from various groups. Other bonding theories like mechanical interlocking [9], diffusion [10], oxide layer breakup [11] and interface amorphization [12] have been proposed; however, these mechanisms were not as popular as ASI and hydrodynamic plasticity.

7.2.1 Adiabatic shear instability

This theory was first proposed by Assadi et al. [7] followed by Grujicic et al. [13] as the bonding mechanism behind cold spray. In this, with the help of finite element modelling, the authors created a single particle and impacted it onto a substrate at different velocities while keeping the substrate and the particle at room temperature. Upon impact, they have observed some abnormal changes in some of the elements in the particle/substrate interface where jetting occurs. They named these elements critical elements (Figure 7.2a), and this abnormal phenomenon was observed only after crossing a certain velocity. If the particle is impacted above this velocity, then in these elements, a sudden rise in temperature and plastic strain with a drop in flow stress is observed (Figure 7.2b), which indicates a localized melting at these critical elements, which can be identified by jetting. This localized melting helps in bonding these particles onto the substrate, and the velocity at which ASI occurs is termed critical velocity (V_{cr}). It is an indication that above this velocity coating deposition occurs. The main assumption in these simulations is that due to low impact time (typically in nanoseconds), high speed and small particle size, it is assumed that the heating is adiabatic. It has been observed from the SEM images that morphologies of the deformed particles also supported the data where jetting in deformed splat morphology of simulation and experiment are identical. It is observed that above the critical velocity deposition occurs; however, typically only a certain percentage of splat undergoes ASI during deposition in real time. These simulations rely strongly on the Lagrangian method, where plastic strain of the element deforms excessively above certain velocities for soft metals, which ultimately aborts the program. Later studies involving the Eulerian method, however, have overcome this drawback and confirmed the ASI as well [14, 15].

Subsequent studies have been performed to understand this bonding mechanism in a closer way, and an empirical equation derived from the experiments was proposed to estimate the critical velocity of the particle (Eq. 7.1) based on the tensile strength, specific heat and operating temperature [16–20]. As mentioned, critical velocity is dependent on the operating temperature as well. With increase in the process temperature, the temperature gain to the inflight particle increases, which further decreases the critical velocity. In real-time deposition, these coatings are generated at a certain process temperature. However, for the critical velocity calculations it is to be noted that particle and substrate temperatures are given as room temperature. For these impact simulations, the Johnson-Cook (JC) plasticity model, which has contributions from strain hardening, strain rate hardening and thermal softening, is used (shown in Eq. 7.2). Even though the JC equation is applicable well up to 10^4 and strain rate in CS is more than 10^6, it is still a widely used model because of its easy availability in the software and because it is an established model and computationally easy. Some scholars have used the modified JC model [21, 22] and other in-house developed models [23, 24] to understand the effect of deviation from JC. Other popular simulation approaches include molecular dynamic simulations [25].

$$v_{cr} = \sqrt{\frac{a\sigma}{\rho} + bC_P\left(T_m - Tp\right)} \tag{7.1}$$

Where σ is the temperature-dependent flow stress, C_p is heat capacity, T_m and T_p are melting and particle temperature of the feedstock, ρ is the density of feedstock, a and b are graph fitting constants.

$$\sigma = \left[A + B\left(\varepsilon_p\right)^n \right]\left[1 + C ln\left(\frac{\dot{\varepsilon}_p}{\dot{\varepsilon}_0}\right) \right]\left[1 - \left(\frac{T - T_i}{T_m - T_i}\right)^m \right] \tag{7.2}$$

Where σ is the flow stress, T_m and T_{ref} are the melting and reference temperature. A and B are uniaxial yield stress and strain hardening coefficient while m, n and c represent thermal softening exponent, strain hardening exponent and strain rate hardening exponent, respectively.

7.2.2 Hydrodynamic plasticity

Recently, Hassani et al. [8] proposed an alternate theory for the bonding mechanism in cold spray, which is known as hydrodynamic plasticity mechanism. In this the authors have proposed that jetting which causes the bonding in CS occurs due to the hydrodynamic plasticity rather than the ASI. They further claimed that ASI is not needed to cause jetting. In order to prove this, single particle impact simulations were run with and without the thermal softening exponent from the JC equation, which is primarily responsible for the melting. It was observed that even without the thermal softening exponent, jetting happens. They also claimed that jetting is attributed more to the pressure release than to ASI. When the particle impacts the substrate, pressure waves emanate from the substrate and the particle. The claim is that this contact edge velocity (V_e) and shock velocity (V_s) are the primary driving factors for causing the jetting. Initially, this edge velocity will be lower than shock velocity, but there is a transition point above which shock velocity will be greater than edge velocity. At this point, the hydrodynamic stresses generated are greater than the flow strength of the material which facilitates the jetting (Figure 7.2c). This theory has received comments [26] and responses [26] from both sides - further proof is needed before disregarding any of the theories. It is to be noted that in both theories, it is given that jetting is the main reason for the particle bonding; however, the theories differ in explaining what constitutes this jetting.

7.2.3 Rebound phenomenon

Rebound phenomenon is not a bonding mechanism like ASI or hydrodynamic plasticity, but it is a deposition behavior, which explains the window of velocity limit of an inflight particle impacting the substrate for successful deposition. As discussed in the ASI, that critical velocity is the minimum required for initial deposition, it is clearly understood that below critical velocity bonding does not occur, which results in no deposition. One may misunderstand that increasing the velocity

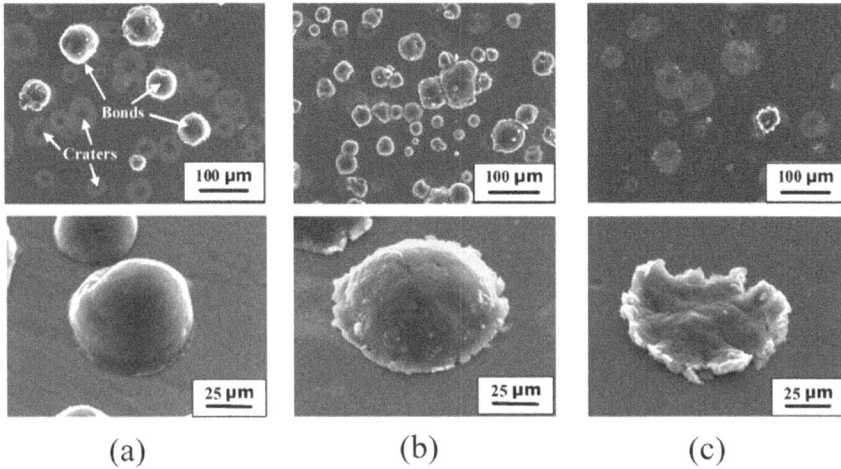

Figure 7.3 SEM micrographs of Al-Si splats attached on substrate sprayed at different velocities (a) 500 m/s, (b) 700 m/s, and (c) 1000 m/s [27].

beyond the critical velocity will further increase the deposition. However, a study done by J. Wu et al. [27] reported that just as there is critical velocity for minimum deposition, there also exists a maximum velocity above which the impacted particle deforms to a maximum limit and rebounds from the substrate, thus also resulting in no deposition (craters in Figure 7.3a, 7.3b, 7.3c). This is known as the rebound phenomenon. It was shown that the bonding of a particle onto a substrate is based on the competition between the adhesion energy and rebound energy [27]. With change in velocity this adhesion energy and rebound energy changes. The initial point of intersection of these energies is termed critical velocity, and the second intersection point is termed rebound velocity. Between these points the adhesion energy will be greater than the rebound energy, which facilitates bonding. Within the parameters of study, it is observed that critical velocity is not much affected with a change in particle size. For a higher particle size, however, the rebound velocity is comparatively lower than the smaller particle size due to higher energies associated with higher mass.

7.3 TYPES OF COLD SPRAY SYSTEM

Even though the principal setup and working mechanisms are the same, cold spray systems are mainly categorized into two types based on the working pressure – low-pressure cold spray system and high-pressure cold spray system. The main difference between the two systems is the working gas; in a low-pressure cold spray the main working gas is compressed air, whereas in a high-pressure cold spray system it is compressed helium (He) or/and nitrogen (N_2).

7.3.1 High-pressure cold spray system

From the name itself it can be understood that these cold spray systems can reach/ use high pressures which are typically up to 50 bar. This high pressure translates into higher gas and particle velocities at the nozzle exit. Typically, helium or nitrogen is used as process gas and powder feeder carrier gas. Due to the higher costs associated with the usage of helium, the cost per product is high for helium-based systems, followed by nitrogen-based systems. To reduce the cost associated, some researchers have tried using nitrogen as the powder feeder carrier gas and helium as the process gas. To further reduce the cost associated with helium usage, helium from the coating chamber has been recycled [28], where the used helium from the process chamber is collected and processed to use again. It reduced the overall cost of the product, but heavy investment is still needed to set up the helium reuse facility. On the other hand, nitrogen is cheaper than helium but costlier than air, and it has the ability to achieve higher pressures up to 50 bar. Velocities achieved and the amount of heat transfer to the particle will differ from the He-based system [29]. Typically, for a given set of process parameters involving pressure and temperature, the velocity attained by the particle is lower in the N_2-based system than in the He-based system, while heat transfer to the particles will be higher in the N_2-based system than in the He-based system. This is due to the low density of helium gas, specific heat ratio and gas constant [30, 31]. Because of this, coatings prepared using He-based systems always exhibit better mechanical properties like hardness and yield strength for the same process parameters of N_2-based systems [29, 31]. One more advantage of high-pressure CS systems is that reaction with oxygen is extremely low in helium and minimal in N_2-based systems, which makes them highly useful in aerospace and medical applications [29, 32, 33].

7.3.2 Low-pressure cold spray system

The cost associated with the usage of gasses like helium and nitrogen led researchers to explore the use of air as the process gas and powder feeder carrier in the CS system. Due to the high density, specific heat ratio and gas constant, the velocities achieved by air-based systems are lower than those of He- and N_2-based systems [20, 29]. Due to the low achievable velocities and use of relatively lower pressures, typically less than 20 bar, these systems are termed low-pressure cold spray systems. As discussed, bonding occurs after reaching the critical velocity, and the critical velocity is different for different material. Sometimes it does not require much velocity to deposit soft materials like aluminum (V_{cr} = 481 m/s), zinc (V_{cr} = 339 m/s), copper (V_{cr} = 452 m/s) and silver (V_{cr} = 375 m/s) [20]. For deposition of these kinds of materials, use of air-based systems are an economically better choice. Some of the researchers have successfully deposited even high-end materials like titanium, Ti6Al4V and super alloys [2, 20, 32]. These materials are very hard and have higher critical velocities. However, due to the extra thermal energy provided in the form of heat transfer from process gas to inflight powder particles, deposition of these materials has become possible [33, 34]. As discussed earlier regarding the bonding mechanism, that critical velocity decreases with increasing particle temperature, higher particle temperatures associated with air-based systems helped in the deposition of these materials. An in-depth explanation of this phenomenon will be provided in subsequent sections.

7.4 PROCESS PARAMETERS

7.4.1 Working gas and process pressure

As discussed in the previous section, compressed gas of air/helium/nitrogen is used in the CS system. In this system, compressed gas has two purposes: first is that the process gas is heated for further expansion and is directed towards the nozzle inlet from the heater assembly, and the second is that the carrier gas acts as a powder feeder which is used to carry the powder from the feeder to the nozzle inlet. At the nozzle, these two gasses mix with each other, so the pressure difference between these two flows need to be adjusted based on the powder flow. Increasing the pressure of process gas increases the velocity of the gas at the nozzle outlet, thereby increasing particle velocities. Helium creates the highest velocities, followed by nitrogen and then air (Table 7.1) [30, 31, 33]. Higher velocities result in higher deposition efficiencies (Figure 7.4a, 7.4b) [19] and therefore better mechanical properties like hardness and yield strength and functional properties, provided that velocities are below the rebound velocity [33]. Gas density, specific heat ratio and gas constant play an important role in achieving higher velocities, which will be discussed later in the nozzle section.

7.4.2 Temperature

A heater is placed inside the cold spray gun assembly, where a typical heating element is made of nichrome. Heaters in high-pressure CS systems are capable of heating up to 1000 °C as process temperature, whereas it is around 600 °C for a low-pressure CS system. This temperature is responsible for the expansion of the gas, which further increases the pressure before entering the nozzle inlet as well as for the heat transfer to the inflight feedstock particle [1, 33]. As discussed earlier regarding the bonding mechanism, temperature also plays an important role in causing ASI along with the particle velocity. Critical velocity, the main parameter which defines the initialization of bonding, is dependent on the temperature of the inflight impacting particle as well. The so-called adiabatic shear instability occurs at lower velocities with an increase in the particle temperature [34], as material can be deformed easily when temperature is applied. This has been tremendously helpful in achieving the coatings intended for inter-splat dependent functional properties like electrical conductivity, corrosion resistance and elastic modulus. Deposition efficiency, electrical conductivity and ductility increases with a rise in process temperature due to thermal softening of particles. However, properties

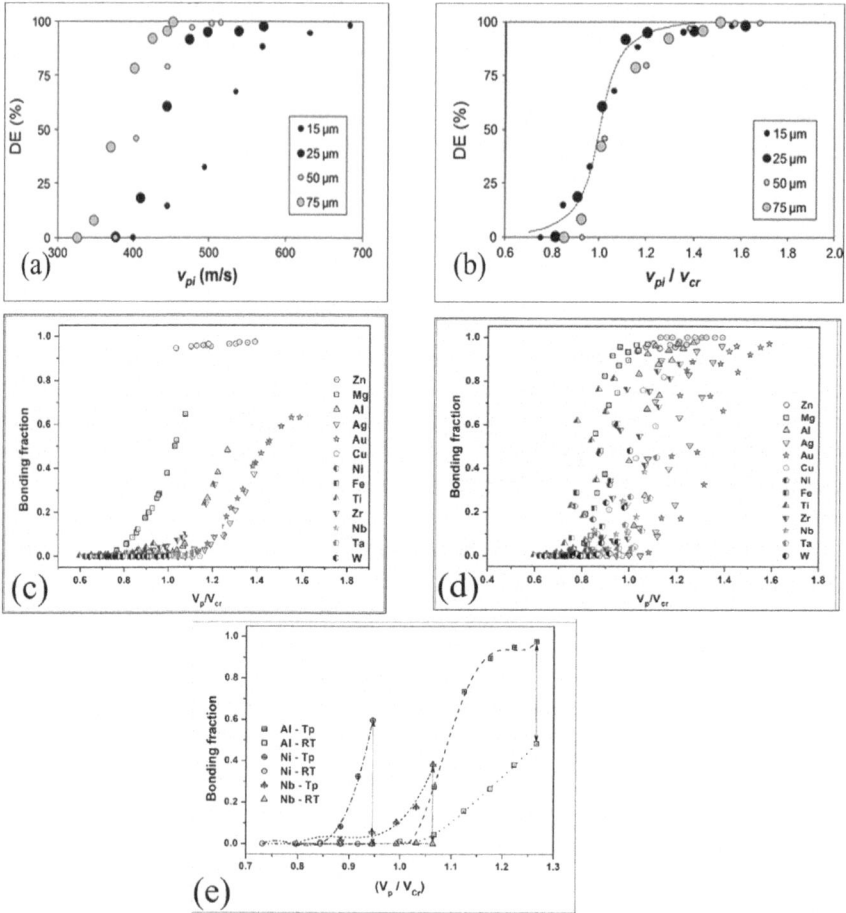

Figure 7.4 Deposition efficiency plotted against (a) particle velocity, (b) V_p/V_{cr} [19]; bonding fraction calculated from peripheral elements in simulation (c) considering particle temperature as room temperature, (d) considering particle temperature and (e) bonding fraction difference for different metals [35].

Table 7.1 Particle velocity table of titanium feedstock for different process parameters [33]

Pressure (MPa)	Temperature (°C)	Gas type	Nozzle type	Velocity (m/s)
3	300	N_2	MOC24	608
3	600	N_2	MOC24	688
0.5	100	He	MOC24	604

such as hardness will be decreased if excessive process temperature is supplied [33, 34, 36]. A proper combination of pressure and temperature needs to be chosen based on the feedstock material and the intended application.

Thermally sensitive materials like titanium and molybdenum tend to form small oxide layers during the travel of the inflight feedstock through the nozzle. The aging of aluminum alloys during inflight travel and nozzle clogging (generally soft/low melting materials) during deposition are some of the problems associated with using a high process temperature [2, 20, 32]. The influence of inflight powder on the deposition was studied using finite element analysis (FEA) by considering powder particle with and without temperature (Figure 7.4c, 7.4d). Correlating the influence temperature with the elastic modulus of the experiment, the researchers concluded that the inflight powder particle temperature effect is significant and contributes to nearly 40% to 60% of bonding (Figure 7.4e) [35].

7.4.3 Feed rate, scan speed, spray angle and standoff distance

Feed rate of the powder feeder in CS is adjusted based on the density and size of the feedstock material as well as desired thickness. Usually for materials with lower density, low feed rate is generally preferred and vice versa for higher density material. Higher feed rates result in more interactions between the inflight particles during the travel in the nozzle, which translates to more velocity loss and ultimately affects the deposition efficiency [38].

Scan speed represents the speed of the robot arm/nozzle transverse movement. Increasing the scan speed results in a lower number of particles accumulating at a particular place on the target substrate during coating deposition, which ultimately decreases the thickness of the coatings achieved per raster. High scan speeds are most useful in generating thin coatings. For repairing/refurbishment of parts and thick coating applications, a slow scan speed is advisable, as a greater number of particles are available to deform and form coating per raster [2, 20].

Spray angle represents the angle between the nozzle and the target substrate and is important for achieving maximum deposition in cold spray coating. A study on the effect of spray angle on deposition efficiency indicated that maximum deposition is achieved when the spray angle is at 90° with the substrate and that deposition efficiency decreases with decreasing the spray angles. No deposition is achieved at angles lower than 30° [39].

Standoff distance is the distance between the nozzle exit and substrate during the spraying of the feedstock. It is observed from the experimental works that the deposition efficiency drastically decreases with an increase in the standoff distance [37]. The microstructure of the coatings revealed that there is increase in porosity with increase in standoff distance, especially for lighter materials like Al and Ti. However, even with lower deposition efficiency and high standoff distance, Cu has obtained dense microstructure due to the tamping effect caused by the higher density of the material (Figure 7.5a, 7.5b). Upon characterization, it is understood

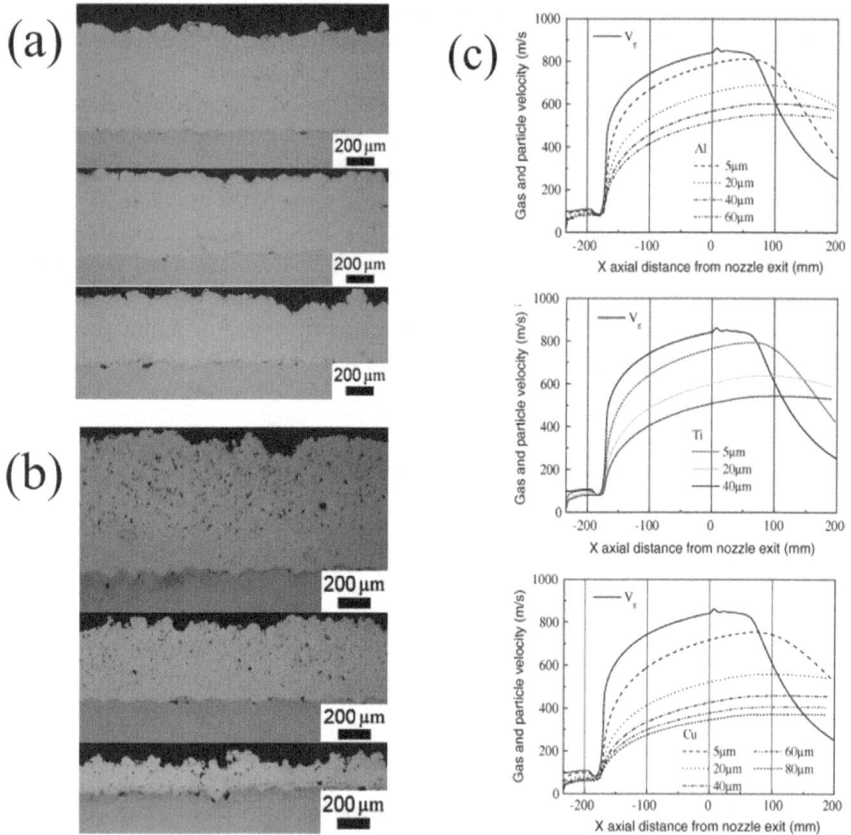

Figure 7.5 Optical microscopic images of coatings deposited at varying standoff distances. Top to bottom (10, 70, 100 mm): (a) Cu coatings; (b) Al coatings; (c) change of gas velocity and particle velocity for different materials with varying standoff distances [37].

that there is no change in hardness for these coatings. This is because there is no change in velocity of the particle within this standoff distance range, and the strain hardening effect due to the impact would be same [37, 40]. Computational fluid dynamics (CFD) analysis of the same has revealed that there is no change in the inflight particle velocity (V_p) and temperature with varying standoff distance, but a drastic decrease in the gas velocity (V_g) has been observed (Figure 7.5c). A reason for low deposition even with no change in particle velocity is the shock wave/bow shock [37, 41]. Bow shock effect is high for low standoff distances and it slowly disappears by increasing the standoff distance. For larger standoff distance, however, gas velocity drastically decreases, and this can affect the inflight particle movement adversely if $V_g < V_p$ [41]. Most articles have suggested that standoff distance between 20 and 30 mm is the best choice in terms of deposition efficiency.

7.4.4 Powder size and morphology

Powder size along with the morphology and density of the feedstock material plays an important role in the successful deposition of feedstock in cold spray. The velocity achieved for larger particle size is less when compared with the smaller particle size for a given process parameter (Figure 7.5c) [37, 41], which ultimately results in lower kinetic energy [36]. Similarly, a material with high density like tantalum and copper always yields lower velocities due to the higher mass when compared with lower density material like aluminum and magnesium for similar process parameters and feedstock size. On the other hand, King et al. [43] observed that extremely small particles of less than 10 μm have yielded lower deformation (less flattening ratio) due to the decrease in velocity caused by the bow shock effect of the nozzle [43]. Usage of extremely small particles also results in flowability issues of powder in the powder feeder. The flowability issue of smaller particles is due to the cohesion of powders caused by forces like Van der Waals, electrostatic and surface tension. Recently Polycontrols, a Canada-based company, developed a vibration-assisted powder feeder which has capability to supply feedstock of very fine size (less than 10 μm) to the cold spray systems without any flowability issue [44]; significant research has yet to be done on these generated coatings. However, it needs to be understood that depositing very fine particles are difficult, as critical velocity increases with decrease in particle size [20].

From the literature reported it is understood that ideal feedstock size for cold spray coatings is anywhere between 15 μm and 50 μm [2, 20]. Particle size highly

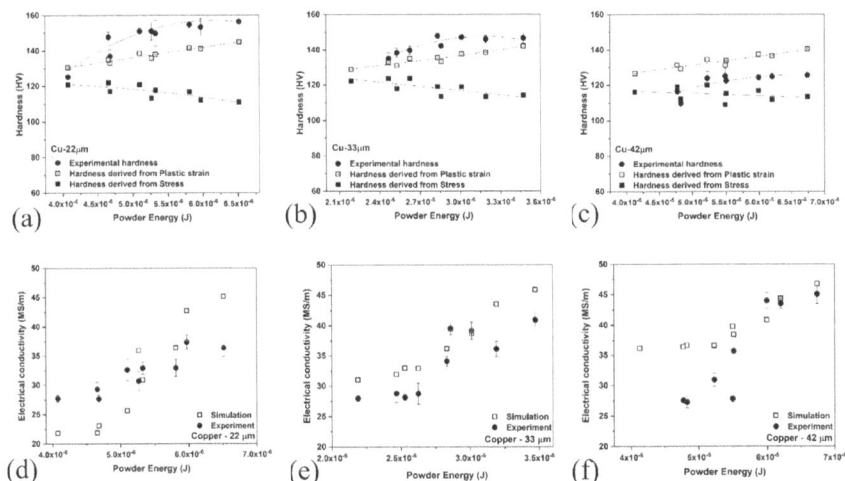

Figure 7.6 Hardness plotted against powder energy (KE + TE) for Cu coatings with average particle size: (a) 22 μm, (b) 33 μm, (c) 42 μm. Electrical conductivity plotted against powder energy (KE + TE) for Cu coatings with average particle size: (d) 22 μm, (e) 33 μm, (f) 42 μm [42].

influences the properties of the coatings, where for a small particle size velocities achieved are higher [45] and the travel time for the feedstock in nozzle is less, therefore lower temperature gains. These particles with higher kinetic energy result in higher strain hardening during particle deformation, and the obtained coatings possess good mechanical properties like high hardness and yield strength. However, there will be a decrease in the overall ductility of the coating due to the high strain hardening. Other functional properties like corrosion resistance and electrical conductivity obtained for these coatings are inferior owing to the losses occurring due to a higher number of splat interfaces.

Kumar et al. [42] have studied the particle size effect on CS copper coatings and from their works it can be concluded that coatings deposited with feedstock of 18 μm average particle size (lower particle size) achieved high hardness of 160 HV, while electrical conductivity achieved is only 36 MS/m (62% of bulk). However, coatings deposited with feedstock of 42 μm average particle size (higher particle sizes) achieved better electrical conductivity, as high as 45 MS/m (77% of bulk), while hardness achieved was only 120 HV (Figure 7.6a, 7.6b, 7.6c, 7.6d, 7.6e, 7.6f) [42]. This increase in the hardness in lower particle size coatings was attributed to the higher grain refinement occurring in smaller size particles compared to the larger size particles (smaller size—higher kinetic energy; more interfaces—more grain refinement places). As already discussed in the temperature section, a larger particle usually achieves lower particle velocities due to the higher mass, which increases the travel time inside the nozzle. This increased travel time helps in gaining more heat from the process gas, and the overall thermal energy of a particle is directly proportional to the mass of the particle and ΔT. This extra thermal energy given helps in achieving better deformation, thereby better bonding and improvement in functional properties like electrical conductivity, corrosion resistance, elastic modulus and ductility [42]. However, due to more thermal softening, hardness achieved is low.

Spherical powder is often chosen for deposition as spherical particles have better flowability; however, non-spherical morphology like tear drop, irregular shaped powders and the like achieves higher velocities due to the low drag force caused inside the nozzle. A study done by Wong et al. [46] revealed the same, where irregular particles achieved better velocities for the same process parameters when compared with the spherical particles due to the low drag coefficient. Hardness achieved for the irregular particle coatings is inferior when compared with the spherical particle coatings because irregular particles impinge on the target/substrate with the sharp corner. So, instead of undergoing plastic deformation, the impacted particle impinges/penetrates (in the previous layer), which results in poor mechanical properties, poor adhesion (jetting phenomenon occurrence sites are fewer) and lower deposition efficiency. The top surfaces of the coating have attracted more porosity, which confirmed the same [46]. An FEA-based impact study of conical and spherical particle simulations revealed the same, in that conical shaped particles stress concentrations are high and jetting was missing, which can result in poor deposition [47].

7.4.5 Substrate

Every metal inherently possesses a small "nm" thickness of oxide layer on top of the surface. Since the bonding mechanism is due to the plastic deformation of impacting particles on the substrate, the state of the substrate is of utmost importance to facilitate bonding. Different approaches like grit blasting [36, 42, 49], polishing [50–52] and preheating of substrates [53] were taken by researchers, who reported interesting results which will be discussed later in this section.

Substrate preparation plays an important role in the adhesion mechanism of coating with the substrate, as the first layer interaction is between the particle and substrate while the interactions of subsequent layers are between the incoming particles and already deposited particles. Without proper adhesion of the first layer, formed coating delaminates easily from the substrate. Substrate condition, such as hard (Figure 7.7a) and soft (Figure 7.7b), also plays an important role in

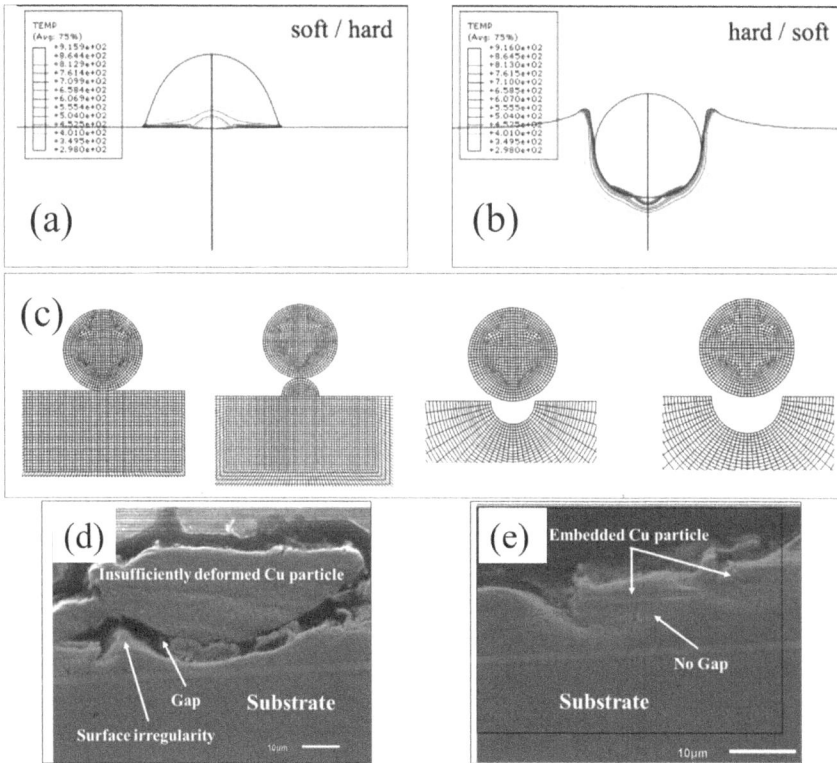

Figure 7.7 Finite element simulation results of particle and substrate combination: (a) soft on hard (Al on steel), (b) hard on soft (Ti on Al) [48]; (c) single particle impact simulations on different roughened substrates [49]; (d) Cu deposited on as received substrate (Ra of substrate 2.5 μm), (e) Cu deposited on mirror polished substrate [50].

the adhesion mechanism. Grit blasting is one of the widely accepted substrate preparation techniques for thermal spray coatings and cold spray [2, 20, 50]. Kumar et al. [49] explained the adhesion mechanism involved behind grit-blasted surfaces and the effect of grit-sized craters on the adhesion mechanism using FEM-based single particle impact simulations (Figure 7.7c). It is reported that frictional energy dissipation increases when a particle impacts on a crater/roughed surface, thereby increasing the interface temperature which facilitates bonding. For this to happen, crater size and particle-substrate type (soft/hard combinations) are some of the important factors that must be considered. Preferably, grit blasted crater size should be lower than the particle size. For the hard material on soft substrate combination, surface roughness doesn't play a large role, as the first layer is mostly due to the interlocking. Some of the recent findings related to substrate effect is that mirror-polished substrates give better adhesion when compared with the roughened substrate [50–52]. Singh et al. [50] deposited Cu particles on different substrates with different roughness, i.e., as received (Ra 2.5 µm), semi-polished (Ra 0.5 µm) and mirror polished (Ra 0.06 µm). SEM analysis of these single splats (Figure 7.7d, 7.7e) revealed that Cu particles deposited on mirror-polished surfaces are better adhered with the substrates with no gap between the substrate and splat, while a gap is observed in splats deposited on roughened substrates. It has been reported that waviness and surface asperities of roughened substrates caused poor bonding. The main differences among these studies are that roughness in these roughened substrates is created using polishing against emery sheets while the grit-blasted substrates are impacted with grit particles at some velocity. The main difference among these studies is the method of roughness generation, where Kumar et al. [49] created roughness using grit blasting while Singh et al. [50] used polishing with different grade emery sheets. A clear picture is yet to be observed in terms of which approach is better, as there is no study done that compares all these effects, which could be of new interest. Another approach tried by researchers is preheating the substrate during the coating process. Early research conducted on this aspect suggests that for soft and low-melting metals like zinc (low critical velocity), deposition efficiency drastically decreased, while an increasing trend is observed for deposition of Al powder (high critical velocity) [53].

As already discussed, it is to be noted that the first layer of deposition is always "particle on substrate" and the subsequent layers are between the "deposited particle and the incoming particle". Critical velocity is measured by impacting the same particle on the same substrate [7], but in reality the substrate may or may not be of the same material, and the deformation behavior of the impacting material on substrate is different for different materials. So, it needs to be realized that the critical velocity changes with substrate as well, but still critical velocity is an important factor to be considered because the bonding of subsequent layers is between the same materials. So there exists two conditions where the first layer critical velocity is based on the substrate used and the subsequent layers critical velocity is based on the same material. For this, a detailed analysis was done by

Bae et al. [48]. The authors observed different trends when studying soft–hard combinations of particle–substrate deposited at their critical velocities. Nevertheless, soft on soft (same material) and hard on hard (same material) materials showed similar results of ASI as what would normally happen. It was the soft on hard and hard on soft combinations which resulted in useful new insights. If a soft particle deposited at its critical velocity on a hard substrate, it was observed that deformation is relatively more in lateral direction, i.e., more flattening and the jetting from the substrate was not observed as it would be observed in a normal impacting ASI scenario (Figure 7.7a). At interfaces, some of the critical elements have experienced high temperatures (near melting), which means that bonding has happened. However, there is no clear temperature transition state as would be observed in ASI (at the critical velocity), which suggests that the ASI would have happened at much earlier velocity itself. So, the critical velocity is much lower when a soft particle impacts onto a hard substrate. Simulations observed for this material combination (soft on hard) at much lower velocities than critical velocity have confirmed this phenomenon. In case of impacting a hard particle on a soft substrate, the hard particle might not deform at all. However, the substrate deforms more and hard particle impinges into the soft substrate, thereby interlocking with the substrate (Figure 7.7b). This study has helped in understanding the deposition of soft metals on hard substrates, hard metals on soft substrates and the effect of substrate material on critical velocity. The same is applicable to the metal coatings generated on polymer/fiber reinforced polymer (FRP) substrates, where bonding of the first layer is due to the interlocking of the incoming particles onto the polymer substrate [54–56]. X. Chen et al. [56] have conducted single particle impact of nano-structured WC-23Co (agglomerated and sintered spherical shape powder) on soft (Q235) and hard (WC-12Co) substrates, thereby observing that the particle is penetrated into the FRP substates and gets interlocked rather than the conventional deformation of feedstock particle which occurs when impacted on a harder substrate [56].

7.5 NOZZLE GEOMETRY

7.5.1 Design and mach number

A De-Laval type nozzle which consists of a convergent–divergent section is used for achieving the required accelerations to the inflight powder particles (Figure 7.8a). The cold spray nozzle is designed for achieving supersonic velocities, so it is rated based on the Mach number a nozzle can achieve. In a convergent–divergent nozzle, at the nozzle entrance a high-pressure gas passes through the convergent section, and its pressure decreases and velocity increases (subsonic) as it reaches the throat section. At the throat section, it achieves Mach 1 (speed of sound), and thereafter from the divergent section there will be a further decrease in pressure and the gas reaches supersonic velocities [16, 58].

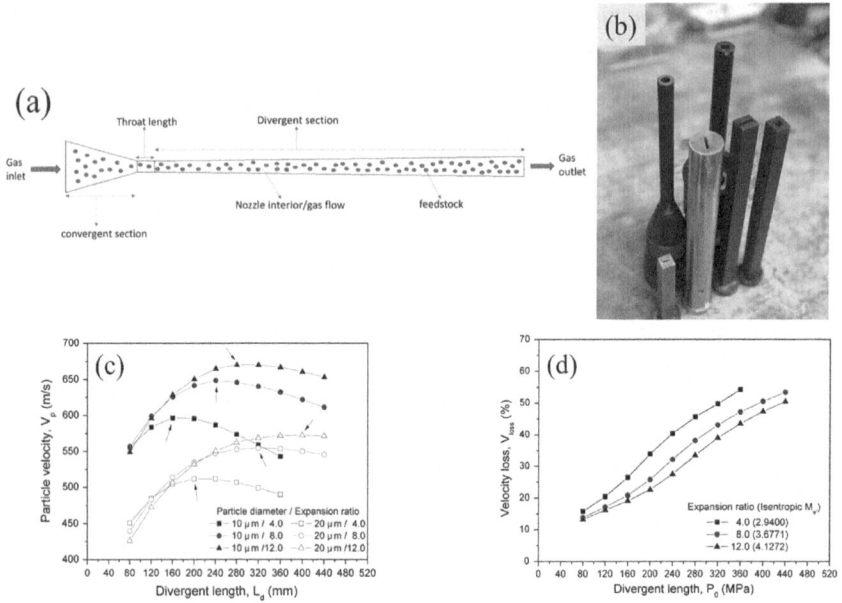

Figure 7.8 (a) Schematic of the computation domain of nozzle CFD model; (b) nozzles with different exit shape [6]; (c) effect of standoff distance on particle velocity; (d) gas velocity loss with increasingly divergent lengths of nozzle [57].

Powder is injected into the nozzle either through axial injection at the entrance of the nozzle [30] or through radial injection in the divergent section [59]. Axial injection is the most common technique for high- and low-pressure CS systems, but some low-pressure CS systems use a radial injection technique for depositing soft/low-melting materials in order to eliminate nozzle clogging at the throat section and for compaction of the system. Nozzle designs are mostly patented by companies, but the objective is the same in that it needs to achieve maximum velocity and is based on the Mach number achievable at the nozzle exit. Mach number depends on the ratio of throat area to the exit area of the nozzle, and it can be calculated from Eq. 7.3, Eq. 7.4 and Eq. 7.5 [16, 58]. Different strategies were employed to achieve this ratio, such as changing the nozzle exit shape to circular exit, rectangular exit and elliptical exit and the throat type to rectangular throat, circular throat and so on (Figure 7.8b). Each has their own advantages and disadvantages; besides this, manufacturability of the nozzle material also has to support these shapes [2, 6, 20].

It is clearly understood that with an increase in pressure, the Mach number achieved at nozzle exit increases, thereby increasing gas and particle velocities. With high pressures it is observed that shock waves and expansion waves form outside the nozzle exit rather than inside the nozzle, which helps in reducing velocity losses [57]. However, along with the process parameters, shock wave formation place is dependent on divergent length as well [57]. Increase in process temperature also increases the particle velocity, but the velocity change is far less than the pressure increase [36].

$$\frac{A}{A^*} = \frac{1}{M}\left[\left(\frac{1}{\gamma+1}\right)\left(1+\frac{\gamma-1}{2}M^2\right)\right]^{\frac{\gamma+1}{\gamma-1}} \qquad (7.3)$$

$$V = M\sqrt{\gamma RT} \qquad (7.4)$$

$$T = \frac{T_0}{1+\frac{\gamma-1}{2}M^2} \qquad (7.5)$$

Where M is the local Mach number, A/A^* represents the expansion ratio of throat area to exit area, γ is the specific heat ratio and V is 1D gas velocity.

7.5.2 Divergent section

Apart from designing the nozzle for high Mach numbers, two other considerations before designing the nozzle are convergent and divergent lengths. These two sections play an important role in particle deposition as they affect the particle velocity and temperature of the inflight particle and ultimately affect the deposition. Particle velocity is estimated by the empirical relation established from the mathematical approach [58] or can be measured experimentally using spray watch or through finite element based CFD simulations. CFD-based simulations are cost effective, and it is easy to understand the effect of changes in the geometry of the nozzle. Using such an approach, Yin et al. [57] studied the effect of divergent length on the Mach number and particle velocity. They observed that with an increase in divergent length, there is a decrease in Mach number and an increase in velocity loss of gas. However, there is an increasing trend for particle velocity with an increase in divergent length due to the reduction in drag force (Figure 7.8c, 7.8d). Even with a reduction of drag forces, after reaching a limit (divergent length) a drastic decrease in particle velocity is observed because of the high velocity loss observed in gas velocity. It is concluded that the optimal divergent length is the result of competition between the drag force and particle acceleration [57].

7.6 POST-PROCESSING OF CS COATINGS

Cold-sprayed coatings are formed due to the heavy plastic deformation of the inflight particles upon impacting with the substrate. So, generally due to heavy plastic deformation, these coatings exhibit low ductility (due to high strain hardening) and the ASI happens only in a fraction of the splat interface due to which other parts of the splat are poorly bonded. These splat interfaces act as nucleation sites for corrosion medium penetration, nucleation sites for rapid crack growth and barriers for electrical conductivity applications. In order to decrease these effects and increase the ductility of the coatings, heat treatments are conducted on cold-sprayed coatings [60, 63]; choice of method is based on the end application, structure and material. Most popular post-processing techniques used vacuum/inert heat treatment of the coatings in a furnace. Other techniques include friction stir processing (FSP), shot peening, *in situ* hammering and laser remelting.

7.6.1 Vacuum/inert heat treatment

This type of heat treatment is one of the most used and common heat treatment processes for CS coatings. However, the main drawback is that the substrate also has to be heat treated along with the coating, which is undesirable when the substrate material is temperature sensitive or for a refurbishment application. Heat treatment studies were done on the cold spray coatings for a long time, and these have helped a great deal in understanding the behavior of cold spray coatings. In the heat treatment process, initially diffusion of pores happens followed by a recrystallization phase and then grain growth (Figure 7.9a) [60]. Most of the studies have conducted heat treatment in a vacuum or inert atmosphere, as the chances of oxidation are high when conducted in a normal atmosphere. A report on heat treatment of CS SS316 steel coatings in different environments, i.e., air, vacuum

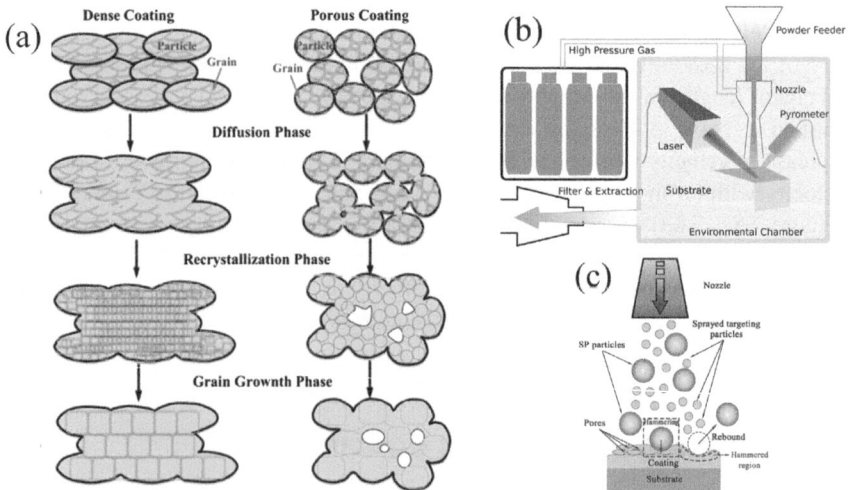

Figure 7.9 (a) Microstructure evolution of coatings in heat treatment [60], (b) schematic representation of laser-assisted CS [61] and (c) schematic representation of *in situ* hammering in CS [62].

and hot isostatic pressing, has shown the ductility and tensile strength increase is less in air-annealed CS coatings compared to vacuum-annealed CS coatings as well as hot isostatic pressing (HIP). Formation of an oxide layer at the splat interface due to the reaction with atmosphere is the reason for this [64].

7.6.2 Friction stir processing

Friction stir processing is an adaptive technology from friction stir welding (FSW), where in FSP a rotating tool with shroud and pin is used as tool which feeds into the workpiece and is moved across the surface to be modified, in contrast to the tool movement on two interfaces to be welded in FSW. Localized melting of the workpiece at the interface happens due to the heat generated from the friction between tool and the workpiece [65, 66]. Reports have shown that due to the high temperatures and instant cooling, dynamic recrystallization occurs in the workpiece material during FSP, resulting in the formation of finer grains and enhancement in properties like hardness. Due to the rotation moment of the tool, homogenization in the material occurs, which helps to distribute the matrix material evenly in cold spray composite coatings. This further helps to improve properties by reducing porosity, properly distributing the matrix phase and homogenizing the grain [66, 67].

7.6.3 Laser remelting

Laser remelting is another post-processing technique used for cold spray coatings. It can be done either during the deposition process itself, which means along with the CS gun movement, or separately as a post-processing technique after the deposition. In the first case a laser source is attached to the CS gun and moves along with it, projecting toward the coating during deposition (Figure 7.9b). In this way, layer by layer the coating is deposited and simultaneously remelted using a laser source [61, 68]. In the second approach, a fully deposited coating is taken and a laser is projected on the coating surface and moves [69]. In both the cases laser power, scan speed and beam diameter decide the depth of laser penetration [61, 68–70]. In most of the cases, like in other heat treatment processes, a decrease in porosity is observed which increases properties like corrosion resistance in Ti coatings [70] and hardness in Al coatings [71]. For metal-ceramic composite coatings, better hardness and uniform distribution of matrix can be observed due to the remelting, but oxidation may occur due to the melting. The effect of laser remelting on these overall properties and optimization is still yet to be explored.

7.6.4 Shot peening and *in situ* hammering

Some researchers have used shot peening as a post-processing technique to increase the properties of the CS coatings [72–74]. Shot peening refers to the bombarding of high-strength particles on a surface to induce compressive stresses and enhance the surface properties. In CS due to the heavy plastic deformation of particles, tensile residual stresses are induced in the coating and compressive residual stresses are induced at the coating-substrate interface. These compressive residual stresses are beneficial for keeping the coating bonded with the substrate, but after reaching a thickness limit the tensile stresses generated in the coatings grow further and delaminate the coating from the substrate. Results from the limited number of studies done for using this post-processing technique reported that shot peening did not induce any compressive residual stresses [73, 74]. However, surface properties like hardness and roughness are enhanced [72]. Even soft material like zinc has exhibited an increase in hardness up to 65 HV through this shot peening post-processing technique [75].

By considering the advantages of shot peening, a new approach named *in situ* hammering was developed, in which large sized hard spherical particles were mixed along with the feedstock powder (Figure 7.9c) and were propelled along with the feedstock and allowed to impact the deposited coating [62, 76]. Due to the higher particle size and mass, these hard particles achieved lower velocities. Upon impact due to their large size, these large sized hard particles will not bond with the coating. Their impact effect increases the compaction, however, and thereby induces the compressive forces and decreases the porosity. In this way, we can increase the properties like hardness and bonding.

7.7 EFFECT OF POST-PROCESSING ON PROPERTIES OF CS COATINGS

7.7.1 Electrical conductivity

One of the important applications of cold spray is electrical conductivity coatings – its ability to deposit materials without phase change provides an opportunity to achieve properties like EC close to their bulk counterpart. Factors like temperature dependence, dislocation density, porosity and grain boundaries are the main reasons for resistivity occurring in a material, which results in loss of electrical conductivity. Phani et al. [77] have tried to quantify these effects for CS copper coatings generated with negligible porosity and showed that for dislocation density to cause significant effect, it needs to be in the order of $10E+16/m^2$ or the hardness increase should be 1000 MPa, which is not observed in their developed coatings. After deducting the resistivity contribution from the grain boundaries, the authors attributed the remaining resistivity to the weak inter-splat bonding (Figure 7.10a, 7.10b). Data from heat-treated coatings also revealed the same, where EC values of the heat-treated coatings achieved close to bulk. This analysis has helped in understanding the effect of inter-splat bonding on properties like electrical conductivity. It is also given that heat treatment of coatings at recrystallization temperatures helps in achieving better inter-splat bonding (provided negligible porosity). Similar kinds of observations are

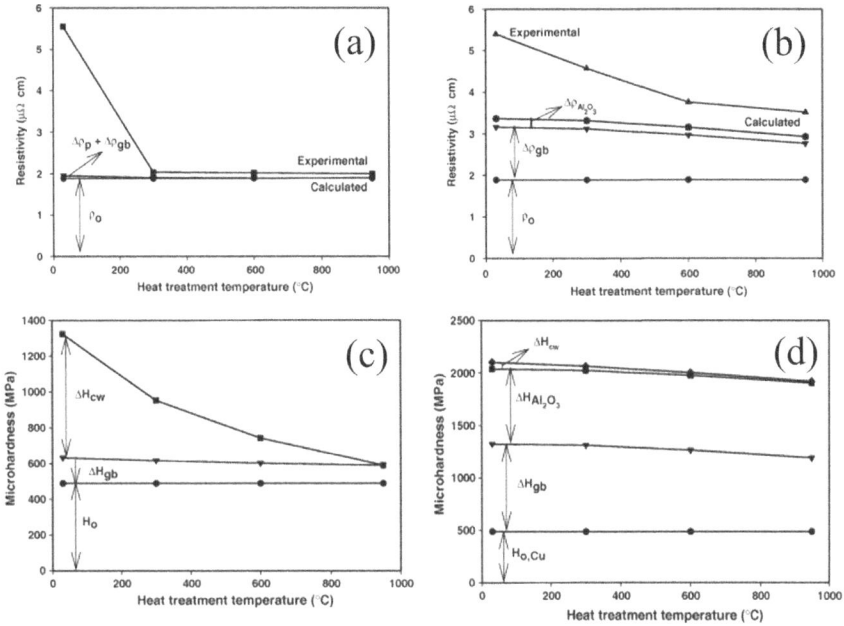

Figure 7.10 Resistivity calculations with annealing temperature for (a) Cu coatings and (b) Cu alumina coatings; hardness contributions from different mechanics with varying annealing temperatures: (c) Cu coatings (d) Cu-alumina coatings [77].

observed from the studies done on heat treatment of cold-sprayed coatings, where in as-sprayed condition the electrical conductivity achieved is low even for dense coatings, but as high as near bulk is observed after heat-treated conditions [52, 77, 78].

In a study done on CS silver coatings where an increase in EC is observed with an increase in the velocity and good EC is obtained in as coated state itself due to the better process conditions, which means that by further heat treatment will result in fractional increase of EC as loss of EC in CS coatings is due to the weak inter-splat bonding. They have conducted heat treatment in air and argon atmospheres and concluded that the change in EC is high in the argon case compared to the air. This is due to the oxidation in air, and further increasing the heat treatment temperatures in the argon case resulted in a decrease of EC. This has been attributed to the no further scope of increasing inter-splat bonding and formation of oxides due to reactions in Ag, which are detected from the XRD results [78]. This once again shows that with the right process parameter combinations, electrical conductivity close to the bulk can be achieved and further post-processing can be avoided. This has been supported from some multi-particle simulations, where interface temperatures are used to estimate EC and the predicted values are well corroborated with the experimental values [42].

7.7.2 Corrosion resistance

Corrosion medium penetration happens primarily because of porosity, weak inter-splat bonding and grain boundaries [79]. In a study done on CS Zn coatings, Naveen et al. [80] observed that a corrosion rate of 24 $\mu A/cm^2$ is obtained for CS Zn coatings in an as-sprayed state, whereas with heat treatment corrosion resistance has improved and corrosion rate is reduced to 4 $\mu A/cm^2$. This is due to the reduction of porosity from 0.5% to 0.25% and improvement of inter-splat bonding, which is clearly visible from the SEM images of an etched cross section. *In situ* hammering of CS Ni coatings has increased the corrosion resistance of generated coatings, where its corrosion rate is almost comparable with the bulk Ni corrosion rate. In this study, AZ31B has been chosen as the substrate; generally Mg alloys have lower E_{corr} and any penetration of corrosion medium through the coating will reach the substrate and start corroding the substrate rapidly due to the lower potential. In all of the studies, because of the densification of coatings, the corrosion medium did not penetrate the substrate even in long hour durations, which is appreciable [62]. The same is observed for Zn coatings where shot peening was adopted as a post-processing technique. Shot peening pressures chosen were between 0.05 MPa to 0.2MPa and with the increase in the shot peening pressure surface roughness decreased, porosity decreased and corrosion rate decreased from 3.5 mm/a to 2.14 mm/a in simulated body fluid [75]. *In situ* hammering-assisted CS Al alloy coatings exhibited an increase in corrosion resistance compared with conventional sprayed coating, in which the corrosion rate decreased by one order because of the decrease of porosity, effectively 12% to 0.2%, due to the hammering effect [81].

A study on corrosion resistance of CS refractory coatings (titanium, tantalum, niobium) in harsh medium (HF) showed that as-sprayed titanium showed better corrosion resistance than did the heat-treated coatings due to the penetration of corrosion medium inside the coatings and the formation of a strong oxide layer, which further stopped the corrosion medium penetration. Other coatings of tantalum and niobium also performed comparatively better in as-coated state when compared with heat treated coatings whose explanation for this not derived. However, in heat-treated coatings corrosion attack was severe on the top surface due to the closure of inter-splat boundaries [82].

7.7.3 Hardness

As already discussed in earlier sections, hardness increases in cold spray coatings with an increase in process pressure or in velocity of the inflight particles. This increase in hardness is attributed to the cold work induced because of the extensive plastic deformation of the inflight particles. So, the total hardness of cold spray coating is a cumulative collection of intrinsic hardness, hardness due to grain size and cold work induced due to the plastic deformation (Figure 7.10c, 7.10d) [77]. A recent study indicated that dislocation densities of around $10E+14/m^2$ are observed in CS coatings compared to $10E+12/m^2$ of the powders; estimations were made from modified Williamson-Hall plot approach and convolutional multiple whole profile (CWMP) fitting [83]. There is a trend of decrease in hardness observed by increasing the heat treatment temperature for CS coatings due to the annihilation of hardening effects. By carefully removing the hardness contribution from grain size by using Hall-Petch relations and intrinsic hardness (material property), one can calculate the hardness contribution from the cold work. Through this one can estimate the contribution of cold work [77]. Other post-processing processes such as shot peening have increased hardness, even in zinc coatings [75] and Al coatings [81]. In cases where FSP is used, a hardness increase was observed in CS AA7075 coatings [84] and metal-ceramic coatings [66, 67]. For metal-ceramic coatings, hardness increased due to the redistribution of ceramic particles all over the coating, improved bonding and decreased porosity. Singh et al. [52] studied the effect of substrate roughness on deposition and properties of Cu particles and the same was studied with the effect of heat treatment, where properties like elastic modulus and hardness increased upon heat treatment. However, the explanation for hardness increase was unclear and has yet to be explored.

7.7.4 Elastic modulus

Elastic modulus (EM) is a material property and can be used as a parameter to talk about the quality of the CS coating, as it is purely based on porosity and crack density [85]. Sundararajan et al. [85] was inspired from the Zimmerman analysis [86] of dependence of elastic modulus on the porosity, Poisson's ratio and crack density and applied the same analogy to understand the EM of CS coatings where crack density is replaced with weak inter-splat bonding. Upon conducting heat treatment at different temperatures, the authors observed that elastic modulus of the coatings is increased and reached near their bulk counter parts. However, coatings with high porosity even after heat treatment have exhibited less increase in EM. So, it is observed that a drop in the elastic modulus of non-porous CS coating is due to poor inter-splat bonding. Vinay et al. [35] applied the same analogy to a multi-particle simulation consisting of 120 particles and estimated the elastic modulus of the coatings through the peripheral elements whose temperatures are greater than the recrystallization temperature. They observed good correlation with the EM obtained through experiment for different materials. It has been observed that with full mechanical interlocking, CS coatings should be able to give 0.7 times of bulk EM, but values achieved below 0.7 times are attributed to the lack of bonding, porosity and pullouts.

7.7.5 Fracture

In a study conducted specifically for fracture toughness, a comparison of cold-sprayed standalone coatings (Al, Cu, Ti, Ni) with the cold-rolled samples of the same material revealed that CS coating fracture toughness achieved was merely

6%–12% for Ti and Cu and 18%–25% for Al and Ni when compared with the cold-rolled samples. Upon analyzing them under SEM, it was observed that fracture was propagated through inter-splat boundaries [87]. From the *in situ* tensile tests of CS Cu thick coatings, it was observed that fracture was occuring through the inter-splat boundaries which was due to the lack of bonding at interfaces. However, when the same was tested after heat treatment, crack propagation hindrance was observed [52, 88]. It was observed that process parameters and annealing temperature were the main factors affecting the crack initiation site where annealing temperatures around 430 °C limited the crack initiation in CS Cu coatings [88]. A fracture toughness test of these samples would give better insights in terms of any improvement in fracture toughness after annealing. White et al. [89] conducted fracture analysis studies using different standard methods (CT and four point bend test) for CS Al alloy by varying the substrate roughness and found that fracture toughness of CT specimen showed a linear trend with substrate roughness, but no correlation was obtained for the four point bend test [89].

7.8 SUMMARY AND SCOPE

In this chapter, the fundamentals of the cold spray process and the detailed influence of process parameters are explored. The De-Laval nozzle, considered to be the central component of the CS system, is discussed in depth, covering design aspects and their impact on inflight particle characteristics. Additionally, its effects on coating characteristics are further examined. The deposition characteristics of some of the commonly sprayed materials like Cu, Zn, Ni, Ti, and Al and their alloys are also discussed in detail. Post-processing, which is often considered an important step in cold spray to improve inter-splat dependent properties as well as to increase ductility, is explored in depth, along with the microstructural changes caused by the different post-processing techniques.

- Constant research interest from researchers around the world has helped in evolving this technology to today's extent, but there is a lot of room for understanding the process fundamentals much more effectively and modifying the system to produce standalone parts much more efficiently, thus making it more economical. One such example of bringing new understandings is the recently proposed alternate bonding mechanism hydrodynamic plasticity (which needs further investigation).
- In the span of two decades cold spray has evolved from metal only coatings to composite (metal-ceramic) coatings, and now steps are being taken to fabricate standalone parts owing to cold spray unique solid state deposition mechanism. Owing to its high deposition rates, cold spray has the ability to produce "parts at need" where other conventional manufacturing processes are not possible to install or usage of conventional 3D printing process is a delayed affair due to low production rate. As discussed, standoff distance

and impact angle are some of the most important factors for achieving high deposition efficiency. Programming the nozzle movement to change the standoff distance according to the thickness of the previously deposited layer is an important step for achieving stable layer thickness in depositing standalone parts. Typically, nozzle outlet dimensions decide the resolution of the layer, but manipulating the nozzle angle to obtain minimum layer resolution is possible due to the change in deposition efficiency with nozzle angle. In order to achieve these, online monitoring of the deposition efficiency can simulate the robot movement before coatings need to be studied and optimized.

- There are many individual modules involved in CS, like process parameters which are given through a user-friendly interface system, robot movement given through a teaching pendant, spray watch setup/CFD simulations used for estimating the velocity and temperature of inflight powder, FEA-based impact simulations for understanding process parameter effect, preprocessing like grit blasting/polishing, inter-processes like machining, online monitoring of thickness deposition and post-processing techniques for property improvements. For performing some of these individual modules, the cold spray process needs to be stopped. Generating a feedback loop and integrating all of these individual modules into a single interconnected module will help to make the CS system a more automated system with minimal human interaction and a highly efficient process which comes under aims of Industry 4.0.

As cold spray is considered an environmentally friendly and sustainable technology because it uses no harmful gasses and no powder is wasted, it can be seen as a "reliable manufacturing technology for future". Happy sustainable spraying!

REFERENCES

[1] A.P. Alkhimov, A.N. Papyrin, U. Vyazemskogo, V.F. Kosarev, N.I. Nesterovich, M.M. Shushpanov, Gas-dynamics spray method for applying a coating, U.S.Patent 5 302 414, 12 Apr 1994.

[2] W. Li, K. Yang, S. Yin, X. Yang, Y. Xu, R. Lupoi, Solid-state additive manufacturing and repairing by cold spraying: A review, Journal of Materials Science and Technology, 34 (2018) 440–457.

[3] R.N. Raoelison, C. Verdy, H. Liao, Cold gas dynamic spray additive manufacturing today: Deposit possibilities, technological solutions and viable applications, Materials and Design, 133 (2017) 266–287.

[4] H. Wu, X. Xie, M. Liu, C. Chen, H. Liao, Y. Zhang, S. Deng, A new approach to simulate coating thickness in cold spray, Surface and Coatings Technology, 382 (2020) 125151.

[5] H. Wu, X. Xie, M. Liu, C. Verdy, Y. Zhang, H. Liao, S. Deng, Stable layer-building strategy to enhance cold-spray-based additive manufacturing, Additive Manufacturing, 35 (2020) 101356.

[6]   H. Wu, S. Liu, Y. Zhang, H. Liao, R.N. Raoelison, S. Deng, New process implementa-
      tion to enhance cold spray-based additive manufacturing, Journal of Thermal Spray
      Technology, 30 (2021) 1284–1293.
[7]   H. Assadi, F. Gärtner, T. Stoltenhoff, H. Kreye, Bonding mechanism in cold gas
      spraying, Acta Materialia, 51 (2003) 4379–4394.
[8]   M. Hassani-Gangaraj, D. Veysset, V.K. Champagne, K.A. Nelson, C.A. Schuh, Adia-
      batic shear instability is not necessary for adhesion in cold spray, Acta Materialia, 158
      (2018) 430–439.
[9]   V.K. Champagne, M.K. West, M. Reza Rokni, T. Curtis, V. Champagne, B. McNally,
      Joining of Cast ZE41A Mg to wrought 6061 Al by the cold spray process and eriction
      stir welding, Journal of Thermal Spray Technology, 25 (2016) 143–159.
[10] S. Guetta, M.H. Berger, F. Borit, V. Guipont, M. Jeandin, M. Boustie, Y. Ichikawa,
      K. Sakaguchi, K. Ogawa, Influence of particle velocity on adhesion of cold-sprayed
      splats, Journal of Thermal Spray Technology, 18 (2009) 331–342.
[11] W.Y. Li, C.J. Li, H. Liao, Significant influence of particle surface oxidation on depo-
      sition efficiency, interface microstructure and adhesive strength of cold-sprayed cop-
      per coatings, Applied Surface Science, 256 (2010) 4953–4958.
[12] K.H. Ko, J.O. Choi, H. Lee, The interfacial restructuring to amorphous: A new adhe-
      sion mechanism of cold-sprayed coatings, Materials Letters, 175 (2016) 13–15.
[13] M. Grujicic, J.R. Saylor, D.E. Beasley, W.S. DeRosset, D. Helfritch, Computational
      analysis of the interfacial bonding between feed-powder particles and the sub-
      strate in the cold-gas dynamic-spray process, Applied Surface Science, 219 (2003)
      211–227.
[14] W.Y. Li, H. Liao, C.J. Li, G. Li, C. Coddet, X. Wang, On high velocity impact of
      micro-sized metallic particles in cold spraying, Applied Surface Science, 253 (2006)
      2852–2862.
[15] A. Fardan, C.C. Berndt, R. Ahmed, Numerical modelling of particle impact and resid-
      ual stresses in cold sprayed coatings: A review, Surface and Coatings Technology, 409
      (2021) 126835.
[16] T. Stoltenhoff, H. Kreye, H.J. Richter, An analysis of the cold spray process and its
      coatings, Journal of Thermal Spray Technology, 11(4) (2002) 542–550.
[17] T. Schmidt, F. Gärtner, H. Assadi, H. Kreye, Development of a generalized parameter
      window for cold spray deposition, Acta Materialia, 54 (2006) 729–742.
[18] T. Schmidt, H. Assadi, F. Gärtner, H. Richter, T. Stoltenhoff, H. Kreye, T. Klassen,
      From particle acceleration to impact and bonding in cold spraying, Journal of Ther-
      mal Spray Technology, 18 (2009) 794–808.
[19] H. Assadi, T. Schmidt, H. Richter, J.O. Kliemann, K. Binder, F. Gärtner, T. Klassen,
      H. Kreye, On parameter selection in cold spraying, Journal of Thermal Spray Tech-
      nology, 20 (2011) 1161–1176.
[20] H. Assadi, H. Kreye, F. Gärtner, T. Klassen, Cold spraying—a materials perspective,
      Acta Materialia, 116 (2016) 382–407.
[21] Q. Chen, A. Alizadeh, W. Xie, X. Wang, V. Champagne, A. Gouldstone, J.H. Lee, S.
      Müftü, High-strain-rate material behavior and adiabatic material instability in impact
      of micron-scale Al-6061 particles, Journal of Thermal Spray Technology, 27 (2018)
      641–653.
[22] R. Chakrabarty, J. Song, A modified Johnson-Cook material model with strain gradi-
      ent plasticity consideration for numerical simulation of cold spray process, Surface
      and Coatings Technology, 397 (2020) 125981.

[23] E. Lin, Q. Chen, O.C. Ozdemir, V.K. Champagne, S. Müftü, Effects of interface bonding on the residual stresses in cold-sprayed Al-6061: A numerical investigation, Journal of Thermal Spray Technology, 28 (2019) 472–483.

[24] Q. Wang, N. Ma, M. Takahashi, X. Luo, C. Li, Development of a material model for predicting extreme deformation and grain refinement during cold spraying, Acta Materialia, 199 (2020) 326–339.

[25] A. Joshi, S. James, Molecular dynamics simulation study of cold spray process, Journal of Manufacturing Processes, 33 (2018) 136–143.

[26] H. Assadi, F. Gärtner, T. Klassen, H. Kreye, Comment on 'Adiabatic shear instability is not necessary for adhesion in cold spray, Scripta Materialia, 162 (2019) 512–514.

[27] J. Wu, H. Fang, S. Yoon, H. Kim, C. Lee, The rebound phenomenon in kinetic spraying deposition, Scripta Materialia, 54 (2006) 665–669.

[28] J. G. Legoux , E. Irissou, S. Desaulniers, J. Bobyn, B. Harvey, W. Wong, E. Gagnon, W. Yue, Characterization and Performance Evaluation of a Helium Recovery System Designed for Cold Spraying. NRC Publ. Arch. 2010; 2: 1–22.

[29] D.L. Guo, D. MacDonald, L. Zhao, B. Jodoin, Cold spray MCrAlY coatings on single-crystal superalloy using nitrogen: Properties and economics, Journal of Thermal Spray Technology, 29 (2020) 1628–1642.

[30] W.Y. Li, C.J. Li, Optimal design of a novel cold spray gun nozzle at a limited space, Journal of Thermal Spray Technology, 14 (2005) 391–396.

[31] G. Bae, S. Kumar, S. Yoon, K. Kang, H. Na, H.J. Kim, C. Lee, Bonding features and associated mechanisms in kinetic sprayed titanium coatings, Acta Materialia, 57 (2009) 5654–5666.

[32] W.Y. Li, C. Zhang, H.T. Wang, X.P. Guo, H.L. Liao, C.J. Li, C. Coddet, Significant influences of metal reactivity and oxide films at particle surfaces on coating microstructure in cold spraying, Applied Surface Science, 253 (2007) 3557–3562.

[33] W. Wong, E. Irissou, A.N. Ryabinin, J.G. Legoux, S. Yue, Influence of helium and nitrogen gases on the properties of cold gas dynamic sprayed pure titanium coatings, Journal of Thermal Spray Technology, 20 (2011) 213–226.

[34] J. Lee, S. Shin, H.J. Kim, C. Lee, Effect of gas temperature on critical velocity and deposition characteristics in kinetic spraying, Applied Surface Science, 253 (2007) 3512–3520.

[35] G. Vinay, S. Kumar, N.M. Chavan, Generalised bonding criteria in cold spraying: Revisiting the influence of in-flight powder temperature, Materials Science and Engineering A, 823 (2021) 141719.

[36] L. Venkatesh, N. M. Chavan, G. Sundararajan, The influence of powder particle velocity and microstructure on the properties of cold sprayed copper coatings, Journal of Thermal Spray Technology, 20 (2011) 1009–1021.

[37] W.Y. Li, C. Zhang, X.P. Guo, G. Zhang, H.L. Liao, C.J. Li, C. Coddet, Effect of standoff distance on coating deposition characteristics in cold spraying, Materials and Design, 29 (2008) 297–304.

[38] M.C. Meyer, S. Yin, K.A. McDonnell, O. Stier, R. Lupoi, Feed rate effect on particulate acceleration in Cold Spray under low stagnation pressure conditions, Surface and Coatings Technology, 304 (2016) 237–245.

[39] C.J. Li, W.Y. Li, Y.Y. Wang, Effect of spray angle on deposition characteristics in cold spraying, thermal spray 2003: Advancing the science & applying the technology, Vol. 2, pp. 91–96, ASM International (2003).

[40] T. Goyal, R.S. Walia, T.S. Sidhu, Effect of parameters on coating density for cold spray process, Materials and Manufacturing Processes, 27 (2012) 193–200.

[41] J. Pattison, S. Celotto, A. Khan, W. O'Neill, Standoff distance and bow shock phenomena in the cold spray process, Surface and Coatings Technology, 202 (2008) 1443–1454.

[42] S. Kumar, B.R. Bodapati, G. Vinay, K. Vamshi Kumar, N.M. Chavan, P. Suresh Babu, A. Jyothirmayi, Estimation of inter-splat bonding and its effect on functional properties of cold sprayed coatings, Surface and Coatings Technology, 420 (2021) 127318.

[43] P.C. King, M. Jahedi, Relationship between particle size and deformation in the cold spray process, Applied Surface Science, 256 (2010) 1735–1738.

[44] www.polycontrols.com

[45] A. Sova, A. Okunkova, S. Grigoriev, I. Smurov, Velocity of the particles accelerated by a cold spray micronozzle: Experimental measurements and numerical simulation, Journal of Thermal Spray Technology, 22 (2013) 75–80.

[46] W. Wong, P. Vo, E. Irissou, A.N. Ryabinin, J.G. Legoux, S. Yue, Effect of particle morphology and size distribution on cold-sprayed pure titanium coatings, Journal of Thermal Spray Technology, 22 (2013) 1140–1153.

[47] L. Palodhi, B. Das, H. Singh, Effect of particle size and morphology on critical velocity and deformation behavior in cold spraying, Journal of Materials Engineering and Performance, 30 (2021) 8276–8288.

[48] G. Bae, Y. Xiong, S. Kumar, K. Kang, C. Lee, General aspects of interface bonding in kinetic sprayed coatings, Acta Materialia, 56 (2008) 4858–4868.

[49] S. Kumar, G. Bae, C. Lee, Influence of substrate roughness on bonding mechanism in cold spray, Surface and Coatings Technology, 304 (2016) 592–605.

[50] S. Singh, H. Singh, R.K. Buddu, Microstructural investigations on bonding mechanisms of cold-sprayed copper with SS316L steel, Surface Engineering, 36 (2020) 1067–1080.

[51] T. Hussain, D.G. McCartney, P.H. Shipway, D. Zhang, Bonding mechanisms in cold spraying: The contributions of metallurgical and mechanical components, Journal of Thermal Spray Technology, 18 (2009) 364–379.

[52] S. Singh, H. Singh, S. Chaudhary, R.K. Buddu, Effect of substrate surface roughness on properties of cold-sprayed copper coatings on SS316L steel, Surface and Coatings Technology, 389 (2020) 125619.

[53] J.G. Legoux, E. Irissou, C. Moreau, Effect of substrate temperature on the formation mechanism of cold-sprayed aluminum, zinc and tin coatings, Journal of Thermal Spray Technology, 16 (2007) 619–626.

[54] P.C. King, A.J. Poole, S. Horne, R. de Nys, S. Gulizia, M.Z. Jahedi, Embedment of copper particles into polymers by cold spray, Surface and Coatings Technology, 216 (2013) 60 67.

[55] R. della Gatta, A. Viscusi, A.S. Perna, A. Caraviello, A. Astarita, Cold spray process for the production of AlSi10Mg coatings on glass fibers reinforced polymers, Materials and Manufacturing Processes, 36 (2021) 106–121.

[56] X. Chen, H.T. Wang, G.C. Ji, X.B. Bai, Z.X. Dong, Deposition behavior of nanostructured WC–23Co particles in cold spraying process, Materials and Manufacturing Processes, 31 (2016) 1507–1513.

[57] S. Yin, M. Zhang, Z. Guo, H. Liao, X. Wang, Numerical investigations on the effect of total pressure and nozzle divergent length on the flow character and particle impact velocity in cold spraying, Surface and Coatings Technology, 232 (2013) 290–297.

[58] A.P. Alkhimov, V.F.Kosarev, S.V.Klinkov, The features of cold spray nozzle design, Journal of Thermal Spray Technology, 10 (2001) 375–381.

[59] V. Varadaraajan, P. Mohanty, Design and optimization of rectangular cold spray noz-zle: Radial injection angle, expansion ratio and traverse speed, Surface and Coatings Technology, 316 (2017) 246–254.

[60] R. Huang, M. Sone, W. Ma, H. Fukanuma, The effects of heat treatment on the mechanical properties of cold-sprayed coatings, Surface and Coatings Technology, 261 (2015) 278–288.

[61] M. Bray, A. Cockburn, W. O'Neill, The laser-assisted cold spray process and deposit characterisation, Surface and Coatings Technology, 203 (2009) 2851–2857.

[62] Y.K. Wei, X.T. Luo, X. Chu, Y. Ge, G.S. Huang, Y.C. Xie, R.Z. Huang, C.J. Li, Ni coatings for corrosion protection of Mg alloys prepared by an in-situ micro-forging assisted cold spray: Effect of powder feedstock characteristics, Corrosion Science, 184 (2021) 109397.

[63] W. Sun, A.W.Y. Tan, K. Wu, S. Yin, X. Yang, I. Marinescu, E. Liu, Post-process treat-ments on supersonic cold sprayed coatings: A review, Coatings, 10 (2020) 123.

[64] S. Yin, J. Cizek, X. Yan, R. Lupoi, Annealing strategies for enhancing mechanical properties of additively manufactured 316L stainless steel deposited by cold spray, Surface and Coatings Technology, 370 (2019) 353–361.

[65] T. Peat, A. Galloway, A. Toumpis, P. McNutt, N. Iqbal, The erosion performance of cold spray deposited metal matrix composite coatings with subsequent friction stir processing, Applied Surface Science, 396 (2017) 1635–1648.

[66] T. Peat, A. Galloway, A. Toumpis, R. Steel, W. Zhu, N. Iqbal, Enhanced erosion per-formance of cold spray co-deposited AISI316 MMCs modified by friction stir pro-cessing, Materials and Design, 120 (2017) 22–35.

[67] K. Yang, W. Li, C. Huang, X. Yang, Y. Xu, Optimization of cold-sprayed AA2024/Al$_2$O$_3$ metal matrix composites via friction stir processing: Effect of rotation speeds, Journal of Materials Science and Technology, 34 (2018) 2167–2177.

[68] E.O. Olakanmi, M. Doyoyo, Laser-assisted cold-sprayed corrosion- and wear-resis-tant coatings: A review, Journal of Thermal Spray Technology, 23 (2014) 765–785.

[69] F. Rubino, A. Astarita, P. Carlone, S. Genna, C. Leone, F. Memola Capece Minutolo, A. Squillace, Selective laser post-treatment on titanium cold spray coatings, Materials and Manufacturing Processes, 31 (2016) 1500–1506.

[70] T. Marrocco, T. Hussain, D.G. McCartney, P.H. Shipway, Corrosion performance of laser posttreated cold sprayed titanium coatings, Journal of Thermal Spray Technol-ogy, 20 (2011) 909–917.

[71] Z. Jing, K. Dejun, Effect of laser remelting on friction-wear behaviors of cold sprayed Al coatings in 3.5% NaCl solution, Materials, 11 (2018) 283.

[72] Q. Zhang, C.J. Li, C.X. Li, G.J. Yang, S.C. Lui, Study of oxidation behavior of nano-structured NiCrAlY bond coatings deposited by cold spraying, Surface and Coatings Technology, 202 (2008) 3378–3384.

[73] R. Ghelichi, S. Bagherifard, I. Fernandez Parienete, M. Guagliano, S. Vezzù, Experi-mental study of shot peening followed by cold spray coating on residual stresses of the treated parts, Structural Durability and Health Monitoring, 6 (2010) 17–29.

[74] A. Moridi, S.M. Hassani-Gangaraj, S. Vezzú, L. Trško, M. Guagliano, Fatigue behav-ior of cold spray coatings: The effect of conventional and severe shot peening as pre-/post-treatment, Surface and Coatings Technology, 283 (2015) 247–254.

[75] H.L. Yao, X.Z. Hu, Z.H. Yi, J. Xia, X.Y. Tu, S. bin Li, B. Yu, M.X. Zhang, X.B. Bai, Q.Y. Chen, H.T. Wang, Microstructure and improved anti-corrosion properties of cold-sprayed Zn coatings fabricated by post shot-peening process, Surface and Coat-ings Technology, 422 (2021) 127557.

[76] H. Zhou, C. Li, C. Bennett, H. Tanvir, C. Li, Numerical analysis of deformation behavior and interface bonding of Ti6Al4V particle after subsequent impact during cold spraying, Journal of Thermal Spray Technology, 30 (2021) 1093–1106.

[77] P. Sudharshan Phani, V. Vishnukanthan, G. Sundararajan, Effect of heat treatment on properties of cold sprayed nanocrystalline copper alumina coatings, Acta Materialia, 55 (2007) 4741–4751.

[78] N.M. Chavan, M. Ramakrishna, P.S. Phani, D.S. Rao, G. Sundararajan, The influence of process parameters and heat treatment on the properties of cold sprayed silver coatings, Surface and Coatings Technology, 205 (2011) 4798–4807.

[79] G. Vinay, N. M. Chavan, S. Kumar, A. Jyothirmayi, B. R. Bodapati, Improved microstructure and properties of cold sprayed zinc coatings in as sprayed condition, Surfaces and Coatings Technology, 438 (2022) 128392.

[80] N.M. Chavan, B. Kiran, A. Jyothirmayi, P.S. Phani, G. Sundararajan, The corrosion behavior of cold sprayed zinc coatings on mild steel substrate, Journal of Thermal Spray Technology, 22 (2013) 463–470.

[81] Y.K. Wei, X.T. Luo, C.X. Li, C.J. Li, Optimization of in-situ shot-peening-assisted cold spraying parameters for full corrosion protection of Mg alloy by fully dense Al-based alloy coating, Journal of Thermal Spray Technology, 26 (2017) 173–183.

[82] S. Kumar, M. Ramakrishna, N.M. Chavan, S. V. Joshi, Correlation of splat state with deposition characteristics of cold sprayed niobium coatings, Acta Materialia, 130 (2017) 177–195.

[83] T. Liu, M.D. Vaudin, J.R. Bunn, T. Ungár, L.N. Brewer, Quantifying dislocation density in Al-Cu coatings produced by cold spray deposition, Acta Materialia, 193 (2020) 115–124.

[84] F. Khodabakhshi, B. Marzbanrad, L.H. Shah, H. Jahed, A.P. Gerlich, Friction-stir processing of a cold sprayed AA7075 coating layer on the AZ31B substrate: Structural homogeneity, microstructures and hardness, Surface and Coatings Technology, 331 (2017) 116–128.

[85] G. Sundararajan, N.M. Chavan, S. Kumar, The elastic modulus of cold spray coatings: Influence of inter-splat boundary cracking, Journal of Thermal Spray Technology, 22 (2013) 1348–1357.

[86] R.W. Zimmerman, The effect of microcracks on the elastic moduli of brittle materials, Journal of Materials Science Letters, 4 (1985) 1457–1460.

[87] O. Kovarik, J. Siegl, J. Cizek, T. Chraska, J. Kondas, Fracture toughness of cold sprayed pure metals, Journal of Thermal Spray Technology, 29 (2020) 147–157.

[88] C. Huang, M. Arseenko, L. Zhao, Y. Xie, A. Elsenberg, W. Li, F. Gartner, A. Simar, T. Klassen, Property prediction and crack growth behavior in cold sprayed Cu deposits, Materials and Design, 206 (2021) 109826.

[89] B.C. White, W.A. Story, L.N. Brewer, J.B. Jordon, Fracture mechanics methods for evaluating the adhesion of cold spray deposits, Engineering Fracture Mechanics, 205 (2019) 57–69.

Chapter 8

Machine learning and additive manufacturing

A case study for quality control and monitoring

Ayush Pratap, Atul Pandey and Neha Sardana

8.1 INTRODUCTION

Additive manufacturing (AM) and machine learning (ML) are two rapidly developing technologies that have the potential to revolutionize the manufacturing industry. AM, also known as 3D printing, allows for creating highly complex geometries and personalized products at a relatively low cost. On the other hand, deep learning is a subfield of machine learning that involves training neural networks to recognize patterns in large datasets, enabling it to identify complex relationships and make predictions [1]. The combination of AM and deep learning has the potential to transform the way products are designed, developed, and manufactured. Deep learning algorithms can be used to optimize the AM process, improve quality control, and reduce the time and cost involved in the product development process. For example, deep learning algorithms can be used to analyze data from sensors that monitor the printing process, enabling the identification of patterns that can help optimize the process parameters.

Similarly, deep learning algorithms can be used to identify defects in AM parts, reducing the time and cost involved in manual inspection [2]. As the technology continues to develop, we will likely see more advanced deep learning algorithms being used to optimize further and automate the additive manufacturing process. The integration of these two technologies has great potential to significantly improve the efficiency and accuracy of the manufacturing process while also enabling the creation of highly complex and personalized products at a relatively low cost [3]. Additive manufacturing has rapidly grown in popularity over the past decade. As a result, there has been a significant increase in the number of industries using this technology. From aerospace to medical, additive manufacturing revolutionizes how products are designed and manufactured. However, the quality control of 3D-printed parts has been a significant concern for many manufacturers. That's why machine learning and deep learning come in. Machine learning and deep learning are subsets of artificial intelligence that enable computers to learn from data and improve their performance over time. These techniques have been applied to quality control in additive manufacturing to improve the accuracy and reliability of 3D-printed parts.

DOI: 10.1201/9781032703046-8

Machine learning is a subset of artificial intelligence that uses algorithms and statistical models to enable computer systems to improve their performance on a specific task by learning from data. In other words, machine learning allows computers to understand and improve without being explicitly programmed. Deep learning is a machine learning subfield that involves using neural networks to learn from data [4]. Neural networks are a set of algorithms that are modeled after the structure and function of the human brain, and they can be used to recognize patterns and make predictions based on input data [5]. Machine learning and deep learning are classified into several categories based on the type of learning involved, such as supervised learning (labeled data), unsupervised learning (unlabeled data), and reinforcement learning (reward-based system) [6].

In the area of materials and manufacturing, machine learning and deep learning can be applied in various ways. For example, machine learning can be used to optimize the manufacturing process by predicting the optimal parameters for producing high-quality products. This can involve analyzing large amounts of data from various sensors and cameras to identify potential defects and adjust the printing parameters in real time, reducing the number of defective parts and improving overall product quality [7, 8]. Deep learning can also be used for predictive maintenance in manufacturing by analyzing data from sensors on machines to predict potential maintenance issues before they occur. This can help reduce the risk of unplanned downtime and improve overall efficiency. Furthermore, machine learning and deep learning can be used for materials design by analyzing data on the properties of different materials and using this information to develop new materials with specific properties. This can involve identifying correlations between other variables and predicting the behavior of new materials based on this data.

In summary, machine learning and deep learning have significant potential for improving efficiency, reducing waste, and improving quality control in the area of materials and manufacturing. As technology evolves, we can expect to see more advanced machine learning and deep learning applications in this field.

8.2 MACHINE LEARNING AND QUALITY CONTROL

Manufacturers integrate discrete manufacturing systems, supply chain management, operation management, optimization process, business intelligence, procurement/order fulfillment, transportation, B2B marketing, and decision technique/modeling to implement Industry 4.0 properly. This Industry 4.0 cannot be fulfilled without the help of machine learning, artificial intelligence (AI), big data, the Internet of Things (IoT), cloud computing, robotics, and automation [9, 10]. When we talk about Industry 4.0, ML algorithms can analyze a large volume of data from different sources, such as sensors and production machines. By analyzing this data, machine learning algorithms can identify patterns that are indicative of quality or defects, which can help manufacturers improve their quality control processes [11, 12]. The basic outline of adopting the machine learning process in additive manufacturing is shown in Figure 8.1. The following are some examples of how these techniques have been used:

a. Defect Detection

AI-based algorithms can analyze the surface of the 3D-printed part and identify any defects, such as cracks, voids, and deformation. By identifying these defects, manufacturers can quickly identify and address issues before they become more significant problems. By doing so, it is possible to increase the general quality of 3D-printed parts and decrease the production of defective parts [8].

b. Quality Prediction

Machine learning algorithms can analyze data from previous 3D-printed parts and identify patterns indicative of quality. By using this information, manufacturers can predict the quality of a new 3D-printed part before it is produced. This can help to identify potential quality issues early on and ensure that the final product meets the required standards [13].

c. Process Optimization

By analyzing data from previous 3D-printed parts, these algorithms can identify patterns that indicate optimal process parameters. By using this information, manufacturers can adjust their processing parameters to improve the quality and consistency of printed parts [12, 14].

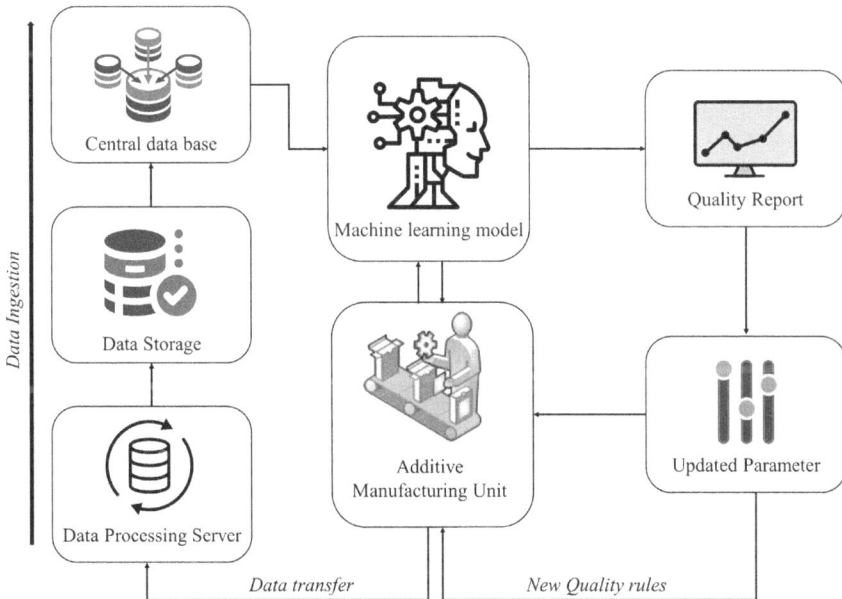

Figure 8.1 Schematic to use machine learning in additive manufacturing.

d. Material Analysis

These algorithms can identify patterns that indicate the quality of the material by analyzing data from previous batches of materials. Manufacturers can use this information to select the best materials for 3D printing and ensure that the final product meets the required standards [15].

e. Predictive Maintenance

Machine learning algorithms can be used to evaluate the data from sensors and other sources to determine the patterns that are indicative of equipment failure. This information can be used to schedule maintenance before equipment fails, reducing downtime and improving overall production efficiency [16, 17].

8.3 ADDITIVE MANUFACTURING IN INDUSTRY 4.0

Additive manufacturing, also known as 3D printing, is a key technology in Industry 4.0, which refers to the fourth industrial revolution characterized by the integration of advanced technologies, such as the Internet of Things, big data, artificial intelligence, and machine learning, to transform the manufacturing industry [18]. Machine learning, in particular, is being used to optimize the additive manufacturing process by leveraging data generated by sensors, cameras, and other sources to improve the quality, speed, and efficiency of 3D printing. For instance, ML algorithms can be used to analyze the performance of 3D printers, predict and prevent defects, and optimize printing parameters such as temperature, speed, and material usage. Moreover, machine learning can also be applied to designing and simulating models by generating and evaluating thousands of design options based on specific criteria and constraints and identifying the most optimal design for a given application [19]. This approach, known as generative design, enables designers and engineers to create more complex and innovative plans that would be difficult or impossible to produce with traditional manufacturing methods. Based on the literature, the various additive manufacturing techniques on which ML has been used are classified in Figure 8.2. One example of an additive manufacturing process that involves machine learning is closed-loop process control, which is used to optimize the 3D printing process in real time by adjusting the printing parameters based on sensor data and machine learning algorithms. In a closed-loop process control system, sensors are placed on the 3D printer to measure variables such as temperature, humidity, and material flow rate. The sensor data is then fed into a machine learning algorithm, which analyzes the data and generates insights about the optimal printing conditions for a particular material or product.

Based on these insights, the machine learning algorithm can adjust the printing parameters, such as the temperature or the speed of the printing process, in real time to achieve the desired output. This iterative data collection, analysis, and adjustment process allows for more precise and efficient additive manufacturing, reducing waste and improving final product quality [20].

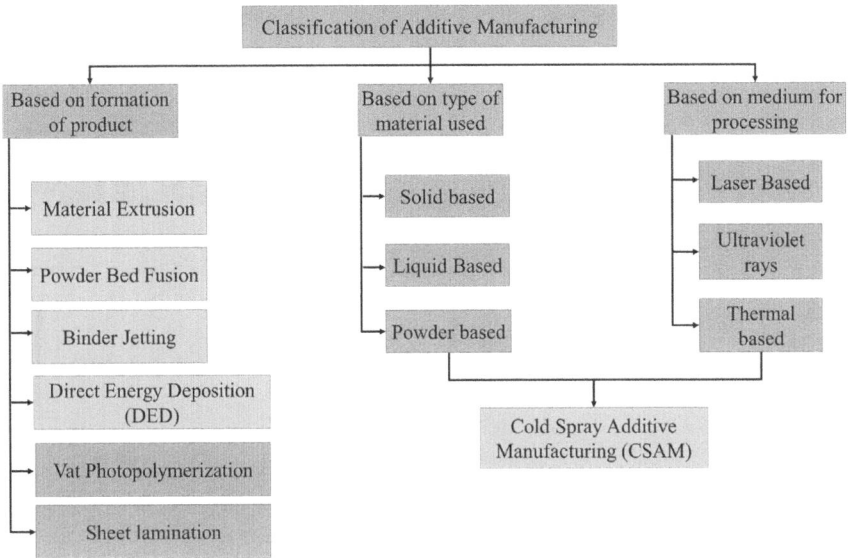

Figure 8.2 Classification of an additive manufacturing technique. The AM technique that uses machine learning extensively is highlighted and discussed.

8.4 CASE STUDIES WITH DEEP LEARNING

8.4.1 Cold spray

Cold spray is a solid-state material deposition process originally developed as a coating technology in the 1980s [21]. It is an additive manufacturing process that involves depositing metallic or non-metallic particles onto a substrate through a supersonic gas jet. The process is sometimes called "cold gas dynamic spraying" or "kinetic metallization." In cold spray, a supersonic gas jet accelerates metal powders to high velocities, typically using nitrogen or helium as the carrier gas. As the particles impact the substrate, they undergo plastic deformation, forming a strong metallurgical bond with the substrate material. This results in the formation of a dense, strongly adherent coating or layer of material on the substrate [22]. It was suggested that the bonding is largely due to mechanical interlocking, where the substrate physically entraps the particles. At high impact velocities, the outer region of the particle that impacts the substrate experiences plastic deformation due to an increase in temperature [23]. The cold spray process is similar to other additive manufacturing techniques, such as powder bed fusion and material extrusion, in that it builds up a 3D part. However, cold spray is unique because it operates at relatively low temperatures, which enables it to deposit a broad range of materials without causing thermal damage or distortion [24]. Cold spray has numerous benefits over other additive manufacturing processes, including the ability to deposit materials with high purity and density and the capacity to cover intricate geometries

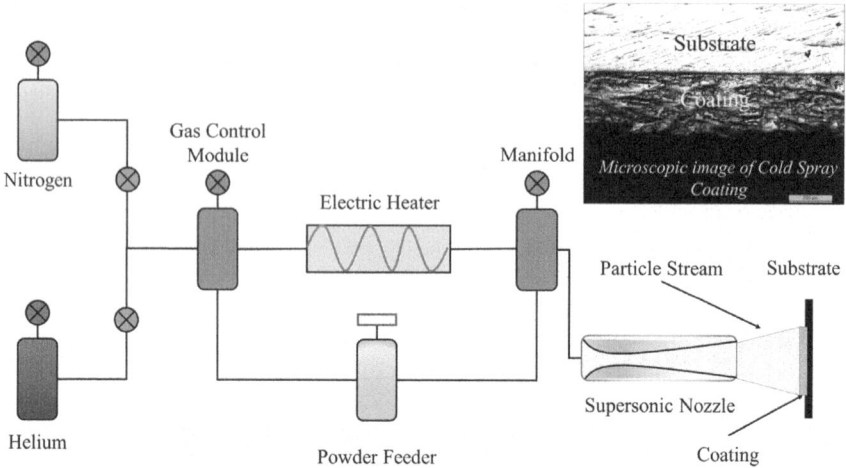

Figure 8.3 Schematic diagram of cold spray coating.

and internal surfaces. It is also a relatively fast process, with up to several kilograms per hour deposition rates possible. In conclusion, cold spray is a promising additive manufacturing technique that is suitable for different applications such as defense, marine, consumer goods, aerospace, and biomedical engineering [25]. The schematic diagram of cold spray coating is shown in Figure 8.3. The cold spray coating of Ti and Al is demonstrated with the microscopic image.

According to a recent study by *Malamousi et al.* [26], most spray shops and manufacturers of thermal spray equipment currently employ sensors for checking thermal spray assets at the product levels. Recently, AI-driven solutions have been needed to merge real-time data, revised data, physical dependency models, and intelligence from many platforms for the optimization of assets and other operations. Another University of Sydney researcher used machine learning to study the track profiles in cold spray processes. From their results, it is evident that with the aid of AI, cold spray track profiles can be monitored [27]. According to *Wang et al.* [28], the critical velocity is predicted using linear and nonlinear artificial neural networks, depending on the outcomes of feature selection. One of the researchers from the University of Barcelona studied the deposition efficiency of the cold spray coating process through machine learning [29]. A summary of various researchers working in cold spray optimization with machine learning is presented in Table 8.1.

8.4.2 Powder bed fusion

The best repeatability and dimensional accuracy in 3D printing of metals production is now provided by the powder bed fusion (PBF) technique, which has attracted significant research funding from academia and industry. Generally, PBF operations use the following steps to fabricate a part: On a machine plate, a coating of powder is spread; in step two, the powder is softened into a precise shape across the specified area; in step three, the plates descend, and a fresh layer is applied to the constructed plate. This procedure is continued until completion [42]. Figure 8.4 depicts a schematic illustration of powder bed fusion.

Table 8.1 Various works use different quality control algorithms in the cold spray process

Quality control	Algorithm	Material	Work
Powder flowability characteristics	Decision tree	Metal powder	[30]
Porosity prediction from processing parameters	Decision tree	Metal/alloy powders	[31]
Optimization of coatings	Semi-analytical experimental based approaches	Simulation-based	[32, 33]
Microstructural defect detection	CNN	Alloys of varying Sn–Ag–Cu compositions	[34, 35]
Track profiling	Self-organizing maps and geodesic-based model	Titanium	[36–38]
Process monitoring and Control	ANN coupled with sensors	Al6061	[39]
Conformance	Computer vision	Metal/alloys	[40, 41]

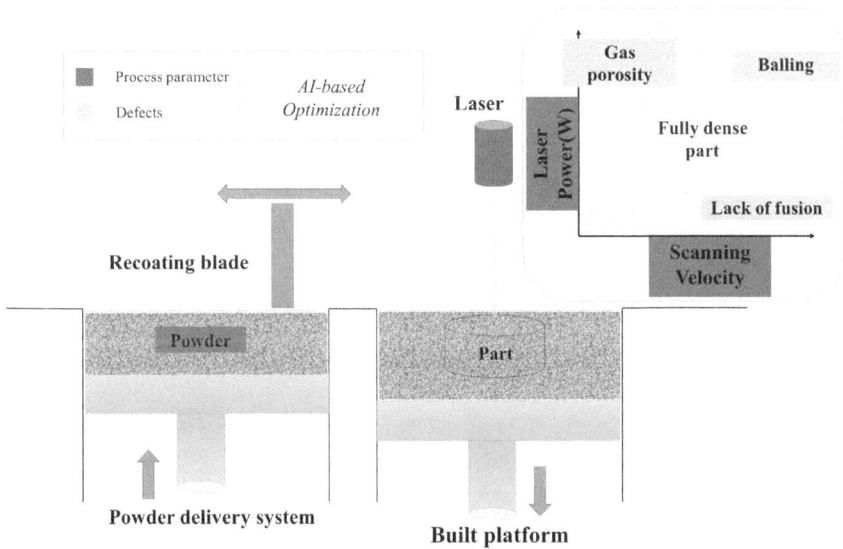

Figure 8.4 Schematic diagram of powder bed fusion.

Flaws in PBF can be divided into three types: powder, process, and defects linked to post-processing. Nevertheless, the investigation of process-related problems is the project's main goal. Many researchers have examined why processing factors, including layer thickness, hatch spacing, scan speed, scan strategy, and power, affect the growth of different defects, namely voids, porosity, and holes [43]. The good superimposed part between melt pools guarantees that all points which have melted once will be used to understand the lack of fusion defect [44]. On the other hand, the keyhole porosity barrier corresponds to pore pinch-off caused by instabilities in deep keyholes [45]. A mixture of fluid flow patterns determines a third boundary and melt pool capillary, known as balling [46].

Deep learning is now being used in many studies where the fingerprint (images) is used to segment and classify the defect associated with the PBF type of additive manufacturing. In recent work, *Ansari et al.* [47] detected small porosity, ensuring the additive manufacturing of metal samples with the help of convolutional neural networks (CNN). *Zhang and Zhao* [48] offer a new process for forecasting visual faults, which is a key factor in deciding whether the laser-based powder bed fusion (LPBF) process can be manufactured. *Pandiyan et al.* [49] constructed a multi-timescale deep learning (DL) model for LPBF monitoring. The hybrid DL architecture was created by combining long short-term memory (LSTM) and convolutional neural networks to detect the flaw. *Chen et al.* [50] leverage the process images captured by PBF equipment to build a convolutional neural network–based detection technique. Powder unevenness, recoated scratches, and powder-spreading flaws are all used in this process. A summary of various researchers working in powder bed fusion optimization with machine learning is presented in Table 8.2.

Table 8.2 Various works use different quality control algorithms in the powder bed fusion process

Quality control	Algorithm	Material	Work
Porosity classification (based on pores size)	Custom model (CNN)	Metal powder	[51]
Internal defect (even, uneven, and porous)	SVM	Carbon steel S30C alloy powder	[52]
Delamination, splatter and ok	Custom model (CNN) with k-fold cross validation	H13 steel	[53]
Plume, melt pool, and spatters	SVM and CNN	Stainless steel 316 L powder	[54]
Normal and defect	Transfer learning	Metal powder	[55]
Debris, recoater streaking, part damage, recoater hopping, super-elevation, incomplete spreading, and ok	Transfer learning	Inconel 625, stainless steel 17–4 PH, Inconel 718, stainless steel 316 L, AlSi10Mg, bronze, and Ti-6Al-4V	[56]

8.4.3 Material extrusion

Material extrusion is a common additive manufacturing process that involves material deposition through a nozzle or extrusion head to create a three-dimensional object. This process is also called fused filament fabrication (FFF) or fused deposition modelling (FDM) [57]. Several types of defects can occur during material extrusion, such as layer shifting, warping, and under-extrusion. These defects can lead to poor part quality, reduced mechanical properties, and even failure. Traditional defect detection and correction methods are often time-consuming and expensive [57]. Machine learning has the potential to address these challenges by enabling automated defect detection and correction in real time. Using machine learning algorithms to analyze data from sensors and cameras during printing defects can be identified and corrected before they become more severe [58, 59]. Figure 8.5 shows the block diagram of the material extrusion process and how ML-based optimization and detection have been done.

One of the most recent works validates closed-loop process parameter modification for robot-based carbon fiber reinforced polymer (CFRP) AM and develops a system for real-time detection of problematic regions. The primary innovation is creating an accurate, real-time deep learning model for the detection of defects, categorization, and evaluation [60]. Another researcher proposes a site-based monitoring system for extrusion-based 3D printers that integrates object identification and computer vision models to identify and repair defects in real time [61].

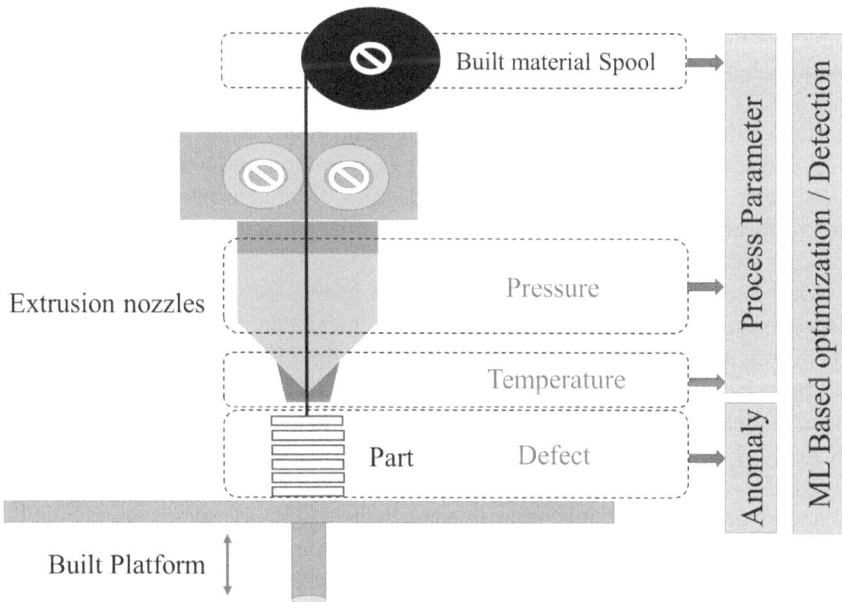

Figure 8.5 Block diagram of the material extrusion process.

Table 8.3 Various works using different algorithms for quality control in the material extrusion process

Quality control	Algorithm	Material	Work
In process failure detection	CNN	Polylactic acid (PLA)	[62, 63]
Dimensional accuracy	NN	Metal	[64]
Void percentage analysis	NN and SVR	Metal	[65]
Surface quality and roughness	Ensemble learning-based approach	Polylactic acid (PLA)	[66]
Bioprinting viability	CNN	Bio ink	[67]
Precision manufacturing	Conditional adversarial networks (CAN)	Acrylonitrile butadiene styrene (ABS)	[68]

A summary of various researchers working in material extrusion optimization with machine learning is presented in Table 8.3.

8.4.4 Binder jetting

Binder jetting, machine learning, and additive manufacturing are three rapidly evolving technologies transforming the manufacturing industry. Binder jetting is a process of additive manufacturing that involves jetting a liquid binder onto a powder bed to create a 3D object. In contrast, machine learning is a branch of artificial intelligence that enables machines to learn from pre-existing data and improve their performance over time. Together, these technologies offer exciting new possibilities for designing, producing, and optimizing complex structures and parts with high precision, speed, and efficiency [69]. In this context, machine learning algorithms can optimize the binder jetting process and improve the final product's quality, reliability, and durability. This integration of binder jetting, machine learning, and additive manufacturing is poised to revolutionize how we design and produce parts and components in various industries, from aerospace and automotive to healthcare and consumer goods [70]. A schematic diagram of the binder jetting process is shown in Figure 8.6.

One researcher used Stylegan3, a type of generative adversarial network, to increase the number of training images by augmentation. Further, they have used YOLOv5, a CNN for detecting objects, to detect the F1 score of 88% [71]. Binder jet parts have a significant nonlinear deformation when they are still in the green state, as stated by another researcher from the University of Pittsburgh in the United States [72].

Binder jetting is a relatively new additive manufacturing process, and like any manufacturing process, it is prone to some defects [73, 74]. Some common faults associated with binder jetting include porosity, warping, surface roughness, inconsistency in layer bonding, and inaccuracies in dimensions.

It is important to note that many of these defects can be minimized or eliminated through proper process control, design optimization, and post-processing techniques. As technology advances, we can expect to see further improvements in the quality and reliability of parts produced through binder jetting. A summary of various researchers working in binder jetting optimization with machine learning is presented in Table 8.4.

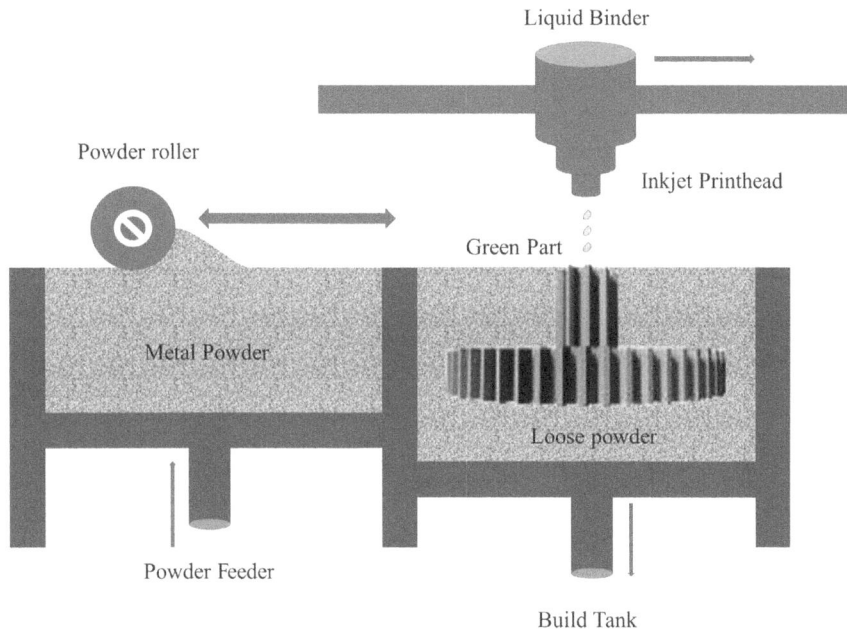

Figure 8.6 Schematic diagram of binder jetting.

Table 8.4 Various works using different algorithms for quality control in binder jetting process

Quality control	Algorithm	Material	Work
Condition monitoring	Deep learning	Sand	[75]
Defect parameter optimization	Weighted k-nearest neighbors' algorithm	Co-Cr-Mo alloy	[76]
Pores evolution	Principal component analysis (PCA)	Copper powder	[77]
Parameters recommendation system	Backward propagation (BP) neural network (NN) algorithm orthogonal experiment Taguchi method	420 stainless steel	[78]
Parametric study	Multivariable linear and Gaussian process regression models	Alumina	[79]
Microstructural analysis	Light GBM classifier	Gas atomized stainless steel	[80]

8.4.5 Direct energy deposition

Direct energy deposition (DED) is an additive manufacturing technique that uses a focused energy source, such as a laser or electron beam, to melt and fuse materials together layer by layer to create complex parts. DED is a versatile technique that can be used to print a wide range of materials, including metals, ceramics, and polymers. It is commonly used in the aerospace, marine, automotive, defense sector, and medical industries [81]. The block diagram of the direct energy deposition process is shown in Figure. 8.7. While DED has many advantages, such as the ability to produce large, complex parts quickly and efficiently, optimizing process parameters to achieve the desired part quality and efficiency can be challenging. This is where machine learning comes in. In the context of DED, machine learning can be used to optimize process parameters, such as excitation source, rate

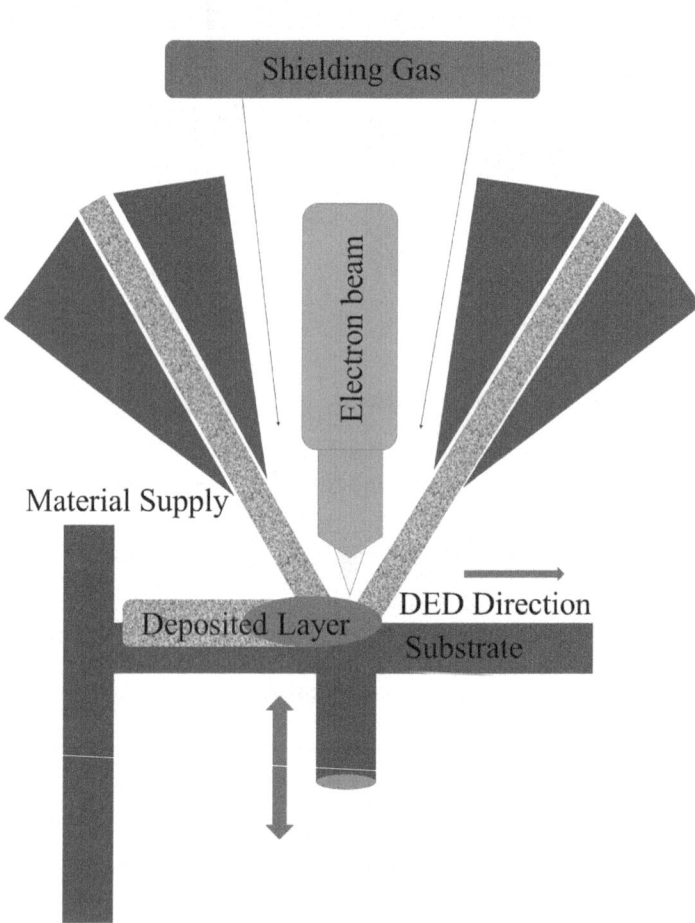

Figure 8.7 Block diagram of direct energy deposition.

of scanning, and powder flow, to improve the quality and efficacy of the printing process [82].

One of the researchers from the Vellore Institute of Technology has demonstrated an ML-based algorithm for pores detection in microstructural pictures of components made of aluminum alloy using a small dataset. The machine learning models led to a mean classification accuracy of about 98% (random forest method) for porosity detection of pore sizes larger than 5 μm [83]. According to another study, anomalous powder feeding brought on by nozzle obstruction or adhesion regularly happens and can result in serious flaws or even cause a process failure. They suggested the unusual powder feeding method based on an AI-based algorithm. The forecast accuracy for the classification performance is up to 99.2%. The findings demonstrate the potential of the deep learning method for DED process defect identification [84]. According to a researcher from the Singapore Institute of Manufacturing Technology, acoustic-based monitoring systems provide additional advantages, including changeable sensor setup and less expensive hardware. The acoustic wave produced by laser-material interactions during the DED process offers details of the intricate physical processes that occur under the surface, including pore development, crack propagation, solidification, and melting [85]. A summary of various researchers working in direct energy deposition process optimization with machine learning is presented in Table 8.5.

The application of machine learning in DED is still in its early stages. Still, it has a great potential to revolutionize the way parts are printed and improve the quality and efficiency of the printing process. As technology advances, we can expect to see more sophisticated machine learning algorithms being developed to optimize the printing process and improve the quality of DED-printed parts.

Table 8.5 Various works using different algorithms for quality control in the direct energy deposition process

Quality control	Algorithm	Material	Work
In situ monitoring of melt pool temperature	XGBoost & LSTM	Nickel-based 718 superalloy	[86]
Selection of effective manufacturing conditions	Random forest (RF) and support vector machine	Titanium alloy	[87]
Microstructural analysis and grain size prediction	NN	Inconel 718	[88, 89]
In situ surface anomaly detection	CNN	SS 316L powder	[90]
Sensitivity analysis of uncertainties	Deep learning (DL)-based surrogate model	Simulation-based	[91]
In situ powder stream fault detection	CNN	Stainless steel 316L	[92]

8.5 FUTURE SCOPE

8.5.1 Digital twin

Digital twin technology is a virtual representation of a physical object or system. This technology allows manufacturers to create a digital model of their product or system, which can be used to simulate and optimize performance. Regarding additive manufacturing, digital twin technology can be used to simulate the 3D printing process and predict the final product's performance [93]. Manufacturers can optimize the additive manufacturing process by creating a digital twin of a 3D-printed part to ensure that the final product fulfills the required qualities and specifications. This can include adjusting the printing parameters such as temperature, speed, and material composition. The digital twin can also be used to simulate different scenarios and predict the final product's performance under various conditions [94]. One of the key benefits of using digital twin technology in additive manufacturing is the ability to reduce the number of defective parts. By simulating the 3D printing process and predicting the final product's performance, manufacturers can identify potential issues before they occur. This can help to reduce the number of failed prints and minimize waste [93]. Digital twin technology can also be used to improve the overall efficiency of the additive manufacturing process. Manufacturers can identify the optimal printing parameters and production schedules by simulating different scenarios to maximize production efficiency [95]. In addition, digital twin technology can be used to improve product design. By creating a digital twin of a product, designers can simulate its performance and adjust the strategy to enhance its functionality and reduce the likelihood of defects. Figure 8.8 shows the integration and future of Industry 4.0 in the digital world.

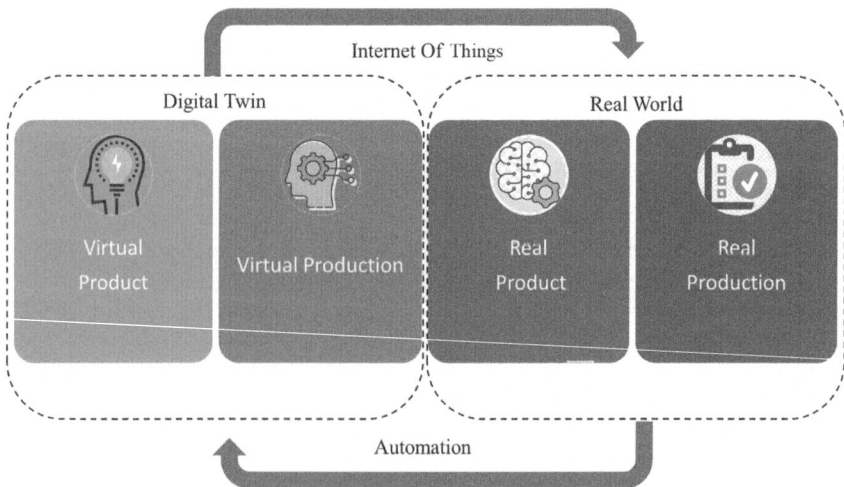

Figure 8.8 Integration of digital with the real world in Industry 4.0.

8.5.2 Internet of things

The Internet of Things refers to the interconnected network of devices that are embedded with sensors, software, and connectivity, allowing them to exchange data and communicate with each other. When it comes to additive manufacturing, the IoT can play an important role in optimizing the 3D printing process and improving the quality of the final product [96]. By integrating IoT devices into the additive manufacturing process, manufacturers can monitor various aspects of the printing process in real time. This can include monitoring the temperature, humidity, and other environmental factors that can affect the quality of the final product. By collecting this data, manufacturers can adjust the printing parameters to ensure that the final product meets the required specifications [97]. The IoT can also be used to improve the overall efficiency of the additive manufacturing process. By collecting data on the performance of various machines and devices, manufacturers can identify potential bottlenecks in the production process and optimize their workflows to improve efficiency [98, 99]. In addition, the IoT can be used to improve the traceability of products throughout the manufacturing process. By attaching sensors to individual parts or products, manufacturers can track their production progress and monitor their real-time performance. This can help to reduce the number of defective products and improve overall quality control.

8.5.3 Automation

Automation and additive manufacturing are closely linked, as automation plays a critical role in optimizing the 3D printing process and improving the overall efficiency of the manufacturing process with the integration of Industry 4.0 [100]. One of the key benefits of using automation in additive manufacturing is the ability to reduce the amount of manual labor required. By automating certain tasks, such as part loading and unloading, manufacturers can improve the speed and accuracy of the printing process while reducing the risk of human error [101]. Automation can also be used to improve quality control in additive manufacturing. By using sensors and cameras to monitor the printing process in real time, manufacturers can identify potential issues before they occur and make adjustments to ensure that the final product meets the required quality and specifications. In addition, automation can be used to optimize the production schedule and reduce the time needed to manufacture a product. By using predictive analytics and machine learning algorithms, manufacturers can identify the optimal printing parameters and production schedule to maximize efficiency and reduce the time required to brings a product to market [102].

8.6 CONCLUSION

In conclusion, the application of machine learning in additive manufacturing has shown significant potential for improving quality control and monitoring in the 3D printing process. Using machine learning algorithms, manufacturers can identify potential defects and adjust the printing parameters in real time, reducing the number of defective parts and improving overall product quality.

The case study highlighted the effectiveness of using machine learning for detecting defects in 3D-printed parts, such as cracks and voids, and adjusting the printing process to improve quality control. By analyzing huge amounts of data from various sensors and cameras, the machine learning algorithm could accurately identify potential issues and recommend adjustments to the printing parameters to ensure that the final product met the required specifications.

Furthermore, the study demonstrated the potential for using machine learning for predictive maintenance in additive manufacturing. By analyzing data from sensors on the 3D printing machines, the algorithm was able to predict potential maintenance issues before they occurred, reducing the risk of unplanned downtime and improving overall efficiency.

Overall, the use of machine learning in additive manufacturing has the potential to revolutionize the way that products are designed and manufactured. Manufacturers can reduce waste, improve efficiency, and produce higher-quality products by optimizing the printing process and improving quality control and monitoring. As technology continues to evolve, we can expect to see even more advanced machine learning applications in additive manufacturing, leading to further advancements in the field.

REFERENCES

[1] L. Meng, B. Mcwilliams, W. Jarosinski, H.-Y. Park, Y.-G. Jung, J. Lee, and J. Zhang, "Machine learning in additive manufacturing: A review," *JOM*, vol. 72, no. 6, pp. 2363–2377, Jun. 2020, doi: 10.1007/S11837-020-04155-Y/FIGURES/11.

[2] F. W. Baumann, A. Sekulla, M. Hassler, B. Himpel, and M. Pfeil, "Trends of machine learning in additive manufacturing," *International Journal of Rapid Manufacturing*, vol. 7, no. 4, p. 310, 2018, doi: 10.1504/IJRAPIDM.2018.095788.

[3] X. Qi, G. Chen, Y. Li, X. Cheng, and C. Li, "Applying neural-network-based machine learning to additive manufacturing: Current applications, challenges, and future perspectives," *Engineering*, vol. 5, no. 4, pp. 721–729, Aug. 2019, doi: 10.1016/J.ENG.2019.04.012.

[4] J. Bell, "What is machine learning?," *Machine Learning and the City*, pp. 207–216, May 2022, doi: 10.1002/9781119815075.CH18.

[5] S. Ray, "A quick review of machine learning algorithms," in *Proceedings of the International Conference on Machine Learning, Big Data, Cloud and Parallel Computing: Trends, Prespectives and Prospects, COMITCon 2019*, pp. 35–39, Feb. 2019, doi: 10.1109/COMITCON.2019.8862451.

[6] A. Pratap and N. Sardana, "Machine learning-based image processing in materials science and engineering: A review," *Materials Today: Proceedings*, vol. 62, no. P14, pp. 7341–7347, Jan. 2022, doi: 10.1016/J.MATPR.2022.01.200.

[7] J. Jiang, Y. Xiong, Z. Zhang, and D. W. Rosen, "Machine learning integrated design for additive manufacturing," *Journal of Intelligent Manufacturing*, vol. 33, no. 4, pp. 1073–1086, Apr. 2022, doi: 10.1007/S10845-020-01715-6/TABLES/7.

[8] R. Li, M. Jin, and V. C. Paquit, "Geometrical defect detection for additive manufacturing with machine learning models," *Materials & Design*, vol. 206, p. 109726, Aug. 2021, doi: 10.1016/J.MATDES.2021.109726.

[9] J. Lee et al., "Review on quality control methods in metal additive manufacturing," *Applied Sciences*, vol. 11, no. 4, p. 1966, Feb. 2021, doi: 10.3390/APP11041966.

[10] H. Kim, Y. Lin, and T. L. B. Tseng, "A review on quality control in additive manufacturing," *Rapid Prototyping Journal*, vol. 24, no. 3, pp. 645–669, 2018, doi: 10.1108/RPJ-03-2017-0048/FULL/PDF.

[11] W. Ren, G. Wen, Z. Zhang, and J. Mazumder, "Quality monitoring in additive manufacturing using emission spectroscopy and unsupervised deep learning," *Materials and Manufacturing Processes*, vol. 37, no. 11, pp. 1339–1346, 2021, doi: 10.1080/10426914.2021.1906891.

[12] F. Imani, R. Chen, E. Diewald, E. Reutzel, and H. Yang, "Deep learning of variant geometry in layerwise imaging profiles for additive manufacturing quality control," *Journal of Manufacturing Science and Engineering, Transactions of the ASME*, vol. 141, no. 11, 2019, doi: 10.1115/1.4044420/956249.

[13] X. Li, X. Jia, Q. Yang, and J. Lee, "Quality analysis in metal additive manufacturing with deep learning," *Journal of Intelligent Manufacturing*, vol. 31, no. 8, pp. 2003–2017, Dec. 2020, doi: 10.1007/S10845-020-01549-2/FIGURES/11.

[14] D. Wu, Y. Wei, and J. Terpenny, "Surface roughness prediction in additive manufacturing using machine learning," in *ASME 2018 13th International Manufacturing Science and Engineering Conference*, MSEC 2018, vol. 3, Sep. 2018, doi: 10.1115/MSEC2018-6501.

[15] M. Parsazadeh, S. Sharma, and N. Dahotre, "Towards the next generation of machine learning models in additive manufacturing: A review of process dependent material evolution," *Progress in Materials Science*, vol. 135, p. 101102, Jun. 2023, doi: 10.1016/J.PMATSCI.2023.101102.

[16] J. Francis and L. Bian, "Deep learning for distortion prediction in laser-based additive manufacturing using big data," *Manufacturing Letters*, vol. 20, pp. 10–14, Apr. 2019, doi: 10.1016/J.MFGLET.2019.02.001.

[17] A. Caggiano, J. Zhang, V. Alfieri, F. Caiazzo, R. Gao, and R. Teti, "Machine learning-based image processing for on-line defect recognition in additive manufacturing," *CIRP Annals*, vol. 68, no. 1, pp. 451–454, Jan. 2019, doi: 10.1016/J.CIRP.2019.03.021.

[18] A. Haleem and M. Javaid, "Additive manufacturing applications in Industry 4.0: A review," vol. 4, no. 4, p. 1930001, Oct. 2019, doi: 10.1142/S2424862219300011.

[19] M. E. H. Korner, M. P. Lambán, J. A. Albajez, J. Santolaria, L. D. C. N. Corrales, and J. Royo, "Systematic literature review: Integration of additive manufacturing and Industry 4.0," *Metals*, vol. 10, no. 8, p. 1061, Aug. 2020, doi: 10.3390/MET10081061.

[20] M. Mehrpouya, A. Dehghanghadikolaei, B. Fotovvati, A. Vosooghnia, S. S. Emamian, and A. Gisario, "The potential of additive manufacturing in the smart factory Industrial 4.0: A review," *Applied Sciences*, vol. 9, no. 18, p. 3865, Sep. 2019, doi: 10.3390/APP9183865.

[21] S. Yin et al., "Cold spray additive manufacturing and repair: Fundamentals and applications," *Additive Manufacturing*, vol. 21, pp. 628–650, May 2018, doi: 10.1016/J.ADDMA.2018.04.017.

[22] W. Li, K. Yang, S. Yin, X. Yang, Y. Xu, and R. Lupoi, "Solid-state additive manufacturing and repairing by cold spraying: A review," *Journal of Materials Science and Technology*, vol. 34, no. 3, pp. 440–457, Mar. 2018, doi: 10.1016/J.JMST.2017.09.015.

[23] H. Assadi, H. Kreye, F. Gärtner, and T. Klassen, "Cold spraying—a materials perspective," *Acta Materialia*, vol. 116, pp. 382–407, Sep. 2016, doi: 10.1016/J. ACTAMAT.2016.06.034.

[24] R. Melentiev, N. Yu, and G. Lubineau, "Polymer metallization via cold spray additive manufacturing: A review of process control, coating qualities, and prospective applications," *Additive Manufacturing*, vol. 48, p. 102459, Dec. 2021, doi: 10.1016/J. ADDMA.2021.102459.

[25] S. Yin et al., "Cold spray additive manufacturing and repair: Fundamentals and applications," *Additive Manufacturing*, vol. 21, pp. 628–650, May 2018, doi: 10.1016/J. ADDMA.2018.04.017.

[26] K. Malamousi, K. Delibasis, B. Allcock, and S. Kamnis, "Digital transformation of thermal and cold spray processes with emphasis on machine learning," *Surface and Coatings Technology*, vol. 433, p. 128138, Mar. 2022, doi: 10.1016/J. SURFCOAT.2022.128138.

[27] D. Ikeuchi, A. Vargas-Uscategui, X. Wu, and P. C. King, "Neural network modelling of track profile in cold spray additive manufacturing," *Materials*, vol. 12, no. 17, p. 2827, Sep. 2019, doi: 10.3390/MA12172827.

[28] Z. Wang, S. Cai, W. Chen, R. A. Ali, and K. Jin, "Analysis of critical velocity of cold spray based on machine learning method with feature selection," *Journal of Thermal Spray Technology*, vol. 30, no. 5, pp. 1213–1225, Jun. 2021, doi: 10.1007/ S11666-021-01198-8/FIGURES/10.

[29] H. Canales, I. G. Cano, and S. Dosta, "Window of deposition description and prediction of deposition efficiency via machine learning techniques in cold spraying," *Surface and Coatings Technology*, vol. 401, p. 126143, Nov. 2020, doi: 10.1016/J. SURFCOAT.2020.126143.

[30] R. Valente et al., "Classifying powder flowability for cold spray additive manufacturing using machine learning," in *Proceedings—2020 IEEE International Conference on Big Data, Big Data 2020*, pp. 2919–2928, Dec. 2020, doi: 10.1109/ BIGDATA50022.2020.9377948.

[31] S. Roy and K. Ravi, "A machine learning based approach for cold spray deposition porosity prediction from processing parameters," in *Thermal Spray 2022: Proceedings from the International Thermal Spray Conference*, vol. 84369, pp. 961–976, May 2022, doi: 10.31399/ASM.CP.ITSC2022P0961.

[32] K. Malamousi, K. Delibasis, B. Allcock, and S. Kamnis, "Digital transformation of thermal and cold spray processes with emphasis on machine learning," *Surface and Coatings Technology*, vol. 433, p. 128138, Mar. 2022, doi: 10.1016/J. SURFCOAT.2022.128138.

[33] M. Tzinava, K. Delibasis, B. Allcock, and S. Kamnis, "A general-purpose spray coating deposition software simulator," *Surface and Coatings Technology*, vol. 399, p. 126148, Oct. 2020, doi: 10.1016/J.SURFCOAT.2020.126148.

[34] A. Chowdhury, E. Kautz, B. Yener, and D. Lewis, "Image driven machine learning methods for microstructure recognition," *Computational Materials Science*, vol. 123, pp. 176–187, Oct. 2016, doi: 10.1016/J.COMMATSCI.2016.05.034.

[35] V. H. C. De Albuquerque, P. C. Cortez, A. R. De Alexandria, and J. M. R. S. Tavares, "A new solution for automatic microstructures analysis from images based on a back-propagation artificial neural network," *Nondestructive Testing and Evaluation*, vol. 23, no. 4, pp. 273–283, Dec. 2008, doi: 10.1080/10589750802258986.

[36] D. Ikeuchi, A. Vargas-Uscategui, X. Wu, and P. C. King, "Data-efficient neural network for track profile modelling in cold spray additive manufacturing," *Applied Sciences (Switzerland)*, vol. 11, no. 4, pp. 1–12, Feb. 2021, doi: 10.3390/app11041654.

[37] M. Tzinava, K. Delibasis, and S. Kamnis, "Self-organizing maps for optimized robotic trajectory planning applied to surface coating," *IFIP Advances in Information and Communication Technology*, vol. 627, pp. 196–206, 2021, doi: 10.1007/978-3-030-79150-6_16/COVER.

[38] Z. Cai, X. Liang, B. Chen, C. Zeng, C. Chen, Z. Mu, and T. Chen, "A geodesic-based robot trajectory planning approach for cold spray applications," *Journal of Thermal Spray Technology*, vol. 28, no. 5, pp. 939–945, Jun. 2019, doi: 10.1007/S11666-019-00871-3/FIGURES/5.

[39] H. Koivuluoto, J. Larjo, D. Marini, G. Pulci, and F. Marra, "Cold-sprayed Al6061 coatings: Online spray monitoring and influence of process parameters on coating properties," *Coatings*, vol. 10, no. 4, p. 348, Apr. 2020, doi: 10.3390/COATINGS10040348.

[40] A. N. Avanaki, "Exact global histogram specification optimized for structural similarity," *Optical Review*, vol. 16, no. 6, pp. 613–621, Dec. 2009, doi: 10.1007/S10043-009-0119-Z/METRICS.

[41] Z. Wang, A. C. Bovik, H. R. Sheikh, and E. P. Simoncelli, "Image quality assessment: From error visibility to structural similarity," *IEEE Transactions on Image Processing*, vol. 13, no. 4, pp. 600–612, Apr. 2004, doi: 10.1109/TIP.2003.819861.

[42] N. Tepylo, X. Huang, and P. C. Patnaik, "Laser-based additive manufacturing technologies for aerospace applications," *Advanced Engineering Materials*, vol. 21, no. 11, p. 1900617, Nov. 2019, doi: 10.1002/adem.201900617.

[43] A. Pratap, N. Sardana, S. Utomo, A. John, P. Karthikeyan, and P.-A. Hsiung, "Analysis of defect associated with powder bed fusion with deep learning and explainable AI," in *2023 15th International Conference on Knowledge and Smart Technology (KST)*, IEEE, pp. 1–6, Feb. 2023. doi: 10.1109/KST57286.2023.10086905.

[44] M. Tang, P. C. Pistorius, and J. L. Beuth, "Prediction of lack-of-fusion porosity for powder bed fusion," *Additive Manufacturing*, vol. 14, pp. 39–48, Mar. 2017, doi: 10.1016/J.ADDMA.2016.12.001.

[45] R. Cunningham, C. Zhao, N. Parab, C. Kantzos, J. Pauza, K. Fezzaa, T. Sun, and A. D. Rollett, "Keyhole threshold and morphology in laser melting revealed by ultrahigh-speed x-ray imaging," *Science (1979)*, vol. 363, no. 6429, pp. 849–852, Feb. 2019, doi: 10.1126/SCIENCE.AAV4687/SUPPL_FILE/AAV4687S9.MOV.

[46] D. Gu and Y. Shen, "Balling phenomena during direct laser sintering of multi-component Cu-based metal powder," *Journal of Alloys and Compounds*, vol. 432, no. 1–2, pp. 163–166, Apr. 2007, doi: 10.1016/J.JALLCOM.2006.06.011.

[47] M. A. Ansari, A. Crampton, R. Garrard, B. Cai, and M. Attallah, "A convolutional neural network (CNN) classification to identify the presence of pores in powder bed fusion images," *The International Journal of Advanced Manufacturing Technology*, vol. 120, no. 7–8, pp. 5133–5150, Jun. 2022, doi: 10.1007/s00170-022-08995-7.

[48] Y. Zhang and Y. F. Zhao, "Hybrid sparse convolutional neural networks for predicting manufacturability of visual defects of laser powder bed fusion processes," *Journal of Manufacturing Systems*, vol. 62, pp. 835–845, Jan. 2022, doi: 10.1016/j.jmsy.2021.07.002.

[49] V. Pandiyan, G. Masinelli, N. Claire, T. Le-Quang, M. Hamidi-Nasab, C. de Formanoir, R. Esmaeilzadeh, S. Goel, F. Marone, R. Logé, S. Van Petegem, and K.

Wasmer, "Deep learning-based monitoring of laser powder bed fusion process on variable time-scales using heterogeneous sensing and operando X-ray radiography guidance," *Additive Manufacturing*, vol. 58, p. 103007, Oct. 2022, doi: 10.1016/j.addma.2022.103007.

[50] H.-Y. Chen, C.-C. Lin, M.-H. Horng, L.-K. Chang, J.-H. Hsu, T.-W. Chang, J.-C. Hung, R.-M. Lee, and M.-C. Tsai, "Deep learning applied to defect detection in powder spreading process of magnetic material additive manufacturing," *Materials*, vol. 15, no. 16, p. 5662, Aug. 2022, doi: 10.3390/ma15165662.

[51] M. A. Ansari, A. Crampton, R. Garrard, B. Cai, and M. Attallah, "A Convolutional Neural Network (CNN) classification to identify the presence of pores in powder bed fusion images," *The International Journal of Advanced Manufacturing Technology*, vol. 120, no. 7–8, pp. 5133–5150, Jun. 2022, doi: 10.1007/s00170-022-08995-7.

[52] Y. Gui, K. Aoyagi, H. Bian, and A. Chiba, "Detection, classification and prediction of internal defects from surface morphology data of metal parts fabricated by powder bed fusion type additive manufacturing using an electron beam," *Additive Manufacturing*, vol. 54, p. 102736, Jun. 2022, doi: 10.1016/j.addma.2022.102736.

[53] H. Baumgartl, J. Tomas, R. Buettner, and M. Merkel, "A deep learning-based model for defect detection in laser-powder bed fusion using in-situ thermographic monitoring," *Progress in Additive Manufacturing*, vol. 5, no. 3, pp. 277–285, Sep. 2020, doi: 10.1007/s40964-019-00108-3.

[54] Y. Zhang, G. S. Hong, D. Ye, K. Zhu, and J. Y. H. Fuh, "Extraction and evaluation of melt pool, plume and spatter information for powder-bed fusion AM process monitoring," *Materials & Design*, vol. 156, pp. 458–469, Oct. 2018, doi: 10.1016/j.matdes.2018.07.002.

[55] B. Duman and K. Özsoy, "Toz yatak füzyon birleştirme eklemeli imalatta kusur tespiti için öğrenme aktarımı kullanan derin öğrenme tabanlı bir yaklaşım," *Gazi Üniversitesi Mühendislik-Mimarlık Fakültesi Dergisi*, Jun. 2021, doi: 10.17341/gazimmfd.870436.

[56] L. Scime and J. Beuth, "A multi-scale convolutional neural network for autonomous anomaly detection and classification in a laser powder bed fusion additive manufacturing process," *Additive Manufacturing*, vol. 24, pp. 273–286, Dec. 2018, doi: 10.1016/j.addma.2018.09.034.

[57] C. Suwanpreecha and A. Manonukul, "A review on material extrusion additive manufacturing of metal and how it compares with metal injection moulding," *Metals*, vol. 12, no. 3, p. 429, Feb. 2022, doi: 10.3390/MET12030429.

[58] J. Qin et al., "Research and application of machine learning for additive manufacturing," *Additive Manufacturing*, vol. 52, p. 102691, Apr. 2022, doi: 10.1016/J.ADDMA.2022.102691.

[59] G. Papazetis and G. C. Vosniakos, "Mapping of deposition-stable and defect-free additive manufacturing via material extrusion from minimal experiments," *International Journal of Advanced Manufacturing Technology*, vol. 100, no. 9–12, pp. 2207–2219, Feb. 2019, doi: 10.1007/S00170-018-2820-1/METRICS.

[60] L. Lu, J. Hou, S. Yuan, X. Yao, Y. Li, and J. Zhu, "Deep learning-assisted real-time defect detection and closed-loop adjustment for additive manufacturing of continuous fiber-reinforced polymer composites," *Robotics and Computer-Integrated Manufacturing*, vol. 79, p. 102431, Feb. 2023, doi: 10.1016/J.RCIM.2022.102431.

[61] G. D. Goh, N. M. Bin Hamzah, and W. Y. Yeong, "Anomaly detection in fused filament fabrication using machine learning," https://home.liebertpub.com/3dp, Jan. 2022, doi: 10.1089/3DP.2021.0231.

[62] Z. Zhang, I. Fidan, and M. Allen, "Detection of material extrusion in-process failures via deep learning," *Inventions*, vol. 5, no. 3, p. 25, Jul. 2020, doi: 10.3390/INVENTIONS5030025.

[63] H. Kim, H. Lee, J. S. Kim, and S. H. Ahn, "Image-based failure detection for material extrusion process using a convolutional neural network," *International Journal of Advanced Manufacturing Technology*, vol. 111, no. 5–6, pp. 1291–1302, Nov. 2020, doi: 10.1007/S00170-020-06201-0/FIGURES/11.

[64] Z. Zhang, J. Femi-Oyetoro, I. Fidan, M. Ismail, and M. Allen, "Prediction of dimensional changes of low-cost metal material extrusion fabricated parts using machine learning techniques," *Metals*, vol. 11, no. 5, p. 690, Apr. 2021, doi: 10.3390/MET11050690.

[65] Z. Zhang and I. Fidan, "Machine learning-based void percentage analysis of components fabricated with the low-cost metal material extrusion process," *Materials*, vol. 15, no. 12, p. 4292, Jun. 2022, doi: 10.3390/MA15124292.

[66] Z. Li, Z. Zhang, J. Shi, and D. Wu, "Prediction of surface roughness in extrusion-based additive manufacturing with machine learning," *Robotics and Computer-Integrated Manufacturing*, vol. 57, pp. 488–495, Jun. 2019, doi: 10.1016/J.RCIM.2019.01.004.

[67] A. Malekpour and X. Chen, "Printability and cell viability in extrusion-based bioprinting from experimental, computational, and machine learning views," *Journal of Functional Biomaterials*, vol. 13, no. 2, p. 40, Apr. 2022, doi: 10.3390/JFB13020040.

[68] L. Li, R. McGuan, R. Isaac, P. Kavehpour, and R. Candler, "Improving precision of material extrusion 3D printing by in-situ monitoring & predicting 3D geometric deviation using conditional adversarial networks," *Additive Manufacturing*, vol. 38, p. 101695, Feb. 2021, doi: 10.1016/J.ADDMA.2020.101695.

[69] M. Ziaee and N. B. Crane, "Binder jetting: A review of process, materials, and methods," *Additive Manufacturing*, vol. 28, pp. 781–801, Aug. 2019, doi: 10.1016/J.ADDMA.2019.05.031.

[70] A. Mostafaei, A. M. Elliott, J. E. Barnes, F. Li, W. Tan, C. L. Cramer, P. Nandwana, and M. Chmielus, "Binder jet 3D printing—Process parameters, materials, properties, modeling, and challenges," *Progress in Materials Science*, vol. 119, p. 100707, Jun. 2021, doi: 10.1016/J.PMATSCI.2020.100707.

[71] N. Satterlee, E. Torresani, E. Olevsky, and J. S. Kang, "Automatic detection and characterization of porosities in cross-section images of metal parts produced by binder jetting using machine learning and image augmentation," *Journal of Intelligent Manufacturing*, pp. 1–23, Apr. 2023, doi: 10.1007/S10845-023-02100-9/FIGURES/21.

[72] B. J. Paudel, H. Deng, and A. C. To, "A physics-based data-driven distortion compensation model for sintered binder jet parts considering size effects," *Additive Manufacturing*, vol. 68, p. 103517, Apr. 2023, doi: 10.1016/J.ADDMA.2023.103517.

[73] F. Dini, S. A. Ghaffari, J. Jafar, R. Hamidreza, and S. Marjan, "A review of binder jet process parameters; powder, binder, printing and sintering condition," *Metal Powder Report*, vol. 75, no. 2, pp. 95–100, Nov. 2021, doi: 10.1016/J.MPRP.2019.05.001.

[74] M. Jamalkhani, M. Asherloo, O. Gurlekce, T. Ho, M. Heim, D. Nelson, and A. Mosta-faei, "Deciphering microstructure-defect-property relationships of vacuum-sintered binder jetted fine 316 L austenitic stainless steel powder," *Additive Manufacturing*, vol. 59, p. 103133, Nov. 2022, doi: 10.1016/J.ADDMA.2022.103133.

[75] D. Gunther, M. F. Pirehgalin, I. Weis, and B. Vogel-Heuser, "Condition monitoring for the Binder Jetting AM-process with machine learning approaches," in *Proceedings—2020 IEEE Conference on Industrial Cyberphysical Systems, ICPS 2020*, pp. 417–420, Jun. 2020, doi: 10.1109/ICPS48405.2020.9274716.

[76] R. Onler, A. S. Koca, B. Kirim, and E. Soylemez, "Multi-objective optimization of binder jet additive manufacturing of Co-Cr-Mo using machine learning," *International Journal of Advanced Manufacturing Technology*, vol. 119, no. 1–2, pp. 1091–1108, Mar. 2022, doi: 10.1007/S00170-021-08183-Z/FIGURES/12.

[77] Y. Zhu, Z. Wu, W. D. Hartley, J. M. Sietins, C. B. Williams, and H. Z. Yu, "Unravel-ing pore evolution in post-processing of binder jetting materials: X-ray computed tomography, computer vision, and machine learning," *Additive Manufacturing*, vol. 34, p. 101183, Aug. 2020, doi: 10.1016/J.ADDMA.2020.101183.

[78] H. Chen and Y. F. Zhao, "Learning algorithm based modeling and process parameters recommendation system for binder jetting additive manufacturing process," *Proceed-ings of the ASME Design Engineering Technical Conference*, vol. 1A-2015, Jan. 2016, doi: 10.1115/DETC2015-47627.

[79] E. Mendoza Jimenez et al., "Parametric analysis to quantify process input influence on the printed densities of binder jetted alumina ceramics," *Additive Manufacturing*, vol. 30, p. 100864, Dec. 2019, doi: 10.1016/J.ADDMA.2019.100864.

[80] S. Bafaluy Ojea, J. Torrents-Barrena, M. T. Pérez-Prado, R. Muñoz Moreno, and F. Sket, "Binder jet green parts microstructure: Advanced quantitative analysis," *Jour-nal of Materials Research and Technology*, vol. 23, pp. 3974–3986, Mar. 2023, doi: 10.1016/J.JMRT.2023.02.051.

[81] I. Gibson, D. Rosen, and B. Stucker, "Directed energy deposition processes," *Addi-tive Manufacturing Technologies*, pp. 245–268, 2015, doi: 10.1007/978-1-4939-2113-3_10.

[82] D. G. Ahn, "Directed Energy Deposition (DED) Process: State of the art," *Interna-tional Journal of Precision Engineering and Manufacturing-Green Technology*, vol. 8, no. 2, pp. 703–742, Feb. 2021, doi: 10.1007/S40684-020-00302-7.

[83] P. K. Nalajam and R. V, "Microstructural porosity segmentation using machine learn-ing techniques in wire-based direct energy deposition of AA6061," *Micron*, vol. 151, p. 103161, Dec. 2021, doi: 10.1016/J.MICRON.2021.103161.

[84] H. Lee, W. Heogh, J. Yang, J. Yoon, J. Park, S. Ji, and H. Lee, "Deep learning for in-situ powder stream fault detection in directed energy deposition process," *Jour-nal of Manufacturing Systems*, vol. 62, pp. 575–587, Jan. 2022, doi: 10.1016/J. JMSY.2022.01.013.

[85] L. Chen, X. Yao, and S. K. Moon, "In-situ acoustic monitoring of direct energy depo-sition process with deep learning-assisted signal denoising," *Materials Today: Pro-ceedings*, vol. 70, pp. 136–142, Jan. 2022, doi: 10.1016/J.MATPR.2022.09.008.

[86] Z. Zhang, Z. Liu, and D. Wu, "Prediction of melt pool temperature in directed energy deposition using machine learning," *Additive Manufacturing*, vol. 37, p. 101692, Jan. 2021, doi: 10.1016/J.ADDMA.2020.101692.

[87] J. S. Lim, W. J. Oh, C. M. Lee, and D. H. Kim, "Selection of effective manufacturing conditions for directed energy deposition process using machine learning methods," *Scientific Reports*, vol. 11, no. 1, pp. 1–13, Dec. 2021, doi: 10.1038/s41598-021-03622-z.

[88] D. Kats, Z. Wang, Z. Gan, W. K. Liu, G. J. Wagner, and Y. Lian, "A physics-informed machine learning method for predicting grain structure characteristics in directed energy deposition," *Computational Materials Science*, vol. 202, p. 110958, Feb. 2022, doi: 10.1016/J.COMMATSCI.2021.110958.

[89] J. Li, M. Sage, X. Guan, M. Brochu, and Y. F. Zhao, "Machine learning-enabled competitive grain growth behavior study in directed energy deposition fabricated Ti6Al4V," *JOM*, vol. 72, no. 1, pp. 458–464, Jan. 2020, doi: 10.1007/S11837-019-03917-7/FIGURES/6.

[90] F. Kaji, H. Nguyen-Huu, A. Budhwani, J. A. Narayanan, M. Zimny, and E. Toyserkani, "A deep-learning-based in-situ surface anomaly detection methodology for laser directed energy deposition via powder feeding," *Journal of Manufacturing Processes*, vol. 81, pp. 624–637, Sep. 2022, doi: 10.1016/J.JMAPRO.2022.06.046.

[91] T. Q. D. Pham, T. V. Hoang, X. V. Tran, S. Fetni, L. Duchêne, H. S. Tran, and A. M. Habraken, "Characterization, propagation, and sensitivity analysis of uncertainties in the directed energy deposition process using a deep learning-based surrogate model," *Probabilistic Engineering Mechanics*, vol. 69, p. 103297, Jul. 2022, doi: 10.1016/J.PROBENGMECH.2022.103297.

[92] H. Lee, W. Heogh, J. Yang, J. Yoon, J. Park, S. Ji, and H. Lee, "Deep learning for in-situ powder stream fault detection in directed energy deposition process," *Journal of Manufacturing Systems*, vol. 62, pp. 575–587, Jan. 2022, doi: 10.1016/J.JMSY.2022.01.013.

[93] D. Guo, S. Ling, and H. Li, "A framework for personalized production based on digital twin, blockchain and additive manufacturing in the context of Industry 4.0," in *IEEE International Conference on Automation Science and Engineering*, vol. 2020, pp. 1181–1186, Aug. 2020, doi: 10.1109/CASE48305.2020.9216732.

[94] C. Liu, L. Le Roux, C. Körner, O. Tabaste, F. Lacan, and S. Bigot, "Digital twin-enabled collaborative data management for metal additive manufacturing systems," *Journal of Manufacturing Systems*, vol. 62, pp. 857–874, Jan. 2022, doi: 10.1016/J.JMSY.2020.05.010.

[95] Y. Cai, Y. Wang, and M. Burnett, "Using augmented reality to build digital twin for reconfigurable additive manufacturing system," *Journal of Manufacturing Systems*, vol. 56, pp. 598–604, Jul. 2020, doi: 10.1016/J.JMSY.2020.04.005.

[96] A. Suresh, R. Udendhran, and G. Yamini, "Internet of Things and additive manufacturing: Toward intelligent production systems in industry 4.0," *EAI/Springer Innovations in Communication and Computing*, pp. 73–89, 2020, doi: 10.1007/978-3-030-32530-5_5/COVER.

[97] C. Yang, W. Shen, and X. Wang, "The Internet of Things in manufacturing: Key issues and potential applications," *IEEE Systems, Man, and Cybernetics Magazine*, vol. 4, no. 1, pp. 6–15, Jan. 2018, doi: 10.1109/MSMC.2017.2702391.

[98] R. Y. Zhong and W. Ge, "Internet of Things enabled manufacturing: A review," *International Journal of Agile Systems and Management*, vol. 11, no. 2, pp. 126–154, 2018, doi: 10.1504/IJASM.2018.092545.

[99] Z. Bi, G. Wang, and L. Da Xu, "A visualization platform for Internet of Things in manufacturing applications," *Internet Research*, vol. 26, no. 2, pp. 377–401, Apr. 2016, doi: 10.1108/INTR-02-2014-0043/FULL/PDF.

[100] R. Ashima, A. Haleem, S. Bahl, M. Javaid, S. K. Mahla, and S. Singh, "Automation and manufacturing of smart materials in additive manufacturing technologies using Internet of Things towards the adoption of industry 4.0," *Materials Today: Proceedings*, vol. 45, pp. 5081–5088, Jan. 2021, doi: 10.1016/J.MATPR.2021.01.583.

[101] C. Xia, Z. Pan, J. Polden, H. Li, Y. Xu, S. Chen, and Y. Zhang, "A review on wire arc additive manufacturing: Monitoring, control and a framework of automated system," *Journal of Manufacturing Systems*, vol. 57, pp. 31–45, Oct. 2020, doi: 10.1016/J.JMSY.2020.08.008.

[102] P. Becker, C. Eichmann, A. Roennau, and R. Dillmann, "Automation of post-processing in additive manufacturing with industrial robots," in *IEEE International Conference on Automation Science and Engineering*, vol. 2020, pp. 1578–1583, Aug. 2020, doi: 10.1109/CASE48305.2020.9216955.

Chapter 9

An insight into applications of laser in modern era

Rahul Nair, Ravi Kant and Hema Gurung

9.1 INTRODUCTION

A laser is a source of a very intense, monochromatic and unidirectional beam of light. Monochromatic means that the beam is of a single wavelength, and coherence means the beam is in the same phase. The transition of electrons from higher to lower energy states is random in ordinary light with respect to time. This is the cause of getting different wavelength, frequency, energy and phase in the emitted photons. But due to the stimulated emission in a laser, a chain reaction is started in which one atom's radiation energizes another atom in succession until all stimulated atoms come to their original state. This provides a coherent monochromaticity and directionality to the laser beam. Because of these properties, the laser beam appears intensely focused and can concentrate a large amount of energy in a small area. Modern-day manufacturing, especially Industry 4.0, relies heavily on laser and sensor technology. Implementing lasers in the manufacturing sector has laid a path for improved, rapid and hassle-free product manufacturing. Nowadays, producing solar cells, batteries and fuel cells needs laser technology to execute them. Even microchips or semiconductors that are made today are manufactured with the help of optical lithography, which is a type of laser technology.

Laser, being flexible and able to adjust to changes in a material like thickness, shape, orientation and reflectivity, makes it a perfect tool for Industry 4.0. Processes like laser-based additive manufacturing are becoming crucial components of Industry 4.0 as they employ several sensors for continuous workflow monitoring and control. This chapter focuses on the application of lasers in three major domains: academics, industry and the medical sector. The following sections also discuss the implementation of lasers in modern-day sectors.

9.2 LASER IN SPECTROSCOPY

The study of emission and absorption of light by matter is termed spectroscopy. Spectroscopy is the analysis and measuring of a specific wavelength, and it is generally utilized for the spectroscopic examination of materials. At a glance, the

DOI: 10.1201/9781032703046-9

working principle of spectroscopy is akin to that of a prism, which disperses light into the colours of the rainbow. Indeed, earlier spectroscopy was carried out using photographic plates and prisms [1]. Due to the high sensitivity of spectrographic techniques, their usage is very prominent in the detection of a single atom and its isotopes. Spectroscopy exploits the basic fundamentals of physics for determining the absorption, emission or scattering of light and its optical phase changes. In the mid-infrared region, atoms have a strong and narrow absorption line compared to their rotational and vibrational mode [2]. This property helps to determine a wide variety of elements with high accuracy and sensitivity. The basic application of this can be seen in finding tiny elements in air pollution [3]. Even the change in optical phase concept has been used for spectroscopy analysis. The intended interaction occurs at one side of the interferometer, and the phase change can be detected at the output of the interferometer. These phase changes resulting from absorption or dispersion associated with narrow spectral features are used in frequency modulation spectroscopy. When a matter is exposed to heating or simple exposure to sunlight, it has the tendency to emit or scatter light with some specific characteristic features. This technique of scattering of light is widely used in satellites, where a spectrometer records the scattered light from the earth's surface to differentiate between sea, flora, fauna and rocky surfaces. This scattering phenomenon is also utilized in astronomy, where optical spectra of light from galaxies and stars have been used to determine movements, speed, temperature and chemical compositions.

The modern-day spectral analysis uses lasers as a tool for the illumination of specimens. Absorbed light will heat the atoms and excite them, resulting in the emission of fluorescent light. The utilization of laser beams as the light source in spectroscopy is termed laser spectroscopy. The divergent properties of lasers, such as exhibition of temporal and spatial coherence, wavelength tunability, high optical powers and the generation of ultrashort pulses, have opened many new doors for spectroscopy. The incorporation of lasers into spectroscopy has revitalized traditional spectroscopic analysis. This is primarily because lasers offer brightness, directionality and spectral purity compared to other light sources. The utilization of laser in traditional spectroscopy processes like Raman spectroscopy, absorption spectroscopy and fluorescence spectroscopy has helped to increase the accuracy and sensitivity of spectroscopy analysis. The recent developments helped in analyzing new techniques for the measurement of wavelength with exceptional precision, and these are mainly based on informetric devices. Laser spectroscopy has become an important technique in medicine, science and industry where high sensitivity and accuracy are required.

9.2.1 Laser induced-breakdown spectroscopy

Laser-induced-breakdown spectroscopy (LIBS), commonly termed laser-induced plasma spectroscopy (LIPS) or laser spark spectroscopy (LSS), is a type of atomic emission spectroscopy. LIBS generates a plasma that is produced with the help of a low-level pulsed laser on the surface of the sample, which vaporizes a small amount of material. This spectroscopy exploits the phenomenon of emission of spectra when the excited atom returns to its ground state for the analysis of the specimen's elements. The generated laser focuses on the targeted specimen via a focusing lens, and the plasma emits a white light which is then collected by a second lens. The secondary lens sends the captured light to a spectrometer through an optical fiber to detect elements. The spectrometer then disseminates the emitted light from the electrons into signals that a detector will further record. These recorded signals are then digitized and displayed through an electronic medium [4]. The process seems simple in an overview, but the physical and chemical processes involved in LIBS are not that straightforward. The lifespan of the plasma beam is itself a complicated process. Inverse bremsstrahlung is a mechanism that predominates in the absorption of incident lasers, and this is accompanied by a collision between electron, ion and molecule. Inverse bremsstrahlung is a process in which electrons in the electric field of the atom will absorb the photon, resulting in a gain in energy of electrons and an increase in temperature through an ionized collision [5]. As soon as the plasma is generated, the continuum and ionic spectra become visible. The "white light", known as a continuum, contains all the spectral information and the ions ejected from electrons. Assessment of LIBS uncovers quick subjective data about a specimen's structure. The data provided by LIBS gives an insight into the chemical fingerprint of material in solid, air or liquid samples. A detailed working diagram of LIBS is demonstrated in Figure 9.1. The mechanism of excitation of a particular energy level in discrete atoms is a highly complex process and depends on the interaction between atoms, molecules and thermodynamic equilibrium. As soon as the laser pulse is ceased, plasma decay starts at

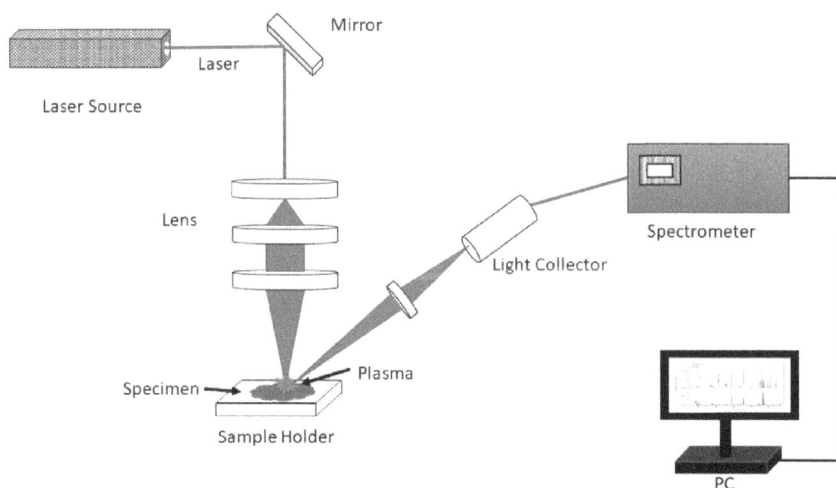

Figure 9.1 Detailed working diagram of LIBS.

a very fast rate, about one to several microseconds. This rate will depend on the deposition of laser energy. Most LIBS investigations use a frequency of 10 Hz or greater for repetitive plasmas.

LIBS has shown its compatibility with various matter, such as solids, liquids, gases, aerosols and slurries, which are difficult to investigate with the conventionally used spectroscopy processes like inductively coupled plasma (ICP). While ICP is mainly confined to research centre use, LIBS is particularly valuable in field applications. Over the years, LIBS has garnered attention in various sectors for its ability to detect impurities in metals, analyze the deterioration of metals used in nuclear reactors and identify soil contamination. Nowadays, its usage is very significant in detecting elemental components of rocks and soils on Mars; it was also able to analyze toxic compounds like anthrax.

Studies in diverse fields have been undertaken to improve the accuracy of the machine and overcome the limitations that restrict its use. Fu et al. [6] studied the plasma evolution of the laser on titanium alloy. Three high-speed cameras were used to capture plasma evolution, and it was found that the plasma and surrounding gases interact drastically during the earlier stage of the development. Earlier stage plasma experiences morphological fluctuation with an increased time in delay, resulting in the transformation of stable plasma to a fluctuating plasma and therefore inaccurate results. Liu et al. [7] provided a review work in which the analysis of the plastics was presented using LIBS. The specific advantages such as discrimination and quantitative and qualitative analysis of plastics helped in the fast detection of toxic particles in food containers and toys. Yuan et al. [8] provided a single-sample calibration method (SSC-LIBS) based on the Lomakin-Scherbe formula. The experimental results showed that SSC-LIBS had enhanced the accuracy of element determination as compared to the multi-point calibration LIBS. Wang et al. [9] reported the effects of preheating samples in femtosecond LIBS. Results indicated that the assistance of preheating improved the diagnosing accuracy of femtosecond LIBS. Zang et al. [10] developed a new LIBS spectrum data treatment method for identifying the concentration of the minor element in steel with the help of machine learning. This approach resulted in identifying the non-metal elements in the steel.

Advantages of LIBS

1. It is a rapid, portable and real-time analysis technique.
2. Being a non-destructive analysis technique, a very small amount of sample is used for analysis.
3. The necessity of sample preparation is almost negligible; contamination is much less.
4. Sample analysis is versatile, i.e. liquid, aerosols, solids, gases, slurry.
5. Multi-elemental analysis can be done.
6. Analysis of hard materials which are difficult to dissolve, like semiconductors and ceramics, can be done easily.

Disadvantages of LIBS

1. Generally, detection limits are not as accurate as conventional solution techniques.
2. The input cost is fairly high and the system is quite complex to use.
3. The interference effect is quite large.
4. Damage to human vision is a big risk when working with high-intensity pulsed laser.
5. Precision obtained by LIBS is low.

9.2.2 Photoacoustic spectroscopy

The term photoacoustic defines the generation of acoustic waves in a sample through the absorption of photons. Measuring the change in optical power in the light beam is the method for determining wavelength-dependent absorption in the sample. The way photoacoustic spectroscopy (PAS) works is quite simple. A high-intensity beam collides with the targeted specimen, which excites the ground state electron population to the higher state. The excited electrons will return to their ground state in radiative and nonradiative pathways. The nonradiative part will eventually produce heat in the restricted district of the excitation light and create a pressure wave that proliferates away from the source. The cyclic excitation is transformed into a cyclic variation of temperature in the targeted specimen by a radiationless mitigation process. Properties like refractive index and pressure are altered during this process, which gives rise to acoustic disturbances that can be monitored by appropriate detectors. These detectors will then emit signals which can further be altered, amplified and averaged to produce significant data regarding the specimen's energy transfer, composition and other valuable information [11]. PAS consists of some critical components: (a) source of the emission of light, (b) specimen chamber, (c) detector for acoustic signals and (d) signal processing apparatus [12].

The generation of a high-intensity beam in photoacoustic spectroscopy is currently executed with the help of a monochromator/lamp or a laser beam. Laser, a monochromatic high intensity beam, has a huge advantage over the traditional monochromator/lamp, the use of which dominates in photoacoustic spectroscopy. The laser has an accurately parallel and cylindrical beam which is suitable for the current resonators used in the photoacoustic cell [13]. Instead of a continuous wave, a pulsed laser source is used because of its potential to generate a high-intensity beam. This high-intensity beam can be utilized in various ways, such as the detection of thermal diffusivity, investigation of phase transitions, and the non-destructive analysis of materials by thermal wave imaging [14]. PAS is utilized when the surface of the solid material is not reflective and can provide optical information of the mass material itself. PAS can also study opaque materials for their optical properties. This proficient PAS method can be utilized to examine materials like semiconductors, metallic frameworks and insulators that cannot be understood promptly by traditional reflection or absorption procedures. Freidlin et al. [15] demonstrated dual-comb spectroscopy for PA measurements. This approach helped to overcome the limitation of a separate image at each wavelength of interest, which causes errors during the sample change between the image acquired at different wavelengths. Krivoshein et al. [16] showed an FTIR-PAS-based approach for identifying and assessing soil components. The new approach showed higher sensitivity in the mid-wave range of the mid-IR range as compared to the traditional one. Lv et al. [17] proposed a radial cavity quartz-enhanced technique for photoacoustic spectroscopy. The experiment results showed that the new optimized radial cavity helped improve photoacoustic spectroscopy's detection sensitivity by greater than one order of magnitude.

Advantages of PAS

1. Pulsed laser beams can detect even a minute portion of the targeted gas, on the order of parts per trillion.
2. Compared to the photothermal process, the photoacoustic signal has the ability to measure the absorption spectrum, which is very useful for transparent samples in which the absorption of light is significantly less.
3. Absorption spectrum can be calculated by dividing signal spectrum by light intensity spectrum; this plays a significant role in the detection of a microscopic particle of any targeted specimens.

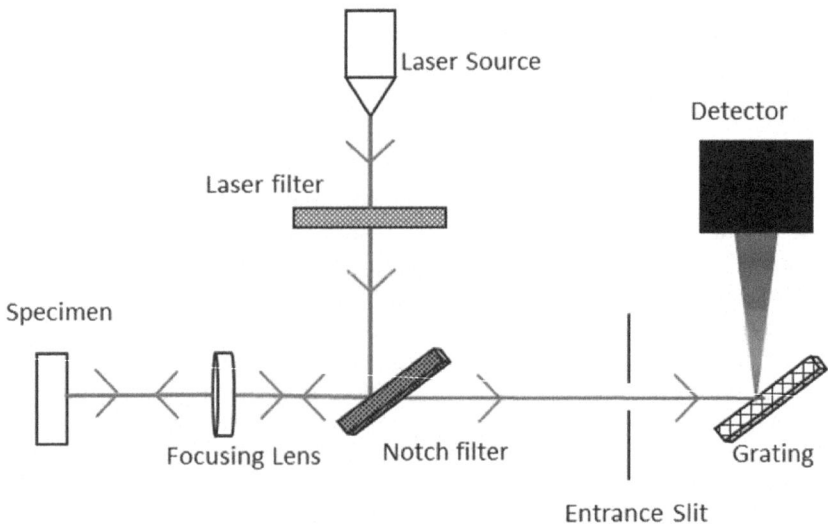

Figure 9.2 Working diagram of Raman spectroscopy.

Disadvantages of PAS

1. Limited tuning range of the laser source.
2. The initial setup and equipment cost is quite high.
3. The bandwidth of the laser is not very broad; the targeted molecule must absorb some light from the source for the analysis.

9.2.3 Raman spectroscopy

Raman spectroscopy is a non-destructive characterization technique that utilizes the inelastic scattering phenomenon to identify the vibrational fingerprint of the sample through which physical properties, molecular interaction, phase, chemical structure and crystallinity can be determined. A wide range of materials can be examined through this process, varying from physical state samples like vapours, liquids and solids to surface layer or microscale samples. In this, the sample will interact with a monochromatic laser source, and the interaction of the photons with the molecules of the sample leads to the rise of scattered light. During the interaction, if the frequency of the scattered light is different from that of incident light, then it will give rise to the Raman spectrum. Most scattered light will have the same frequency as monochromatic light, which does not provide any useful information, and this is termed as Rayleigh scattering [18]. Only a small amount of scattered light will differ (roughly around 0.0000001%) from the source's original frequency, which will make up Raman scatter. When the frequency of scattered light is lower than that of incident radiation, a Stokes line will appear, indicating that the electrons have absorbed energy. But if the frequency is higher than that of incident light, then the anti-Stokes line will appear in the Raman spectrum, indicating that the electron has released energy [19]. In conventional Raman spectroscopy, Stokes bands are measured. This is because the Stokes bands involve the transition of electrons from low to high energy levels, resulting in a stronger band than the anti-Stokes band. Usually, anti-Stokes bands are measured for the fluorescent sample, where the Stokes band experiences disturbance due to fluorescence [20]. A Raman spectrometer includes three essential components: detector, excitation source and sampling apparatus. A typical working diagram of Raman spectroscopy is shown in Figure 9.2.

The identification of appropriate laser wavelength is a critical task depending on the usage. Three categories of laser sources are available for Raman spectroscopy.

1. Diode-pumped single-longitudinal mode laser: The wavelength range in these lasers varies from 300 nm to 1064 nm. The single-longitudinal-mode (SLM) laser contains narrow linewidths and low noise, making it ideal for usage in Raman spectroscopy. The spectral purity provided by these lasers is very high, normally greater than 60 dB side-mode suppression ratio.
2. Single-mode diode laser: The power output from the single-mode laser is less than that of the multimode diode laser, but it provides high power

intensity because of the narrow active area, making it very suitable for high-speed applications like Raman spectroscopy. Generally, the wavelength of these lasers ranges from 785 nm to 1064 nm. The single-mode suppression ratio is limited by the sideband emission to nearly 50 dB, generally obtained some picometers away from the main peak.

3. Volume Bragg-grating (VBG) frequency-stabilized diode lasers: A VBG is a grating element in which the reflective index receives a periodic modulation throughout the photosensitive material. It permits the multiple multimode laser diodes to lock simultaneously, making it useful at any wavelength or power level. This element helps the laser to achieve narrow-linewidth emission at a wavelength that is generally not possible at distributed Bragg reflection or distributed feedback. Compared to the other diode lasers, the side-mode suppression ratio is limited to about 40–50 dB to the main peak, but with the help of VBG it can be improved to 60–70 dB.

Due to its predominant advantages over the other spectroscopy processes, Raman spectroscopy is widely used for various characterization techniques. Technological advancement is also helping Raman spectroscopy to overcome its limitations and widen its field area. Berghian-Grosan and Magdas [21] developed a new method by combining Raman spectroscopy and machine learning for the authentication of edible oils. The proposed approach detected the adulteration as well as the magnitude of it in the oils. Lu et al. [22] proposed a method that incorporates artificial intelligence into Raman spectroscopy to analyze the microbes at a single cell level. The combination of convolution neural network (ConvNet) and Raman spectroscopy helped to categorize the cells according to their spectral features. Lu et al. [23] developed a novel method by combining serum Raman spectroscopy and multiscale convolution independent circulation neural network along with multiscale fusion convolution independent circulation neural network for the diagnosis of hepatitis B virus (HBV). The experimental results demonstrated that the prediction sensitivity, specificity and accuracy were increased by incorporating the developed method, resulting in effective diagnosis of healthy and HBV patients.

Advantages of Raman spectroscopy

1. The major advantage of Raman spectroscopy is that no specific type of sample preparation is needed.
2. It is non-destructive testing, and it cannot be interfered with by water.
3. Raman spectra can be quickly acquired in a matter of seconds.
4. A wide variety of organic and inorganic samples can be detected by Raman analysis, and samples can be analyzed even through glass or polymer packing.
5. A large region (50 cm^{-1} to 400 cm^{-1}) can be covered in a single recording.

Disadvantages of Raman spectroscopy

1. It cannot be used for metal or alloys.
2. The intense heat produced by the laser could destroy the sample or cover the Raman spectrum.
3. As the Raman effect is very weak, a sophisticated instrument is required.

9.3 LASER SENSORS

A laser sensor records the measurement value with the help of laser technology. It converts the physically measured value to an analogue electrical signal. The difference in material properties and variation in environmental conditions make it difficult for certain technology to work. Certain equipment encounters challenges when it comes to accurately analyzing moving particles in a dusty environment, while others struggle with precisely measuring the liquid levels. The laser provides a contactless measurement, which finds its way into various modern applications [24]. Most of the laser sensors are based on the triangulation principle; the term triangulation is defined as the calculation of distance measurement with the help of angular estimation [25]. A laser spot is projected from the sensor to the targeted object, and when this reflected light falls on the receiving element at an angle (depending on the distance between the sensor and targeted object), the sensor calculates the distance at which the measured object is placed. These sensors can be joined with either point or line lasers. A point laser produces a focused small circle of light on the surface of the specimen, which is intended to transfer the point from one surface to another. In contrast, a line laser produces vertical and horizontal lines with the help of LED diodes. The sensors integrated with line lasers are commonly used for indoor applications and have the plumb down and up potential. The modern-day models use a pulsing light technology that operates with a light detector, which aids to expand their application in outdoor environments.

According to the laser sensor type, there are four types available: complementary metal-oxide semiconductors (CMOS), photoelectric sensors, charge-coupled devices (CCD) and position-sensitive detectors (PSD). The laser sensors used in industries help to determine or analyze the dimensions of various materials travelling at high speed or to monitor the distances of the material irrespective of the range of colours and the ambient background light. Some common laser sensors used in industries include laser displacement, time of flight, laser photoelectric and laser grid sensors. A detailed explanation of the aforementioned sensors is provided in the following sections.

9.3.1 Laser displacement sensor

The laser displacement sensor (LDS) calculates the distance between the sensor and the targeted object by using different elements and then converting it into a mathematical value of distance. Some of the most common sensor types are ultrasonic displacement, optical displacement and linear proximity. The lens collects the laser beams and makes them parallel, which is then emitted to the targeted specimen. The receiver detects the reflected light from the targeted object, which is then detected by the image sensor through the lens. The distance D can be calculated by the following formula:

$$D = \frac{L \times S}{I}$$

where D is the distance between the laser and the targeted object. The distance between the lenses is denoted by L, and S is the distance between lens and the image sensor. The term I is the current from the image sensor. A detailed diagram is depicted in Figure 9.3. The head of the sensor is mounted in such a way that the angle of the emitted light and the angle of the received light are the same [26].

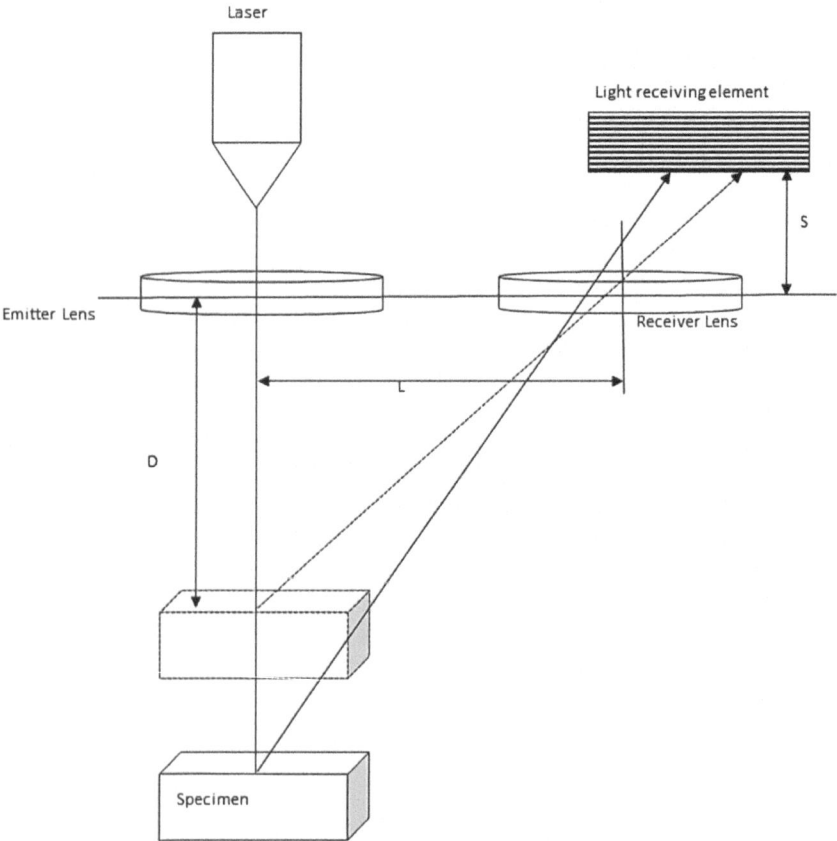

Figure 9.3 Schematic of laser displacement sensing.

Laser displacement sensing is applied extensively in modern industries. Laser distance sensors are commonly used in the automobile industry to detect the complex geometrical shapes of a single component. Contacting sensors could not accurately detect these complex shapes, whereas non-contacting sensors like laser sensors provide high accuracy. A laser displacement transducer is a commonly used tool in automobile industries to control the dimensions of the headlight component. It helps to eliminate the inconsistencies in the earlier stage so that a smooth production could be acquired and delays could be avoided in the production line. Displacement sensors like laser displacement-level measurement sensors are useful in the chemical, food and plastics industries.

Smith and Zheng [27] developed a multi-laser displacement sensor for the digitizing of unknown sculpture surfaces. This newly developed technique was used to measure the curvatures, tangents and changes in surface displacement. The results showed that it was very useful in detecting the arc length of the sample, with minimal error. Sun and Li [28] demonstrated a novel error-compensation model which can reduce the error in the incident angle of the laser displacement sensor. The designed model helped to increase the efficiency of the surface detection of an aero-engine blade. Giri and Kharkovsky [29] reported a measurement technique with the help of the laser displacement sensor to detect the crack propagation that occurred in the concrete specimens. Sandak and Tanaka [30] compared the roughness of a wood specimen obtained from the laser displacement sensor with the contact type stylus. The experimental results showed that the profile obtained by the LDS is much more accurate than that of the contact type stylus.

9.3.2 Time-of-flight sensor

Time-of-flight (ToF) sensors are used to measure the distance travelled between two points by the photons in terms of time. Time-of-flight sensors use various types of carriers; light and sound are the most commonly used carriers. Light-based ToF sensors are better than ultrasound-based sensors. They provide faster readings, greater accuracy and extended range with low power consumption, compact size and less weight. Infrared light proves to be much more beneficial than other light mediums because of its ability to distinguish from the ambient light and maintain fewer signal disturbances. These abilities make them cost-effective sensors considering their size and weight. The working diagram of a ToF sensor is shown in Figure 9.4.

A tiny laser is fitted inside the ToF sensors, emitting IR light at the targeted specimen. The light is then reflected from the targeted objects and returned to the ToF sensor [31]. There are two ways in which the ToF determines the distance and the depth – by time pulses and phase shift of an amplitude-modulated wave. The timed pulse illuminates the specimen with a laser and then measures the reflected light with a scanner. The speed of light is used to calculate the distance of the object accurately. A digital 3D representation is created with the help of the time taken by the laser to return and its wavelength. This technology is now widely used

Laser Source   S

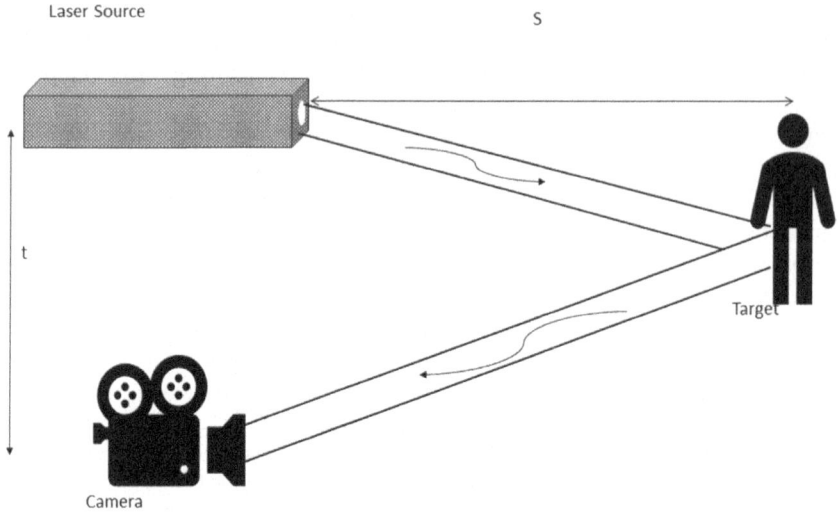

Camera

Figure 9.4 Working diagram of time-of-flight sensor.

in phone cameras [32]. The distance (D) between the sensor and the specimen can be calculated by the given formula:

$$D = \frac{S \times T}{2}$$

where S is the speed of the light and T is the time of flight. A timer is activated when the laser beam exits the sensor. It records the time the laser beam returns to the receiver sensor after the interaction with the object. As the speed of light is constant, subtracting these two times will help to calculate the time of flight.

Modern-day technology uses a continuous wave in ToF to determine the phase shift in the reflected light to identify the distance travelled and the depth. The amplitude modulation will create a sinusoidal form of the light source. The frequency will be pre-known, allowing the detector to analyze the phase shift of the reflected light. Kumar et al. [33] proposed a ToF sensor ring device for collision avoidance in the industrial robotic workspace. The simulation results showed that the ToF sensor ring device was better than conventionally available collision avoidance setups in terms of productivity, performance and safety. Langmann et al. [34] experimentally compared the depth measurements of several commercially available setups with the ToF sensor. Experimental results indicated that the ToF sensor more accurately predicted the depth than did the commercial sensors. Moreover, a monocular and fully registered 2D/3D camera was also proposed to benefit machine vision applications. Tsuji and Kohama [35] suggested a proximity skin sensor that uses a ToF sensor, which can be connected to the surface of human cooperative robots for safety. The proposed prototype was installed at the head of the robot arm and can identify the position and distance of an object without making any physical contact.

9.3.3 Photoelectric sensor

Photoelectric sensors are one of the most commonly found sensors in our everyday life. They are helpful in controlling elevators, managing the opening and closing of grocery store gates, opening sink faucets with a wave of a hand and much more. A photoelectric sensor identifies the change in light intensity to detect the difference in surface conditions and objects through the optical property. The emitted light and the target detection method depend on the sensor type that has been used. These sensors consist of an amplifier, a receiver, a light source and a signal converter. The emitted light is either interrupted or reflected by the target object, thereby altering the intensity of the light that reaches the receiver. The receiver identifies this change and converts it into an electrical output [36]. These sensors have more advantages than previously available sensors had. The sensing range of these photoelectric sensors can surpass the ultrasonic, inductive, magnetic and capacitive technologies. Their small size ratio to sensing range makes them an ideal sensor for almost any application.

Photoelectric sensors provide three types of target detection: retro-reflective, diffused and thru-beam. The diffused mode sensing, also known as proximity mode, contains the transmitter and receiver in the same housing. The transmitter emits the laser beam, which strikes the specimen; after striking, the beam will be reflected at different angles. At the same time, a portion of the reflected beam is captured at the receiver end, which helps to detect the specimen. Since a significant amount energy is lost due to the specimen's ability to reflect light, the use of diffused mode in shorter ranges is limited. The major advantage of using such a type of mode is that it limits the requirement of any secondary device like a separate receiver or a reflector. The retro-reflective mode of photoelectric sensing is typically used for longer sensing ranges due to the increased efficiency of the reflector. These types of sensors may or may not contain polarization filters. A polarization filter will permit only a certain phase angle of light to the receiver. This helps the sensor in detecting shiny objects as the targeted specimen. The thru-beam mode, also known as the opposed mode, is the final type of detection of the photoelectric sensor in this list. Thru-beam uses two housings separately, one for the receiver and the other for a transmitter. The light emitted from the transistor will be aimed at the receiver end and when a target interrupts this beam, the receiver's output is activated. Thru-beam mode is the most efficient compared to the other two modes and provides the longest sensing range for the photoelectric sensors. The commonly available mode includes receiver housing and transmitter housing. The laser beam will be placed between these two housings. Another widely used type is a slot and fork type sensor that contains both receiver and transmitter in one housing, therefore no alignment is required.

Zhang and Zhang [37] proposed a design method for a laser-based photoelectric detection sensor that showed better detection stability and higher detection ability than the conventionally used photoelectric detectors. Xin et al. [38] demonstrated an intelligent transplanting system for the potter pepper seedling. This system contained a laser-based photoelectric sensor for the detection of the seedlings. Lv and Luo [39] showed an intelligent vehicle system design which was based on the infrared photoelectric sensor. The data showed that the proposed vehicle could move at high speed and stability on a straight road.

9.3.4 Laser light grid sensor

A laser light sensor uses an emitter that sends parallel beams from a laser light barrier, and then these beams will radiate on a line sensor at the receiver's end. When the object comes in contact with these beams, it will obstruct the laser light barrier, due to which the laser beam from the emitter will not reach the receiver. This obstructed portion of beam is then used to measure the desired details about the specimen. These laser curtains help the obstructing specimen to create a shadow on the receiver side. Both receiver and transmitter can be used in a complete form like a U-shape mounting or may be used as separate components [40]. A line sensor inside the laser grid contains hundreds of light-sensitive cells known as pixels. A high-intensity signal is generated when the laser light falls on the pixels. There is also a chance that a pixel is partially exposed; in these cases, a half intensity signal will be emitted. A sensor processes this signal and then converts it into a video signal used to set a threshold value. Whenever this video signal intersects the threshold value, it demonstrates an edge of the specimen in the laser curtain. One edge will help to identify the position and two edges help to measure the diameter; with three or four edges, we can identify the distance between two objects. The schematic of a laser grid sensor is presented in Figure 9.5.

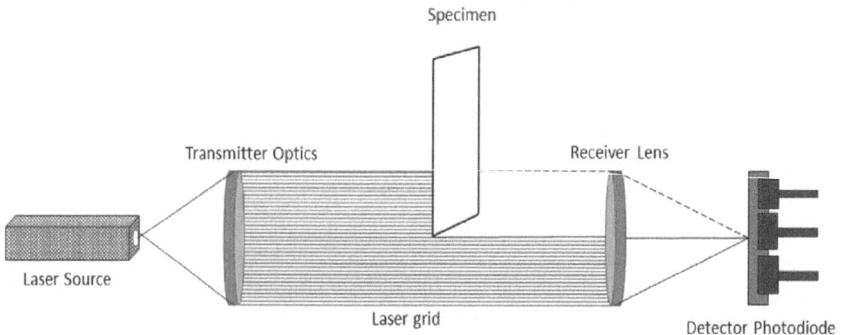

Figure 9.5 Schematic of laser grid sensor.

Laser light grid sensors are very useful for measuring distances or positions between objects. This sensor can commonly be seen in the automotive industry to detect the diameter of the moulds used to produce turbo. The turbo will pass between the receiver and transmitter to control the diameter. The interruption caused by this between the laser curtains aids the sensor in determining the diameter, helping to maintain the quality of the product. Jia et al. [41] proposed a 3D reconstruction method using fusion analysis of an image taken from an object using orthogonal structure grid laser illumination. The grid laser system enables the full view of the specimen in a single shot. This newly developed model was executed on industrial welds, and it was found that an error less than ± 0.09 was yielded for the weld areas. Zang et al. [42] provided a vision sensing method for multi-layer and multi-pass welding using a laser light grid sensor. An algorithm based on grid intersection extraction was designed to determine the earlier weld beads to foresee the next weld path position of the coming layer. The experimental results demonstrated that this method effectively measured a multi-layer multi-scan weld seam. Ma et al. [43] proposed a novel method to determine the obstacles and the parking spaces by utilizing laser device reorganization and visual sensor technology. The laser transmitter produced a laser grid, which changes accordingly to the conditions encountered on the ground. A camera detects this structure and uses it as the area of interest for image processing. The results showed that the proposed method was able to identify the size of the obstacles and parking spaces even in the surroundings where the obstacle and background had the same colour.

9.4 LASERS IN THE BIOMEDICAL SECTOR

Laser technology is being rapidly adopted in modern industries. The ability of lasers to work in a controlled atmosphere and with a wide variety of materials makes them useful for automobile, clothing, aerospace and biomedical industries. Their distinct characteristics like excellent hemostatic effect, low postoperative complications and less pain during the procedure make them an essential tool in the care and intervention sector. The ability of lasers to concentrate energy in a tiny area makes them an optimized tool for cutting through tissues and for precise surgeries. Laser-based diagnostic devices are also quickly making their way into biomedical imaging and research. Medical procedures like dentistry, corneal decrement, wound healing, cosmetic surgeries and nerve stimulation are using lasers for precise and painless procedures [44]. Nowadays, diode lasers have been combined with nanoparticles for use in drug delivery, cancer therapy, diagnostics of cancer cells and biosensing. Lasers in the visible region (430–630 nm), that have absorption in colour-dyes and blood, are currently being used in tattoo removal, phototherapy of oral cancer and retina decrement. The IR region lasers that have a wavelength of 750–1200 nm can penetrate tissue deeper than the visible laser, making them ideal candidates for

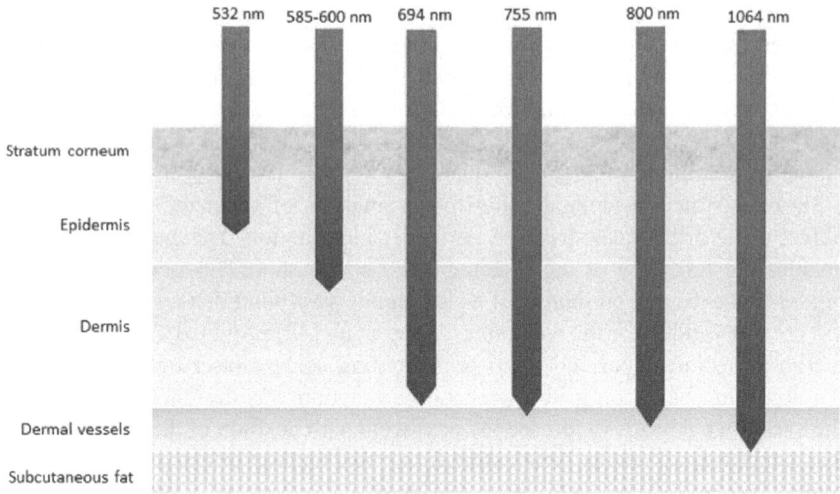

Figure 9.6 Penetration depth of different wavelength lasers in skin.

nano-gold mediated cancer therapy and hair removal [45]. On the other hand, mid-IR lasers with wavelengths of 1.9–3 μm and 9.3–10.6 μm with powerful absorption in tissue and water make them useful for soft and hard ablation type procedures.

The interaction between the laser and tissue involves the transfer of photon energy to the tissues through the process of absorption. The heating of the tissues by the medium of the laser energy should be between 50 °C and 100 °C; otherwise, there will be a disordering of bio-molecules and proteins, which is called photocoagulation. In the region where the laser has operated, a colour change in tissue happens with a loss in the mechanical integrity of that area. This leads to the dying of the tissue in the photocoagulation area, which can be removed or pulled off later. Laser-assisted photocoagulation helps to treat tumours and retinal disorders [46]. There are techniques in which high power density lasers are used, in which the laser heats the tissue above 100 °C, leading to the boiling and evaporation of the water inside it. The change of the tissue into a gaseous state during boiling is called photo-vaporization. This process completely removes the tissue from the surface, which is a suitable option for bloodless incision, skin rejuvenation or resurfacing. Many high-power lasers such as excimer are in the ultraviolet range and can split the chemical bonds without locally heating the tissues. This type of interaction gives rise to photochemical ablation. This process results in a clean incision with a minimal heat-affected zone near the incision wall. The interaction of different wavelength-based lasers with skin levels is shown in Figure 9.6.

9.4.1 Laser in dermatology

Leon Goldman, known as the "father of laser in medicine", introduced the use of laser technology in dermatology in 1963, laying the foundation for unimaginable technological development and innovative clinical potential. Goldman reported the effect of using Maiman's laser for selective destruction of cutaneous pigment structures such as black hairs [47]. He also explained the usage of the ruby laser and Q switched-based devices for pigment lesions and tattoo removal. Mester et al. [48] identified the beneficial aspects of low-energy laser on the hair growth of rats in 1966 and further decided to use the same process to heal pressure ulcers. Finlay et al. [49] used a laser to activate photosensitive substances, resulting in blinding and destroying selective cancer cells. This process is further termed photodynamic therapy. The 1980s and 1990s showed a rapid increase in the use of lasers in dermatology such as laser rejuvenation, laser resurfacing and laser hair removal. The effect of the laser beam on the skin tissue is presented in Figure 9.7.

The laser-tissue interaction is divided into three categories: photodisruption, photothermal and photochemical. Among these, the common type of interaction is photothermal. In this, the tissue absorbs the light which increases the temperature of the region. Depending on the heat generated, different stages of tissue damage occur. Hyperthermia occurs when the laser heat denatures the cellular proteins,

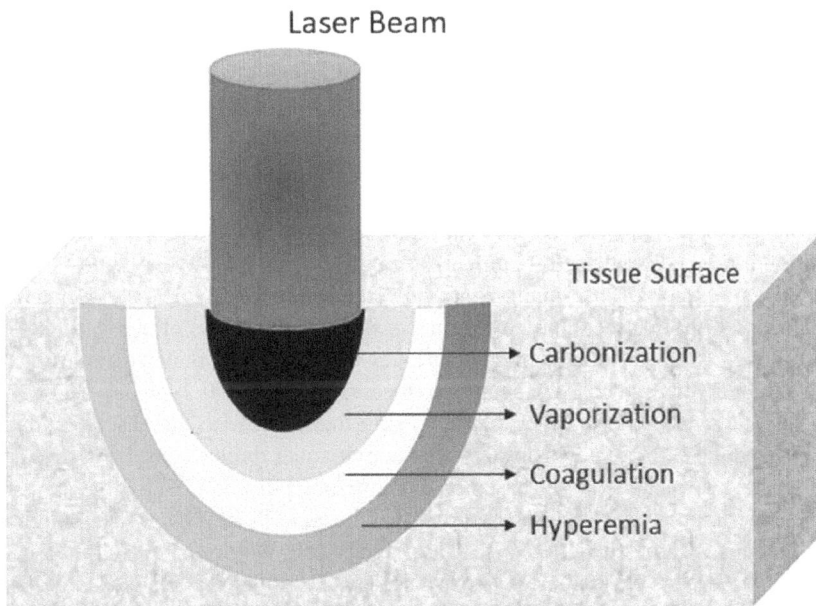

Figure 9.7 Various effects of laser beam interaction with skin tissues.

Table 9.1 Effect of laser-tissue interaction at different temperatures

Temperature (°C)	Thermal effect
37	
> 42	Hyperthermia
> 60	Coagulation
100	Vaporization
> 100	Carbonization
> 300	Melting

which may result in permanent tissue damage. Adding additional heat will result in tissue vaporization, and most surgeons aim to acquire this process. Diffusion allows the cells to absorb energy from the vaporized tissue during the interaction of laser and tissue. Further heating will lead to the carbonization of the tissue which changes the optical properties of the laser-interacted tissue, resulting in an increase in reflectivity, energy absorption and scattering of light. During this phase, the surgeon is doing detrimental tissue damage, and at this point the surgeon should avoid further lasing. Further laser interaction will result in incandescence in which the temperature of the tissue will rise to 350 °C and a visualization of a spark may occur [50]. As the laser-tissue interaction area rises to this temperature, the chances of thermal injury increase, resulting in worse cracking of wound, infection and delay in wound healing. Table 9.1 demonstrates the effect of laser-tissue interaction at different temperatures.

Surgical lasers are the most commonly used lasers among doctors, especially the CO_2 (carbon dioxide) laser. The use of these lasers is common, mainly due to their wavelength (1060 nm), making them beneficial for skin and mucosal diseases. Modern dermatology uses a wide range of laser equipment that can treat cutaneous diseases with safety and efficiency compared to the time of Goldman. Fernández et al. [51] used a combination of laser (405 nm and 639 nm) and ozone (40 ppm or 60 ppm) based treatment for rapid therapy for onychomycosis vitro. This combination of laser and ozone showed a synergistic effect as compared to the ozone-only treatment, and experiments showed that it was effective for seven out of eight studied fungal pathogens. Kislevitz et al. [52] conducted experiments for skin rejuvenation on 12 patients who received single 1470/2940 nm laser treatment. The study reported that the facial skin treatment with this laser treatment improved the skin by increasing vascular activity over three weeks. Silvestri et al. [53] used a Q-switched laser based on a picosecond pulse to treat hypomelanosis. The experimental investigation reported that the superficial melanosis and lighter phototypes responded better to the laser treatment. Wollina and Goldman [54] showed the usage of dual lasers in dermatology. The investigation used two diode lasers with wavelengths of 980 nm and 1470 nm for skin vascular lesions; this resulted in better treatment results and higher patient satisfaction.

9.4.2 Laser in ophthalmology

The development of a functioning ruby laser in 1960 with a wavelength of 694 nm has helped the field of ophthalmic lasers to evolve rapidly. The research in retinal light photocoagulation was started in 1940 by Meyer-Schwickerath with the solar coagulation technique, and soon he developed xenon arc photocoagulation in 1950. Even though the selectivity and precision were good, the method had many limitations. Retinal light photocoagulation was widely recognized but not incorporated into clinical practice because of the difficulty of the technique. Soon after this invention, it was clear that lasers would be better for creating light than xenon-based flash lamps. Rapid research was started on the use of lasers in the ophthalmology field, and many positive outcomes have been observed over the years.

Normally, retinal photocoagulation involves the use of a laser pulse with a varying duration between 10 and 200 ms. The most common laser used in photocoagulation is yellow semiconductor laser (577 nm) and Nd:YAG (532 nm). Laser beam interaction with the cornea helps to reshape its focus and generate a channel inside the eye to release the intraocular pressure of glaucoma [55]. Haemoglobin, retinal pigment epithelium (RPE) and choroid are the primary elements that absorb the laser energy. In the RPE, heat is absorbed by the irradiation of a laser, resulting in the diffusion of choroid into the retina and causing coagulation of photoreceptors and inner retina. Applying a 100 ms pulse laser causes the heat to diffuse up to 200 mm, thus smoothening the edge and expanding the coagulated zone above the laser spot, known as "thermal blooming". Utilizing a shorter pulse for heat diffusion and lesser spot can avoid or minimize retinal damage [56]. The acute retinal lesion diameter increases logarithmically with the increase in pulse duration and increases linearly with the increase in laser power.

In contrast, threshold power needs to be increased with a decrease in the pulse duration to create a retinal lesion. A duration of 2 ms or less is not feasible to create a visible lesion of moderate grade without damaging the retina; therefore, a pulse duration in this range should be avoided to generate visible lesions. The proliferation and RPE migration help reinstate the continuity of retinal pigment epithelium monolayer within one week compared to the damage caused by full thickness with 100 ms laser pulses. Initially, the damaged zone was filled up with glia (a connecting tissue of the nervous system) in the photoreceptor layer. After some time, photoreceptors move to the damaged zone from the adjacent retina, decreasing the size of the damaged zone. Small lesions with no damage in the retinal area can be refilled by photoreceptors and help to rewire the bipolar cells with time. This will help to restore the retinal structure and avoid the neuronal loss which is accompanied by long-duration retinal burns.

In modern-day eye treatments, lasers are increasingly being used due to their precision and ability to perform the required treatment quickly. Modern surgeries such as laser-assisted *in situ* keratomileusis are very beneficial for patients who require a correction in their vision. The vision is corrected either immediately or the day after surgery with almost no pain. The major benefit of using this type of surgery is that no badges or stitches are required and as the patient ages, more adjustments could be made to correct the vision. During the past decade, a rapid increase in research has been observed on the utilization of lasers in ophthalmology. De Cillà et al. [57] demonstrated a proper mechanism of subthreshold micro-pulse laser (SMPL) and used this mechanism for the retina, restoring the oxidant/antioxidant balance and counteracting the programmed form of cell in the eyes of mice. Subramaniam et al. [58] reported that micro-pulse transscleral cyclophotocoagulation was a better alternative to the traditional glaucoma filtration surgery for intraocular pressure reduction in keratoplasty eyes. A series of experiments were conducted on patients. The results showed that the patients who received subthreshold micro-pulse laser surgery showed better results than those who received glaucoma filtration surgery.

9.4.3 Laser in dentistry

Theodore Maiman was the first to introduce lasers in dentistry in 1960, and since then, there has been extensive research in the field of dental practice. There are two types of lasers, based on the source: hard lasers and soft lasers. In hard lasers, Er:YAG, CO2 and Nd: YAG are commonly used as a source for the generation of the laser beam. Semiconductor-based diode devices are used for soft tissue treatments because they are cheap and compact. This makes them widely used in low-level laser therapies and biostimulation applications. The laser being efficient, specific, low cost and comfortable to use makes it superior to the traditional techniques used in dental practices.

An argon laser with a wavelength of 488 nm and 514 nm is common in dentistry because it is poorly absorbed by the dentin and enamel [59]. Due to this poor absorption during sculpting and cutting, there will be minimal interaction between the laser and the hard tissue, resulting in minimal damage to the tooth surface. During the interaction between the argon laser and the decayed tooth, the decayed area changes its colour to dark orange-red, which helps to discriminate it from the healthy surrounding area. Er: YAG laser proved efficient and safe to execute the treatment without patient discomfort and tooth structure damage. Laser irradiation over the tissue enables the operator (clinician) to collect a signal from bacteria present in the infected dentin. The control over the pulse laser helps to obtain an automated-based decay removal. The diode and Nd:YAG are suitable for soft tissue operations. Photodynamic therapy, which is a photochemical reaction based on the laser, is suitable for treating the malignancies of the oral mucosa. Photodynamic therapy helps to kill the tumour cells and encourages the generation of anti-tumour immunity by activating T lymphocytes and macrophages. Photodynamic therapy is useful in carcinoma *in situ* and squamous cell carcinoma treatments in dentistry.

Recently, the use of lasers in dentistry has increased drastically, mainly because of the easy-to-control parameters and patient-specific treatments. Researchers are continuing to make laser-assisted dentistry treatments more feasible for a wide range of problems in the oral areas. Garrocho-Rangel et al. [60] conducted a literature survey for better treatment of ankyloglossia. The survey reported that laser-based surgery treatments for lingual frenectomy showed much efficiency and comfortability with the pediatric dentist and child as compared to the conventional blade-based methods. Dobrzański and Dobrzański [61] conducted a review of the current trends and technical advancements in dental prosthetics. The authors reported a digital approach in clinical procedures and technological development in implant preparation with the help of additive manufacturing, computer-aided design, selective laser sintering and other manufacturing procedures.

9.5 LASERS IN THE MANUFACTURING AND DEFENCE SECTOR

Over the years, humans have witnessed the role of technology in the evolution of the manufacturing industry. We have seen rapid growth in this industry from the Stone Age to the Electric Age. The Electric Age marked the latest stage of maturity, and now it's the Optical Age that is poised to bloom. The last few decades have shown that optical technology has become a vital tool in manufacturing. Laser, being the frontline in this optics technology, showed that due to its concentrated heat source, it could join, form and build various products irrespective of the material type used. Industries are adopting laser technology to perform various tasks such as the modification of surface roughness, cutting metals, measuring part dimensions and welding metals with polymers [62]. They have been prevalent in metals and EV industries for over two decades. Laser machining can design exquisite features in a product that would be impossible or difficult to produce using conventional equipment. The laser cutting machine is becoming a go-to cutting tool for the metal and plastic industrial sectors as it produces a clean cut without producing any burs, therefore eliminating the need for additional finishing steps. Lasers are one of the key technologies of Industry 4.0, and research is ongoing to use them more efficiently in manufacturing processes [63]. In 2018, the National Institute of Standards and Technology constructed a laser in which the pulse lasted a quadrillionth of a second, which is 100 times faster than traditional ultrafast lasers. A group of scientists in Germany are working on the integration of tiny lasers with silicon chips to improve processing speed. Artificial intelligence (AI) is also being used to develop smart lasers that can understand the material and calculate the processing time accordingly. Strumpf, a German manufacturer, is working on the development of a laser that utilizes AI technology to understand the best welding points for manufacturing copper coils for the automobile industry. In the current manufacturing sector, many companies are accepting additive manufacturing, AI technology, Industry 4.0 and sensor technology, and laser technologies will be playing an ever-expanding part in the modern era of manufacturing. The utilization of lasers in modern industries is discussed in the following sections.

9.5.1 Laser in welding

In the welding industry, lasers provide deep and narrow weld lengths at high speed, making them an optimal tool for high-volume productions. In sectors such as automotive, aeronautics and marine where rapid production rates are required, lasers are being integrated with robotic systems for smoother and higher-quality productions [64]. A laser generates two types of beams depending on the application – continuous and pulsed beams. In laser welding, a high-power intensity beam is generated, providing a small heat-affected area with high heating and cooling rates. Generally, the penetration depth is directly dependent on the power supplied by the generator, but there are instances where it depends on the location of focal length; that is, more penetration will be achieved when the focal length is below the material surface. Laser welding can produce deep weld beads in a vacuum environment, giving it the advantage over traditional arc and plasma welding technology [65].

There are two types of welding available in the current industry: direct laser welding and laser transmission welding. Direct laser welding joins the material through heat generated by the laser at the outer surface of the specimen. This type of laser technique is compatible with various materials, including metals, alloys, ceramics and plastics. Direct laser welding has proved to be an efficient tool for joining metal and polymer joints (metal-polymer). Wang et al. [66] studied the effect of laser power and velocity on the weld strength of 5182 Al alloy and glass fiber reinforced PA66 plate. The tensile strength showed that the weld strength increases at first but then decreases with an increase in velocity and laser power. Zou et al. [67] analyzed the joining mechanism of Ti-6Al-4V alloy and PEEK-CFRTP by direct laser welding. The results showed that the shear strength could reach near 56.3 kN/mm. Characterization showed that "anchor-shaped" structures provided the mechanical bonding effect. Zhou et al. [68] developed a novel ultrasonic vibration assisted laser welding process to eliminate the defects that occur in the laser welding of dissimilar metals for nuclear reactor pump end sealing. The assistance of ultrasonic vibration has decreased the width of the unmixed zone with the homogenized distribution of macro elements because of the enhanced molten convention. Laser transmission welding is an innovative, non-contact, highly flexible, fast and automated joining process that provides consistently high-quality products [69]. In the current thermoplastics industry, laser transmission welding has established its dominance over conventional welding methods [70].

Kumar et al. [71] investigated the parameter optimization and mathematical modelling of wobble laser transmission welding of acrylic and polycarbonate sheets using a low-power laser system. Goyal et al. [72] provided a novel integrated approach to predict the temperature at the interface of the laser transmission welding joint and the absorptivity of the absorber. The developed numerical model was in good agreement with the experimental results with an average error of less than 6%. The selection of optimized process parameters is essential to obtain a high-quality joint in laser transmission welding. Goyal et al.

optimized the process parameters up to 36.32% by applying a genetic algorithm for laser transmission welding of polycarbonate sheet by using iron powder absorber [73].

9.5.2 Laser in forming

Laser forming has become a contemporary tool used to mould metal sheets into the desired shape without heavy die or external force. Laser forming is a thermomechanical process where a controlled defocused beam is used to change the shape of the workpiece [74]. The advantage of using laser forming is that it does not provide any spring back effect, thus can be used to generate complex shapes with optimized irradiation strategies [75–78]. The mechanism utilizes a high-intensity laser to irradiate the workpiece, thus increasing the temperature in the irradiated and surrounding areas. The temperature increase generates thermal stresses in the irradiated region, giving rise to the bending of the sheet [79]. In the current world, where customized products, prototyping and low-volume production-centric industries exist, laser forming is a sustainable tool for these industries. In industries like aerospace and semiconductor, where a small error or change in shape could force the manufacturer to replace the whole working setup, laser-based forming can be used as a correction tool and can save a substantial amount of money [80]. As the laser consumes a high amount of energy to complete a task, it becomes necessary to optimize process parameters before the actual process. This opened a wide area of scope for research in the laser forming field. Researchers have witnessed various numerical and analytical models over decades to predict the bend angle [81–83]. Nair et al. [84] developed an analytical model to predict the bending of mild steel. The experimental results showed good agreement with the analytical results, with an absolute error of 9.51%. Kant and Joshi [85, 86] reported the effect of scan speed on difficult-to-form magnesium M1A alloy material used commonly for tool steel. The experiments provided an insight into process conditions and mechanisms to obtain maximum bend angle with minimum edge effect. Yadav et al. [87] investigated the effect of line energy on edge effect, microstructure, bend angle and mechanical properties of duplex stainless steel used for shipbuilding. Results showed that with constant line energy, the bend angle increased with laser power and scanning speed. Furthermore, Yadav et al. [88–90] explored the effect of different cooling conditions along with line energy and reported that at low line energy, the forced cooling did not affect the bend angle but significantly reduced the cooling time and heat affected zone, whereas it significantly improved the bend angle for both single and multiscan laser bending [91, 92]. Kant and Joshi [93] explored laser bending with the assistance of moving mechanical load and reported that the bend angle could be notably enhanced with the assistance of mechanical load. Pal et al. [94] attempted to enhance the bending angle with the application of electro-magnetic force and reported significant improvement in the bend angle.

9.5.3 Laser in machining

Machining processes such as lathe machining, milling and drilling are some of the most common processes used in manufacturing industries. Hybrid machining is one such operation that has been developed to overcome the flaws of conventional machining operations such as the usage of cutting fluids, poor machinability and high cost. A combination of energy sources such as magnetic power, heat, vibration and pressure fluids can be integrated into a single machine to operate various machining operations simultaneously [95–98]. This integration provides superior machining performance compared to conventional machining. Deswal and Kant [99] investigated the effect of laser and ultrasonic vibration on chips, machining temperature, microstructure and surface roughness during the turning operation of magnesium alloy. The results showed that low tool wear, fine microstructure, low machining forces, low surface roughness, high microhardness and machining temperature were obtained during ultrasonic vibration-laser assisted turning compared to laser-assisted, ultrasonic-vibration assisted and conventional turning. Besides, ultrasonic vibration-laser assisted turning was also utilized to machine aluminium alloys by varying ultrasonic vibration directions [100–102]. Przestacki [103] investigated the influence of laser assisted machining on difficult-to-machine A359/20SiCp composite material used for aerospace. The results demonstrated that the tool wear was significantly lower in laser-assisted turning than in conventional operation.

9.5.4 Laser in the defence sector

Lasers have rapidly made their way into the defence sector in recent years. Security and military operations often require the transfer of timely and secure information from one place to another. In earlier days, effective communication in the military heavily relied on the radio spectrum, which is susceptible to electromagnetic interference and security threats. Another flaw with this spectrum is that it needs to be hard pressed to be synchronized with the bandwidth required for current real-time transfer, high-resolution images and on-air video conferencing [104]. Due to these reasons, research has shifted to the infrared and visible spectrums. Laser is one such technology which is immune to electromagnetic interference. The broad classification of lasers and their increasing maturity have opened many gates for their usage for military purposes. Modern-day laser technologies are changing the warfare model by being utilized in data relays, weather regulators, sensing devices, active lighting, directed energy weapons and target designators. The defence sector is accepting laser weapons due to their various advantages over conventional weapon systems. Firstly, laser weapons can transmit at the speed of light, allowing them to destroy distant targets just after their detection. Secondly, laser deployments can flexibly tune the gradient effect to customize the range of results to disruptive, fatal and non-fatal outcomes.

Finally, laser energy maintains a low profile, offers precise point target selections and provides low collateral damage, making it a valuable tool for covert operations. Although the initial setup cost is high for laser-based systems, they provide cost-effective engagement after deployment [105].

While this field is gaining traction in defence applications, it has not yet reached its full maturity, leaving opportunities for research to address its limitations. Laser, being a high-energy system, requires an effective cooling system which could maintain the temperature during repetitive transmission cycles. Furthermore, improving the accuracy of laser transmission under jamming or during naturally occurring unfavourable situations such as hail storms, sand storms, humidity and heavy smoke are some important areas for research.

9.6 SUMMARY

This chapter provides insight into the utilization of laser technology in the academic, medical, manufacturing and defence sectors. The chapter describes the challenges, advantages, disadvantages and current innovations of each field. Literature showed that lasers have become an essential tool irrespective of field in the modern era. Its monochromatic, coherent, minimum HAZ and flexibility in operation make it a versatile tool for almost every industry. Laser has opened many gates for academicians and researchers to understand material behaviour and its structure. In the current scenario, many material characterizations have been dependent on laser technology, providing a better path of understanding. The medical field has rapidly grown in the last few decades, aiding in the treatment of various diseases and making medical procedures less painful for patients. Laser is one such technology that allowed the medical industry to grow at this speed. Modern-day laser equipment and treatments are providing many patients with a speedy recovery and a permanent solution for their suffering. The manufacturing sector has evolved over the years, providing the foundation for any industry to grow. A strong manufacturing sector represents a country's economic growth and urbanization. Laser have extended their applications to nearly every aspect of production, including cutting, welding, surface finishing, machining and cleaning. The beauty of this technology is that almost no post-processing is required after laser operations, providing cost-effectiveness and high-quality efficiency to industries. The defence sector is also adapting lasers for more secure and covert operations. Many countries have recently developed laser weapons that could destroy distant targets with minimal collateral damage. High-intensity laser weapons are now one of the most prominent weapons in the defence sector. Laser has contributed to the growth of the defence sector by enhancing communication, transmission and tracking.

Still, laser technology is not a fully matured process due to its low suitability for mass production, high energy consumption and cost of operations. Industry 4.0 and cloud storage systems are the next big step in manufacturing, and laser technology has a bright future in the evolving manufacturing sector with these smart technologies. Smart lasers will redefine the roles of factories, customers, surgeons and designers in future manufacturing sectors.

REFERENCES

[1] L. Radziemski and D. Cremers, "A brief history of laser-induced breakdown spectroscopy: From the concept of atoms to LIBS 2012," *Spectrochim. Acta B At. Spectrosc.*, vol. 87, pp. 3–10, 2013, doi: 10.1016/j.sab.2013.05.013.

[2] S. D. Jackson, "Towards high-power mid-infrared emission from a fibre laser," *Nat. Photonics*, vol. 6, no. 7, pp. 423–431, 2012, doi: 10.1038/nphoton.2012.149.

[3] P. L. Hanst, "Infrared spectroscopy and infrared lasers in air pollution research and monitoring," *Appl. Spectrosc.*, vol. 24, no. 2, pp. 161–174, 1970, doi: 10.1366/000370270774371930.

[4] K. Yu, J. Ren, and Y. Zhao, "Principles, developments and applications of laser-induced breakdown spectroscopy in agriculture: A review," *Artif. Intell. Agric.*, vol. 4, pp. 127–139, 2020, doi: 10.1016/j.aiia.2020.07.001.

[5] M. Moll, M. Schlanges, T. Bornath, and V. P. Krainov, "Inverse bremsstrahlung heating beyond the first born approximation for dense plasmas in laser fields," *New J. Phys.*, vol. 14, no. 6, p. 065010, 2012. doi:10.1088/1367-2630/14/6/065010.

[6] Y. T. Fu, W. L. Gu, Z. Y. Hou, S. A. Muhammed, T. Q. Li, Y. Wang, and Z. Wang, "Mechanism of signal uncertainty generation for laser-induced breakdown spectroscopy," *Front. Phys.*, vol. 16, no. 2, 2021, doi: 10.1007/s11467-020-1006-0.

[7] K. Liu, D. Tian, C. Li, Y. Li, G. Yang, and Y. Ding, "A review of laser-induced breakdown spectroscopy for plastic analysis," *TrAC—Trends Anal. Chem.*, vol. 110, pp. 327–334, 2019, doi: 10.1016/j.trac.2018.11.025.

[8] R. Yuan, Y. Tang, Z. Zhu, Z. Hao, J. Li, H. Yu, Y. Yu, L. Guo, X. Zeng, and Y. Lu, "Accuracy improvement of quantitative analysis for major elements in laser-induced breakdown spectroscopy using single-sample calibration," *Anal. Chim. Acta*, vol. 1064, pp. 11–16, 2019, doi: 10.1016/j.aca.2019.02.056.

[9] Q. Wang, A. Chen, H. Qi, S. Li, Y. Jiang, and M. Jin, "Femtosecond laser-induced breakdown spectroscopy of a preheated Cu target," *Opt. Laser Technol.*, vol. 121, no. August 2019, p. 105773, 2020, doi: 10.1016/j.optlastec.2019.105773.

[10] Y. Zhang, C. Sun, L. Gao, Z. Yue, S. Shabbir, W. Xu, M. Wu, and J. Yu, "Determination of minor metal elements in steel using laser-induced breakdown spectroscopy combined with machine learning algorithms," *Spectrochim. Acta Part B At. Spectrosc.*, vol. 166, no. November 2019, p. 105802, 2020, doi: 10.1016/j.sab.2020.105802.

[11] J. F. McClelland, "Photoacoustic spectroscopy," *Anal. Chem.*, vol. 55, no. 1, pp. 89–105, 1983, doi: 10.1021/ac00252a003.

[12] G. A. West, J. J. Barrett, D. R. Siebert, and K. V. Reddy, "Photoacoustic spectroscopy," *Rev. Sci. Instrum.*, vol. 54, no. 7, pp. 797–817, 1983, doi: 10.1063/1.1137483.

[13] A. Miklós, P. Hess, and Z. Bozóki, "Application of acoustic resonators in photoacoustic trace gas analysis and metrology," *Rev. Sci. Instrum.*, vol. 72, no. 4, pp. 1937–1955, 2001, doi: 10.1063/1.1353198.

[14] W. Markus, "Electronic spectroscopy vibrational, rotational & raman properties," *Photoacoust. Spectrosc., Appl.*, pp. 1800–1809, 1997.

[15] J. T. Friedlein, E. Baumann, K. A. Briggman, G. M. Colacion, F. R. Giorgetta, A. M. Goldfain, D. I. Herman, E. V. Hoenig, J. Hwang, N. R. Newbury, E. F. Perez, C. S.

Yung, I. Coddington, and K. C. Cossel, "Dual-comb photoacoustic spectroscopy," *Nat. Commun.*, vol. 11, no. 1, 2020, doi: 10.1038/s41467-020-16917-y.

[16] P. K. Krivoshein, D. S. Volkov, O. B. Rogova, and M. A. Proskurnin, "FTIR photoacoustic spectroscopy for identification and assessment of soil components: Chernozems and their size fractions," *Photoacoustics*, vol. 18, no. December 2019, p. 100162, 2020, doi: 10.1016/j.pacs.2020.100162.

[17] H. Lv, H. Zheng, Y. Liu, Z. Yang, Q. Wu, H. Lin, B. A. Z Montano, W. Zhu, J. Yu, R. Kan, Z. Chen, and F. K. Tittel, "Radial-cavity quartz-enhanced photoacoustic spectroscopy," *Optics Letters,* vol. 46, no. 16, pp. 3917–3920, 2021.

[18] G. S. Bumbrah and R. M. Sharma, "Raman spectroscopy—basic principle, instrumentation and selected applications for the characterization of drugs of abuse," *Egypt. J. Forensic Sci.*, vol. 6, no. 3, pp. 209–215, 2016, doi: 10.1016/j.ejfs.2015.06.001.

[19] M. D. Duncan, J. Reintjes, and T. J. Manuccia, "Scanning coherent anti-Stokes Raman microscope," *Opt. Lett.*, vol. 7, no. 8, p. 350, 1982, doi: 10.1364/ol.7.000350.

[20] R. R. Jones, D. C. Hooper, L. Zhang, D. Wolverson, and V. K. Valev, "Raman techniques: Fundamentals and frontiers," *Nanoscale Research Letters*, vol. 14, no. 1, 2019. doi:10.1186/s11671-019-3039-2.

[21] C. Berghian-Grosan and D. A. Magdas, "Raman spectroscopy and machine-learning for edible oils evaluation," *Talanta*, vol. 218, no. April, p. 121176, 2020, doi: 10.1016/j.talanta.2020.121176.

[22] W. Lu, X. Chen, L. Wang, H. Li, and Y. V. Fu, "Combination of an artificial intelligence approach and laser tweezers Raman spectroscopy for microbial identification," *Anal. Chem.*, vol. 92, no. 9, pp. 6288–6296, 2020, doi: 10.1021/acs.analchem.9b04946.

[23] H. Lu, S. Tian, L. Yu, X. Lv, and S. Chen, "Diagnosis of hepatitis B based on Raman spectroscopy combined with a multiscale convolutional neural network," *Vib. Spectrosc.*, vol. 107, no. November 2019, p. 103038, 2020, doi: 10.1016/j.vibspec.2020.103038.

[24] H. Schwenke, W. Knapp, H. Haitjema, A. Weckenmann, R. Schmitt, and F. Delbressine, "Geometric error measurement and compensation of machines-an update," *CIRP Ann. Manuf. Technol.*, vol. 57, no. 2, pp. 660–675, 2008, doi: 10.1016/j.cirp.2008.09.008.

[25] R. G. Dorsch, G. Häusler, and J. M. Herrmann, "Laser triangulation: fundamental uncertainty in distance measurement," *Appl. Opt.*, vol. 33, no. 7, p. 1306, 1994, doi: 10.1364/ao.33.001306.

[26] J. Sun, J. Zhang, Z. Liu, and G. Zhang, "A vision measurement model of laser displacement sensor and its calibration method," *Opt. Lasers Eng.*, vol. 51, no. 12, pp. 1344–1352, 2013, doi: 10.1016/j.optlaseng.2013.05.009.

[27] K. B. Smith and Y. F. Zheng, "Multi-laser displacement sensor used in accurate digitizing technique," *J. Manuf. Sci. Eng. Trans. ASME*, vol. 116, no. 4, pp. 482–490, 1994, doi: 10.1115/1.2902132.

[28] B. Sun and B. Li, "Laser displacement sensor in the application of aero-engine blade measurement," *IEEE Sens. J.*, vol. 16, no. 5, pp. 1377–1384, 2016, doi: 10.1109/JSEN.2015.2497363.

[29] P. Giri and S. Kharkovsky, "Detection of surface crack in concrete using measurement technique with laser displacement sensor," *IEEE Trans. Instrum. Meas.*, vol. 65, no. 8, pp. 1951–1953, 2016, doi: 10.1109/TIM.2016.2541358.

[30] J. Sandak and C. Tanaka, "Evaluation of surface smoothness by laser displacement sensor 1: Effect of wood species," *J. Wood Sci.*, vol. 49, no. 4, pp. 305–311, 2003, doi: 10.1007/s10086-002-0486-6.

[31] L. Li, "Time-of-flight camera–an introduction," *Texas Instruments—Tech. White Pap.*, no. January, p. 10, 2014.

[32] T. M. A. Zulcaffle, F. Kurugollu, D. Crookes, A. Bouridane, and M. Farid, "Frontal view gait recognition with fusion of depth features from a time of flight camera," *IEEE Trans. Inf. Forensics Secur.*, vol. 14, no. 4, pp. 1067–1082, 2019, doi: 10.1109/TIFS.2018.2870594.

[33] S. Kumar, C. Savur, and F. Sahin, "Dynamic awareness of an industrial robotic arm using time-of-flight laser-ranging sensors," *Proc. 2018 IEEE Int. Conf. Syst. Man, Cybern. SMC 2018*, pp. 2850–2857, 2019, doi: 10.1109/SMC.2018.00485.

[34] B. Langmann, K. Hartmann, and O. Loffeld, "Increasing the accuracy of time-of-flight cameras for machine vision applications," *Comput. Ind.*, vol. 64, no. 9, pp. 1090–1098, 2013, doi: 10.1016/j.compind.2013.06.006.

[35] S. Tsuji and T. Kohama, "Proximity skin sensor using time-of-flight sensor for human collaborative robot," *IEEE Sens. J.*, vol. 19, no. 14, pp. 5859–5864, 2019, doi: 10.1109/JSEN.2019.2905848.

[36] Z. Ruirui and G. Xiao, "The application and development of photoelectric sensor," *Energ. Procedia*, vol. 17, pp. 1304–1308, 2012, doi: 10.1016/j.egypro.2012.02.243.

[37] X. Zhang, H. Li, and S. Zhang, "Design and analysis of laser photoelectric detection sensor," *Microw. Opt. Technol. Lett.*, vol. 63, no. 12, pp. 3092–3099, 2021, doi: 10.1002/mop.33011.

[38] J. Xin, Z. Kaixuan, J. Jiangtao, D. Xinwu, M. Hao, and Q. Zhaomei, "Design and implementation of Intelligent transplanting system based on photoelectric sensor and PLC," *Futur. Gener. Comput. Syst.*, vol. 88, pp. 127–139, 2018, doi: 10.1016/j.future.2018.05.034.

[39] X. Lv and P. Luo, "A design of autonomous tracing in intelligent vehicle based on photoelectric sensor," *Yadian Yu Shengguang/Piezoelectrics Acoustooptics*, vol. 33, no. 6, pp. 939–942, 2011.

[40] R. Singh and K. S. Nagla, "Development of an efficient laser grid mapping technique: P-SLAM," *Int. J. Image Data Fusion*, vol. 10, no. 3, pp. 177–198, 2019, doi: 10.1080/19479832.2019.1625449.

[41] N. Jia, Z. Li, J. Ren, Y. Wang, and L. Yang, "A 3D reconstruction method based on grid laser and gray scale photo for visual inspection of welds," *Opt. Laser Technol.*, vol. 119, no. August 2018, p. 105648, 2019, doi: 10.1016/j.optlastec.2019.105648.

[42] C. Zhang, H. Li, Z. Jin, and H. Gao, "Seam sensing of multi-layer and multi-pass welding based on grid structured laser," *Int. J. Adv. Manuf. Technol.*, vol. 91, no. 1–4, pp. 1103–1110, 2017, doi: 10.1007/s00170-016-9733-7.

[43] S. Ma, Z. Jiang, H. Jiang, M. Han, and C. Li, "Parking space and obstacle detection based on a vision sensor and checkerboard grid laser," *Appl. Sci.*, vol. 10, no. 7, 2020, doi: 10.3390/app10072582.

[44] E. Mester, A. F. Mester, and A. Mester, "The biomedical effects of laser application," *Lasers Surg. Med.*, vol. 5, no. 1, pp. 31–39, 1985, doi: 10.1002/lsm.1900050105.

[45] J.-T. Lin, "Progress of medical lasers: Fundamentals and applications," *Med. Devices Diagnostic Eng.*, vol. 1, no. 2, pp. 36–41, 2016, doi: 10.15761/mdde.1000111.

[46] L. A. Everett and Y. M. Paulus, "Laser therapy in the treatment of diabetic retinopathy and diabetic macular edema," *Curr. Diab. Rep.*, vol. 21, no. 9, 2021, doi: 10.1007/s11892-021-01403-6.

[47] S. Gianfaldoni, G. Tchernev, U. Wollina, M. Fioranelli, M. G. Roccia, R. Gianfaldoni, and T. Lotti, "An overview of laser in dermatology: The past, the present and . . . the future (?)," *Open Access Maced. J. Med. Sci.*, vol. 5, no. 4 Special Issue Global Dermatology, pp. 526–530, 2017, doi: 10.3889/oamjms.2017.130.

[48] E. Mester, A. F. Mester, and A. Mester, "The biomedical effects of laser application," *Lasers Surg. Med.*, vol. 5, no. 1, pp. 31–39, 1985, doi: 10.1002/lsm.1900050105.

[49] J. C. Finlay, K. Cengel, T. M. Busch, and T. C. Zhu, "Photodynamic therapy," *Handb. Biomed. Opt.*, vol. 90, no. 12, pp. 733–750, 2016.

[50] M. Boord, "Laser in dermatology," *Clin. Tech. Small Anim. Pract.*, vol. 21, no. 3, pp. 145–149, 2006, doi: 10.1053/j.ctsap.2006.05.007.

[51] J. Fernández, I. del Valle Fernández, C. J. Villar, and F. Lombó, "Combined laser and ozone therapy for onychomycosis in an in vitro and ex vivo model," *PLoS One*, vol. 16, no. 6, pp. 1–13, 2021, doi: 10.1371/journal.pone.0253979.

[52] M. Kislevitz, K. B. Lu, C. Wamsley, J. Hoopman, J. Kenkel, and Y. Akgul, "Novel use of non-invasive devices and microbiopsies to assess facial skin rejuvenation following laser treatment," *Lasers Surg. Med.*, vol. 52, no. 9, pp. 822–830, 2020, doi: 10.1002/lsm.23233.

[53] M. Silvestri, L. Bennardo, E. Zappia, F. Tamburi, N. Cameli, G. Cannarozzo, and S. P Nistico, "Q-switched 1064/532 nm laser with picosecond pulse to treat benign hyperpigmentations: A single-center retrospective study," *Appl. Sci.*, vol. 11, no. 16, 2021, doi: 10.3390/app11167478.

[54] U. Wollina and A. Goldman, "The dual 980-nm and 1470-nm diode laser for vascular lesions," *Dermatol. Ther.*, vol. 33, no. 4, 2020, doi: 10.1111/dth.13558.

[55] K. P. Thompson and Q. S. Ren, "Therapeutic and diagnostic application of lasers in ophthalmology," *Lasers Med.*, vol. 80, no. 6, pp. 211–245, 2001, doi: 10.1201/9781420040746.ch8.

[56] G. Dorin, "Evolution of retinal laser therapy: Minimum intensity photocoagulation (MIP). Can the laser heal the retina without harming it?," *Semin. Ophthalmol.*, vol. 19, no. 1–2, pp. 62–68, 2004, doi: 10.1080/08820530490884173.

[57] S. De Cillà, D. Vezzola, S. Farruggio, S. Vujosevic, N. Clemente, G. Raina, D. Mary, G. Casini, L. Rossetti, L. Avagliano, C. Martinelli, G. Bulfamante, and E. Grossini, "The subthreshold micropulse laser treatment of the retina restores the oxidant/antioxidant balance and counteracts programmed forms of cell death in the mice eyes," *Acta Ophthalmol.*, vol. 97, no. 4, pp. e559–e567, 2019, doi: 10.1111/aos.13995.

[58] K. Subramaniam, M. O. Price, M. T. Feng, and F. W. Price, "Micropulse transscleral cyclophotocoagulation in keratoplasty eyes," *Cornea*, vol. 38, no. 5, pp. 542–545, 2019, doi: 10.1097/ICO.0000000000001897.

[59] C. Mercer, "Lasers in dentistry: A review. Part 1," *Dent. Update*, vol. 23, no. 2, pp. 74–80, 1996.

[60] A. Garrocho-Rangel, D. Herrera-Badillo, I. Pérez-Alfaro, V. Fierro-Serna, and A. Pozos-Guillén, "Treatment of ankyloglossia with dental laser in paediatric patients: Scoping review and a case report," *Eur. J. Paediatr. Dent.*, vol. 20, no. 2, pp. 155–163, 2019, doi: 10.23804/ejpd.2019.20.02.15.

[61] L. A. Dobrzański and L. B. Dobrzański, "Dentistry 4.0 concept in the design and manufacturing of prosthetic dental restorations," *Processes*, vol. 8, no. 5, 2020, doi: 10.3390/PR8050525.

[62] S. C. Dass, *Interaction of Lasers with Plastics and Other Materials*. Plastics Design Library, 1999.

[63] Y. C. Shin, B. Wu, S. Lei, G. J. Cheng, and Y. Lawrence Yao, "Overview of laser applications in manufacturing and materials processing in recent years," *J. Manuf. Sci. Eng. Trans. ASME*, vol. 142, no. 11, 2020, doi: 10.1115/1.4048397.

[64] L. S. Bass and M. R. Treat, "Laser tissue welding: A comprehensive review of current and future," *Lasers Surg. Med.*, vol. 17, no. 4, pp. 315–349, 1995, doi: 10.1002/lsm.1900170402.

[65] A. Mahrle and E. Beyer, "Hybrid laser beam welding—classification, characteristics, and applications," *J. Laser Appl.*, vol. 18, no. 3, pp. 169–180, 2006, doi: 10.2351/1.2227012.

[66] C. Wang, G. Zhang, Q. Zhu, H. Yang, C. Yang, and Y. Liu, "Mechanism and numerical simulation analysis of laser welding 5182 aluminum alloy/PA66 based on surface texture treatment," *Opt. Laser Technol.*, vol. 153, 2022, doi: 10.1016/j.optlastec.2022.108273.

[67] P. Zou, H. Zhang, M. Lei, and D. Cheng, "Formation mechanism of direct laser welding joint between CFRTP and Ti-6Al-4V," *Mater. Lett.*, vol. 325, no. April, p. 132707, 2022, doi: 10.1016/j.matlet.2022.132707.

[68] S. Zhou, B. Wang, D. Wu, G. Ma, G. Yang, and W. Wei, "Follow-up ultrasonic vibration assisted laser welding dissimilar metals for nuclear reactor pump can end sealing," *Nucl. Mater. Energy*, vol. 27, no. March, p. 100975, 2021, doi: 10.1016/j.nme.2021.100975.

[69] D.K. Goyal, R. Yadav, and R. Kant, "Study of temperature field considering gradient volumetric heat absorption in transparent sheet during laser transmission welding (LTW)," *Lasers Eng.*, vol. 53, pp. 215–229, 2022.

[70] D.K. Goyal and R. Kant, "Mechanism of bonding during laser transmission welding using EIP absorber," *Mater. Manuf. Process.*, vol. 38, pp. 485–493, 2023. https://doi.org/10.1080/10426914.2023.2165676.

[71] D. Kumar, S. Paitandi, A. S. Kuar, and D. Bose, "Experimental investigation on laser transmission welding of polycarbonate and acrylic," *Mach. Learn. Appl. Non-Conventional Mach. Process.*, no. January, pp. 160–180, 2021, doi: 10.4018/978-1-7998-3624-7.ch010.

[72] D. K. Goyal, R. Yadav, and R. Kant, "An integrated hybrid methodology for estimation of absorptivity and interface temperature in laser transmission welding," *Int. J. Adv. Manuf. Technol.*, pp. 3771–3786, 2022, doi: 10.1007/s00170-022-09536-y.

[73] D. K. Goyal, R. Yadav, R. Kant, "Laser transmission welding of polycarbonate sheets using electrolytic iron powder absorber," *Opt. Laser Technol.*, vol. 161, p. 109165, 2023. doi: 10.1016/j.optlastec.2023.109165.

[74] Y. Shi, Z. Yao, H. Shen, and J. Hu, "Research on the mechanisms of laser forming for the metal plate," *Int. J. Mach. Tools Manuf.*, vol. 46, no. 12–13, pp. 1689–1697, 2006, doi: 10.1016/j.ijmachtools.2005.09.016.

[75] A. Gisario, M. Barletta, C. Conti, and S. Guarino, "Springback control in sheet metal bending by laser-assisted bending: Experimental analysis, empirical and neural network modelling," *Opt. Lasers Eng.*, vol. 49, no. 12, pp. 1372–1383, 2011, doi: 10.1016/j.optlaseng.2011.07.010.

[76] R. Kant, P. M. Bhuyan, and S. N. Joshi, "Experimental studies on TGM and BM dominated curvilinear laser bending of aluminum alloy sheets," *Lasers Based Manuf*, no. June 2016, pp. 69–91, 2015.

[77] R. Kant and S. N. Joshi, "Numerical modeling and experimental validation of curvilinear laser bending of magnesium alloy sheets," *Proc. Inst. Mech. Eng. Part B J. Eng. Manuf.*, vol. 228, no. 9, pp. 1036–1047, 2014, doi: 10.1177/0954405413506419.

[78] R. Kant and S. N. Joshi, "Numerical investigations into influence of scanning path curvature on deformation behavior during curvilinear laser bending of magnesium sheets," *J. Therm. Stress.*, vol. 41, no. 3, pp. 313–330, 2018, doi: 10.1080/01495739.2017.1403298.

[79] U. S. Dixit, S. N. Joshi, and R. Kant, "Laser forming systems: A review," *Int. J. Mechatronics Manuf. Syst.*, vol. 8, pp. 160–205, 2015.

[80] J. Magee, K. G. Watkins, and W. M. Steen, "Advances in laser forming," *J. Laser Appl.*, vol. 10, no. 6, pp. 235–246, 1998, doi: 10.2351/1.521859.

[81] R. Kant, S. N. Joshi, and U. S. Dixit, "An integrated FEM-ANN model for laser bending process with inverse estimation of absorptivity," *Mech. Adv. Mater. Mod. Process.*, vol. 1, no. 1, pp. 1–12, 2015, doi: 10.1186/s40759-015-0006-1.

[82] P. M. Bhuyan, R. Kant, and S. N. Joshi, "Experimental investigation on laser bending of metal sheets using parabolic irradiations," In *Proceedings of the 5th International and 26th All India Manufacturing Technology, Design and Research Conference AIMTDR 2014*, no. Aimtdr, pp. 1–6, 2014.

[83] S. N. Joshi and Ravi Kant, "Numerical simulation of multi-pass laser bending processes using finite element method," *Second International Conference on Intelligent Robotics, Automation and Manufacturing (IRAM) 2013*, no. December 2013, pp. 1–7, 2013.

[84] R. Nair, R. Kant, R. Yadav and H. Gurung "Experimentally validated analytical modelling of the laser bending of low carbon steel sheets," *Lasers Eng.*, vol. 53, pp. 253–265, 2022.

[85] R. Kant and S. N. Joshi, "Thermo-mechanical studies on bending mechanism, bend angle and edge effect during multi-scan laser bending of magnesium M1A alloy sheets," *J. Manuf. Process.*, vol. 23, pp. 135–148, 2016, doi: 10.1016/j.jmapro.2016.05.017.

[86] R. Kant and S. N. Joshi, "Numerical and experimental studies on the laser bending of magnesium M1A alloy," *Lasers Eng.*, vol. 35, no. 1–4, pp. 39–62, 2016.

[87] R. Yadav, D. K. Goyal, and R. Kant, "A comprehensive study on the effect of line energy during laser bending of duplex stainless steel," *Opt. Laser Technol.*, vol. 151, no. September 2021, p. 108025, 2022, doi: 10.1016/j.optlastec.2022.108025.

[88] R. Yadav, D. K. Goyal, and R. Kant, "Enhancing process competency by forced cooling in laser bending process," *J. Therm. Stress.*, vol. 45, no. 8, pp. 617–629, 2022, doi: 10.1080/01495739.2022.2103057.

[89] R. Yadav, D. K. Goyal, and R. Kant, "Multi-scan laser bending of duplex stainless steel under different cooling conditions," *CIRP J. Manuf. Sci. Technol.*, vol. 39, pp. 345–358, 2022, doi: 10.1016/j.cirpj.2022.10.002.

[90] R. Yadav, D. K. Goyal, and R. Kant, "An experimental study of forced cooling in single scan laser bending," in Advances in Forming, Machining and Automation, U. S. Dixit, Uday S., Kanthababu M., Babu A. Ramesh, Ed. Singapore: Springer, 2022, pp. 13–21.

[91] R. Yadav and R. Kant, "Effectiveness of forced cooling during laser bending of duplex-2205," *Mater. Manuf. Process.*, vol. 38, no. 5, pp. 598–607, 2022, doi: 10.1080/10426914.2022.2146717.

[92] R. Yadav and R. Kant, "Improving bend angle by using forced cooling when laser bending Al sheet," *Lasers Eng.*, vol. 55, pp. 75–85, 2023.

[93] R. Kant and S. N. Joshi, "Finite element simulation of laser assisted bending with moving mechanical load," *Int. J. Mechatronics Manuf. Syst.*, vol. 6, no. 4, pp. 351–366, 2013, doi: 10.1504/IJMMS.2013.057128.

[94] Y. Pal, B. Singh, R. Kant, and R. Yadav, "Enhancing the bend angle and mechanical properties of mild steel using fiber laser bending technique under the influence of electromagnetic force," *Opt. Lasers Eng.*, vol. 168, no. March, p. 107631, 2023, doi: 10.1016/j.optlaseng.2023.107631.

[95] N. Deswal and R. Kant, "Machinability analysis during laser assisted turning of aluminium 3003 alloy," *Lasers Manuf. Mater. Process.*, vol. 9, no. 1, pp. 56–71, 2022, doi: 10.1007/s40516-022-00163-9.

[96] N. Deswal and R. Kant, "Laser-assisted turning of aluminium 3003 alloy," *Lasers Eng.*, vol. 53, no. 3–4, pp. 197–214, 2022.

[97] N. Deswal and R. Kant, "Experimental investigation on magnesium az31b alloy during ultrasonic vibration assisted turning process," *Mater. Manuf. Process.*, vol. 37, no. 15, pp. 1708–1714, 2022, doi: 10.1080/10426914.2022.2039701.

[98] N. Deswal and R. Kant, "FE analysis of ultrasonic vibration assisted turning of magnesium az31b alloy," *Mater. Today. Proc.*, vol. 62, no. 14, pp. 7473–7479, 2022, doi: 10.1016/j.matpr.2022.03.507.

[99] N. Deswal and R. Kant, "Synergistic effect of ultrasonic vibration and laser energy during hybrid turning operation in magnesium alloy," *Int. J. Adv. Manuf. Technol.*, vol. 121, no. 1–2, pp. 857–876, 2022, doi: 10.1007/s00170-022-09384-w.

[100] N. Deswal and R. Kant, "Hybrid turning process by interacting ultrasonic vibration and laser energies," *Mater. Manuf. Process.*, vol. 38, no. 5, pp. 570–576, 2022, doi: 10.1080/10426914.2022.2065014.

[101] N. Deswal and R. Kant, "A study on ultrasonic vibration and laser assisted turning of aluminium alloy," in Advances in Forming, Machining and Automation, Lecture Notes on Multidisciplinary Industrial Engineering. Singapore: Springer, 2023, doi: 10.1007/978-981-19-3866-5_14.

[102] N. Deswal and R. Kant, "Radial ultrasonic-vibration and laser assisted turning of al3003 alloy," *Mater. Res. Express*, vol. 10, no. 4, p. 044004, 2023, doi: 10.1088/2053-1591/acce24.

[103] D. Przestacki, "Conventional and laser assisted machining of composite A359/20SiCp," *Procedia CIRP*, vol. 14, pp. 229–233, 2014, doi: 10.1016/j.procir.2014.03.029.

[104] S. Affan Ahmed, M. Mohsin, and S. M. Zubair Ali, "Survey and technological analysis of laser and its defense applications," *Def. Technol.*, vol. 17, no. 2, pp. 583–592, 2021, doi: 10.1016/j.dt.2020.02.012.

[105] T. A. K. Al-Aish, "Design and analysis the fiber laser weapon system FLWS," *Adv. Phys. Theor. Appl.*, vol. 47, pp. 59–69, 2015.

Chapter 10

Utilization of ultrasonic vibration and laser energies during sustainable machining

Neeraj Deswal and Ravi Kant

10.1 INTRODUCTION

The major contributors to energy consumption in the world are industries, transportation, and construction. Industries contribute towards the urbanization, modernization, and economic growth of a country. However, the development of industries has led to the consumption of a great amount of energy and fossil fuels. They also generate a significant amount of carbon dioxide and other greenhouse gas emissions [1]. These emissions have substantially impacted human health and environmental pollution. Industry has accounted for about 32% of the world's energy consumption [2]. Additionally, fossil fuels comprise approximately 80% of the overall world energy supply [3]. Furthermore, the industrial sector is the primary source of greenhouse gas emissions, accounting for approximately 29% of carbon dioxide emissions related to electricity production [4]. Manufacturing operations consume approximately 40% of the overall energy and about 25% of the global natural resources [5]. The manufacturing sector utilizes numerous processes during production, and the machining process is the most commonly utilized in the manufacturing industries. Machining processes are among the most widely utilized manufacturing methods in various industries, including automotive, aviation, marine, and more [6]. It is an energy-consuming and waste-generating process, apart from creating environmental pollution and occupational health hazards [7].

10.2 CHALLENGES DURING MACHINING PROCESS

In conventional machining processes, heat is generated in the cutting zone as a result of the interaction between the cutting tool and the workpiece material. This heat generation presents several challenges, including large machining forces, rapid cutting tool wear, reduced surface quality, an impact on the mechanical properties of the workpiece, increased processing time for finishing a product, and higher production costs due to frequent cutting tool replacement [8]. To overcome heat-generated issues, wet machining is utilized to minimize heat generation, achieve

DOI: 10.1201/9781032703046-10

better surface quality, and enhance cutting tool life [9]. Hence, cutting fluids are used significantly during machining operation in the manufacturing industries. Energy usage and cutting fluids (also known as metalworking fluid (MWF)) are major sources of higher manufacturing costs, waste generation, and environmental deterioration [10]. Energy consumption is related to various adverse impacts because of the usage of non-renewable resources, resulting in harm to the ecosystem, habitat alteration, increased emissions, intensified land usage, and higher risks to human health. Moreover, higher energy utilization from conventional fossil fuel sources has increased the generation of more greenhouse gases, such as carbon dioxide, sulphur dioxide, and nitrogen oxide, to the atmosphere [11]. The cutting fluids are toxic and non-biodegradable because they are produced by mineral oils, and some additives contribute to the depletion of the ozone layer [12]. They are also one of the primary health threats on the machine shop floor [13]. Additionally, cutting fluids break down chemically due to the high cutting temperature of the machining operation, resulting in an adverse impact on human health both externally via skin and internally due to inhalation [14]. The major factors contributing to higher manufacturing costs are power consumption, maintenance of the lubrication system, and disposal and cleaning of the cutting fluids [15].

10.3 NEED FOR SUSTAINABLE MACHINING PROCESS

The 1972 Stockholm UN summit on the human environment has emphasized the need for sustainable development methodologies to resolve the global environmental challenges, such as ecological system degradation, resource depletion, waste minimization, and environmental pollution [16]. Moreover, as per ISO-14000 standards, manufacturing industries should limit their adverse activities to reduce environmental pollution [17]. Therefore, to overcome the ever-increasing challenges of natural resource consumption and environmental pollution, countries are required to implement a sustainable development strategy for better product quality and productivity. Sustainable manufacturing is mainly focused on economically feasible machining processes which can minimize environmental hazards by conserving natural resources and energy. Sustainable manufacturing can be defined as the development of machining processes for the manufacturing of components/products in such a manner that it would minimize adverse environmental impacts, minimize waste, conserve energy, conserve natural resources, and be economically feasible and safer for the operator, community, and consumers [8]. Hence, the primary goals of sustainable manufacturing processes are related to minimal energy requirement, minimal environmental pollution, minimal carbon emissions, increased resource utilization, and better machinability characteristics (such as lower machining force, higher tool life, better surface finish, and more) [18]. The significant functional elements of sustainable manufacturing are sustainable products, processes, and systems [19]. The concept behind sustainable manufacturing is presented in Figure 10.1 [20].

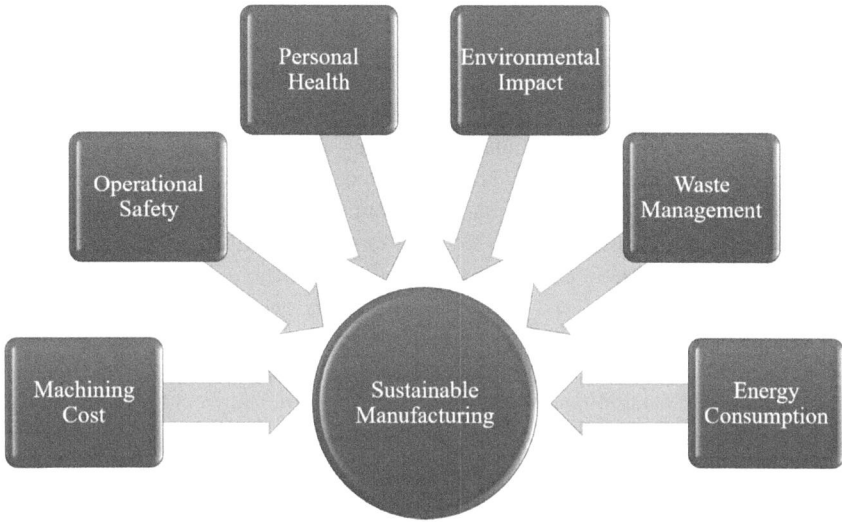

Figure 10.1 A layout of sustainable manufacturing.

Industries and researchers are exploring various techniques to enhance the machining methods in terms of energy efficiency, eco-friendliness, and sustainability [18]. Therefore, various sustainable machining techniques have been developed to minimize energy consumption and environmental pollution. Dry machining, minimum quantity lubrication, cryogenic machining, hybrid machining, and the like are some of the sustainable machining methods employed. Hence, these machining methods are discussed in the next sections with more emphasis on hybrid machining processes.

10.4 DRY MACHINING

A machining process that uses no cutting fluids is termed dry machining. By eliminating the costs associated with cutting fluid procurement, management, disposal, and machine tools equipped with subsystems for cutting fluids, significant reductions in capital investments can be achieved. Furthermore, the environment impacts related to cutting fluids are also eliminated. Hence, the main functions of cutting fluids during conventional machining processes, such as lubrication, chip removal, and lower heat generation, have to be accomplished by some other means in the dry machining process [21, 22]. Modifications to cutting tool design, development of cutting tool materials, and changes in the machine tool have been carried out to dissipate heat during the dry machining operation. Thermoelectric cooling, undercooling by providing coolant under the cutting tool, and self-lubricating tools by inserting solid lubricant into the rake and flank face of the cemented carbide tool have been developed to modify the cutting tool design [16]. The coatings on the cutting tool material provide enhanced thermal insulation, hardness, and self-lubrication. The cutting tool materials should be more heat resistant and generate less heat [23]. The monolayer coating could suffer early delamination or crack. Therefore, preferred is a multilayer coating which combines various coating materials for enhanced tool life. The various coating materials which have been used to obtain better tool life are titanium aluminum nitride (TiAlN) for better heat resistance, titanium carbonitride (TiCN) for enhanced toughness, titanium nitride (TiN) for better adhesion characteristics, and molybdenum disulfide (MoS_2) for self-lubrication [24]. The machine tool design has been modified primarily to remove and dispose of chips. Conveyor, suction pump, compressed air, tilt table to tilt the part from vertical to horizontal, vertical inversion fixture, liquid flushing at the end of the machining process, and vacuum shroud surrounding the tool are some of the modifications to machine tool design [16]. However, dry machining is preferable at higher cutting speed to avoid premature failure of the cutting tool. Moreover, residual stresses are found to be comparable with wet machining, which can hamper the surface integrity in the machining process.

10.5 MINIMUM QUANTITY LUBRICATION

Minimum quantity lubrication (MQL) is an innovative technique in which a minimal amount of cutting fluid is sprayed with a flow rate of 10–500 ml/h in the cutting zone during the machining operation [20]. It is also known as the near dry machining (NDM) technique in which synthetic esters, fatty alcohols, emulsions, gaseous substances, oil-water mixture, and so on are used most commonly as a cooling lubricating fluid during the machining operation. These lubricating fluids are injected in a small quantity to the machining zone with or without the assistance of a transporting medium. A pump equipped with a precision-metered droplet is used to deliver the cooling fluids at a rapid succession at the cutting zone. In certain cases, high-pressure (usually 4–6 bar) compressed air is used to deliver atomized fine droplets to the machining region in the form of aerosol spray [19]. However, the cooling and lubricating capabilities of the aerosol spray are considered poor compared to conventional cutting fluids. Fatty alcohols, synthetic esters, vegetable oils, and the like have provided better cooling and lubrication characteristics as well as better machining performance [25]. In certain cases, oil is used as the lubricant in the air-based MQL technique, but its cooling capabilities are limited. It poses a significant challenge in the machining operation, especially when low thermal conductivity materials are machined. Hence, carbon dioxide at a supercritical state can dissolve many lubricating oils and provide adequate lubrication in the cutting zone, especially when machining low thermal conductivity materials. The supercritical carbon dioxide provides deeper penetration of the lubricating fluids in the cutting zone and uniform oil coating on the surface of the workpiece and cutting tool [26]. The MQL technique provides better machining performance compared to conventional machining, but further studies are required to obtain the optimal design of external nozzles and to deliver the optimal amount of lubricating fluids in the cutting zone.

10.6 CRYOGENIC MACHINING

Cryogenic machining is another effective method of cooling the cutting region during the machining process. In this technique, a cryogenic cooling lubricating fluid is used as a coolant to provide lubrication in the machining zone and lower the temperature at the chip-tool interface [27]. Cryogenic machining is significantly different from the conventional cutting fluids technique and MQL technique. In other cooling techniques, the cooling fluids flush the chips from the machining zone after breaking. However, in cryogenic machining, cooling promotes brittle fracture of chips and chemical stability to the cutting tool [16]. The benefits of cryogenic machining include reduced tool wear, improved surface finish, lower temperatures at the chip-tool interface, increased production rates, absence of harmful emissions, and environmental friendliness. Cryogenic cooling can be applied in several ways, such as workpiece pre-cooling, spraying at the cutting tool edge, indirect cryogenic cooling, cooling at the cutting tool back surface, cooling at the chip-tool interface, cooling by dual nozzle system, and direct cryogenic treatment of cutting tools [19]. Moreover, cryogenic machining employs various cooling mediums, including liquid nitrogen, air, water, steam, vapor, refrigerated carbon dioxide, and refrigerated compressed air. Liquid nitrogen is most commonly used as a cooling medium in cryogenic machining due to its various unique properties, such as abundance in the environment, non-combustibility, non-corrosive properties, no harmful emissions production, no chemical reaction with the cutting tool or the workpiece, no residual oil on chips, safer to the operator, and elimination of coolant disposal cost. However, additional equipment costs are associated with the cryogenic machining setup, and liquid nitrogen is costly and not reusable because it evaporates in the atmosphere. Apart from that, due to the cryogenic cooling, the temperature at the workpiece is decreased. It becomes harder and results in higher cutting forces compared to the conventional machining process, which is not desirable. Also, the cooling effect is reduced significantly at higher cutting speeds due to the lower penetration of the cryogenic cooling at the cutting tool tip [16].

10.7 HYBRID MACHINING PROCESSES

The hybrid machining (HM) process is defined according to the International Academy for Production Engineering (CIRP): "Hybrid manufacturing process is based on the simultaneous and controlled interaction of process mechanisms and/ or energy sources/tools having a significant effect on the process performance."

In the definition of the HM process, "simultaneous and controlled interaction" reveals that the interaction of various processes or energies should be at the same time and at the same impact/processing/machining zone [28, 29]. Generally, in manufacturing technology, the "hybrid" term is used when a combination of different kinds of technologies is associated with the manufacturing of specified parts. HM reveals a relationship between various process energies or assistance of a particular process via different process energies or by some other means. So, hybrid can have various meanings in the manufacturing process [30]:

1. Combination of various energies acting at a similar time and same zone.
2. Combination of various process steps which are produced separately earlier.
3. Combination of various machines on a single machine platform.
4. Combination of various products with hybrid function.

The principal idea in developing hybrid processes is to gain significant advantages from the developed process and minimize the disadvantages in a single process, especially in a cost effective and eco-friendly manner [31]. The combination of two or more processes in HM should have significant benefits, such as reduction in process chains which leads to minimization of clamping, referencing, alignment at several workplaces, ability to precisely handle dimensionally complex products, processing of materials effectively, potential benefits on the machinability, and substantial improvement in surface integrity. Several energies have been utilized during the HM process, such as ultrasonic vibration [32], thermal [33], electric field [34], and magnetic field [35].

The ultrasonic vibration HM process is one of the most commonly utilized processes due to its frequent separation of the cutting tool with the workpiece surface during the machining operation. The thermal HM process is used widely due to the softening of the workpiece material by means of various heat sources such as laser, induction, plasma, and gas torch prior to the machining operation. The electric field HM process is used to reduce the flow stress of the workpiece material which eases the machining process. The magnetic field HM process is used to increase the machining temperature and reduce machine-tool vibrations, resulting in significant improvement in machining. Among the various energies utilized so far, thermal and ultrasonic vibration energy are the most commonly applied during the HM process. Therefore, the application of thermal and ultrasonic vibration energy is discussed here.

10.7.1 Thermally assisted machining process

The thermally assisted machining (TAM) process is a hybrid machining process in which an external heat source is utilized to preheat the workpiece material in the cutting region before the machining operation [36]. The yield strength, hardness, and strain hardening of the workpiece material are decreased, leading to thermal softening of the material, which in turn facilitates the material removal process by the cutting tool [37]. The basic idea of the TAM process is to rapidly heat the workpiece material, leading to thermal softening, while ensuring that the chips efficiently remove the maximum heated material from the cutting zone without causing thermal damage to the cutting tool and machined surface. The applied heat source should fulfill some basic conditions [38, 39]:

1. Rapid and controlled heating should be supplied by the source for smooth operation.
2. Heating should be restricted to a small region to prevent any other part damage.
3. The machined surface should be prevented from overheating to restrict any metallurgical changes.
4. Sufficient intensity should be supplied by the heat source for quick and localized heating.
5. The source should be cost-effective, safe to the operator, and environmentally friendly.

Several external heat sources are utilized during the TAM process, such as electric current, furnace, resistance, gas flame, electron beam, induction, plasma, and laser beam [40–45]. A non-consumable electrode is used to transfer the current in the workpiece material, which raises workpiece temperature using the electric current heat source. A furnace is used to preheat the workpiece material in the furnace heat source. An electric current is passed through the workpiece to the cutting tool during resistance heating. The liquefied petroleum gas or oxy-acetylene gas is used to heat the workpiece material when employing a gas flame heat source. A high-energy electron beam is applied to the workpiece surface for thermal softening of the material when using an electron beam heat source. Induction coils are used like an electric source that induces a high-frequency alternating electric current to preheat the workpiece material when using an induction heat source. A stream of ionized gas is used to increase the workpiece temperature, ensuring thermal softening of the material when employing a plasma heat source. A laser beam is used to preheat the workpiece material prior to the machining operation in the laser heat source process. Although all of the heat sources have benefits, they also have some limitations. Electric current requires stabilization equipment and arc ignition; furnace possesses insufficient energy and temperature is lost during machining; resistance is unable to maintain uniform temperature and requires an insulating machine tool; gas flame possesses poor controllability and low power intensity; electron beam requires a vacuum, works only on metals, and is expensive; induction possesses low power intensity; plasma requires eye protection for the operator; and laser requires high capital cost. A laser beam is the preferred external heat source due to its easy controllability, rapid and localized heating, higher power intensity on small areas, and a smaller heat affected zone (HAZ) compared to the aforementioned heat sources [46]. Therefore, the laser heat source is discussed in the present study.

Turning is one of the most widely utilized techniques for fabricating parts among the various machining processes [47]. The turning operation is performed on the lathe machine, with the workpiece material securely held in place. A cutting tool is used to remove the unwanted material from the workpiece material and obtain the desired shape and size. The turning process is controlled by cutting speed, feed rate, and depth of cut [48]. Cutting speed is defined as the speed at which the workpiece rotates with respect to the cutting tool. Feed rate is defined as the distance covered by the cutting tool during one revolution of the workpiece. Depth of cut is defined as the amount of material removed during each revolution of the workpiece material. Cutting speed, feed rate, and depth of cut are generally expressed in m/min, mm/rev, and mm, respectively [49].

10.7.2 Laser assisted turning process

The laser assisted turning (LAT) process is a hybrid machining process in which a high-intensity defocused laser beam is utilized to preheat the workpiece material usually above the recrystallization temperature to facilitate material removal through a conventional cutting tool. The yield strength and hardness of the material are decreased due to the preheating of the workpiece material, which ensures thermal softening of the workpiece material [50, 51]. The thermal softening of the material leads to enhanced machining performance such as lower machining forces, higher tool life, higher material removal rate (MRR), lower surface roughness, and lower residual stresses [52, 53]. The LAT operation has been investigated for various materials, including iron [54], steel [55], ceramics [56], waspaloy [57], titanium alloys [58], nickel alloys [59], metal matrix composites [60], and single-crystal silicon [61]. LAT is mainly used in various industries, including marine, automotive, aviation, biomedical, and more [62, 63]. A schematic of the LAT process is shown in Figure 10.2.

The utilization of laser heat source in thermally assisted machining areas was initiated in the late 1970s. Initially, it was explored to machine aerospace alloys such as titanium and nickel alloy. With the advancement in materials, it has been utilized to machine various materials, as mentioned earlier. Various types of lasers have been utilized in laser assisted machining processes, such as CO_2, Nd: YAG, Excimer, high power diode, and fiber laser [64–66]. Fiber lasers have excellent energy efficiency, easier beam transfer, better beam profile controllability, and so on compared to other lasers [67]. The laser unit consists of a laser generator, laser source, and laser head. The generator is used to create the laser and transfer it to the source. The source carries the laser through an optical fiber cable to the laser head. The laser cutting head is connected to the output of the laser source through an optical fiber cable. The laser cutting head is positioned in such a manner that the laser beam is always focused on the workpiece surface but not on the cutting tool. Furthermore, compressed air is generally used through the laser nozzle to protect the focusing lens from extreme heat and debris generated during the experiments. The performance of laser heat source depends on various laser parameters, such

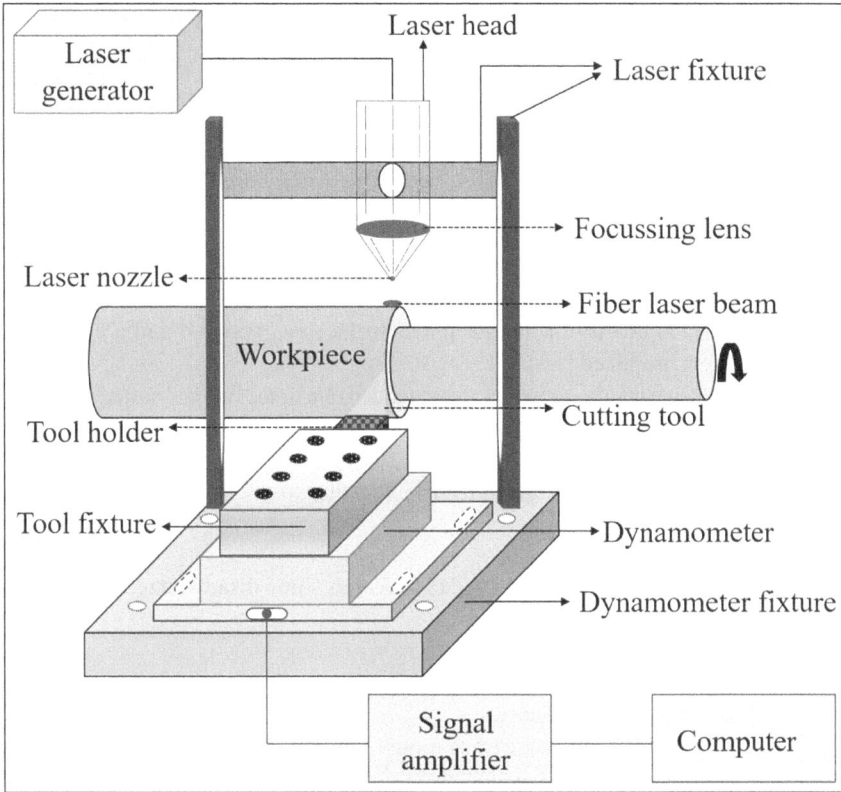

Figure 10.2 A schematic of the LAT process.

as laser power, laser diameter, laser approach angle, the distance between laser nozzle and workpiece surface, and distance between laser beam spot and cutting tool edge, and also on the machining parameters, i.e., cutting speed, feed rate, and depth of cut.

Advantages and disadvantages of the LAT process

The LAT process has various advantages over the conventional turning (CT) process. The important advantages are discussed as follows [68–73]:

1. The laser beam is focused precisely, which eventually creates rapid, localized, controlled heating and minimizes HAZ on the workpiece surface.
2. Laser is operated easily in air and does not require any other media such as a vacuum in electron beam and gas atmosphere in plasma.
3. The laser beam can be applied to any optically reachable location with the required intensity.

4. The machining time is reduced, which significantly reduces the production cost and enhances productivity.
5. The average machining forces are reduced compared to CT due to thermal softening of the workpiece material which eases machining.
6. The machine power consumption is reduced due to the low load on the cutting tool.
7. The MRR is increased because of the lower machining forces and power consumption.
8. Tool wear, tool breakage, and chatter generation are decreased due to the reduction in machining forces and power consumption.
9. The surface quality of the machined surface is improved and a crack-free surface is produced.
10. Compressive residual stresses are generated due to the thermal softening effect.
11. The metallurgical properties are enhanced due to the better machining performance.
12. The process is environmentally friendly due to the elimination of cutting fluids.

Despite its many advantages, the LAT process has some disadvantages, which are described as follows:

1. The initial investment for the laser machine is high, leading to an overall increase in the process cost.
2. External fixtures are required to mount the laser head on the machine prior to machining operation, which also raises the installation cost.
3. The highly reflective workpiece surfaces are difficult to process; however, coatings on the surface can resolve this issue.
4. The laser heating can deteriorate the workpiece surface and even melt the surface if laser parameters are not controlled properly.
5. The laser can harm the operator if proper precautions are not followed, such as using goggles, gloves, and other safety measures.

10.7.3 Ultrasonic vibration assisted turning process

The ultrasonic vibration assisted turning (UVAT) process is a hybrid machining process in which high frequency–low amplitude vibration is provided to the tool. The contact between tool and workpiece occurs for a specific period and after that it disengages and the process continues engaging and disengaging [74]. Due to the periodic separation, machining performance experiences significant enhancements, including lower machining forces, longer tool life, better surface finish, lower segmented chips, better machine stability, no burr formation, and lower built-up edge (BUE) [75, 76]. UVAT has been applied on a variety of materials such as soft metals and alloys [77], hard to machine steels [78], difficult to machine alloys [79], amorphous materials [80], single-crystal materials [81], sintered alloys

[82], ceramics [83], metal matrix composites [84], and fiber-reinforced plastic composites [85]. UVAT is utilized in various industries like electronics, petro-chemical, automotive, power generation, aerospace, biomedical, and more. [86]. A schematic of the UVAT process is presented in Figure 10.3.

The vibration system consists of a frequency generator, transducer, and booster. The vibration is provided by an ultrasonic generator that utilizes a piezoelectric or magneto-restrictive transducer to generate ultrasonic motion of low amplitude and high frequency. The frequency generator is employed to generate an electric signal, and these signals are converted into ultrasonic vibration by the transducer. The booster is used to enhance the ultrasonic vibration amplitude given by the transducer and propagate these vibrations on the cutting tool. The cutting tool is clamped to the bottom of the booster by a standard screw [87]. The cutting tool

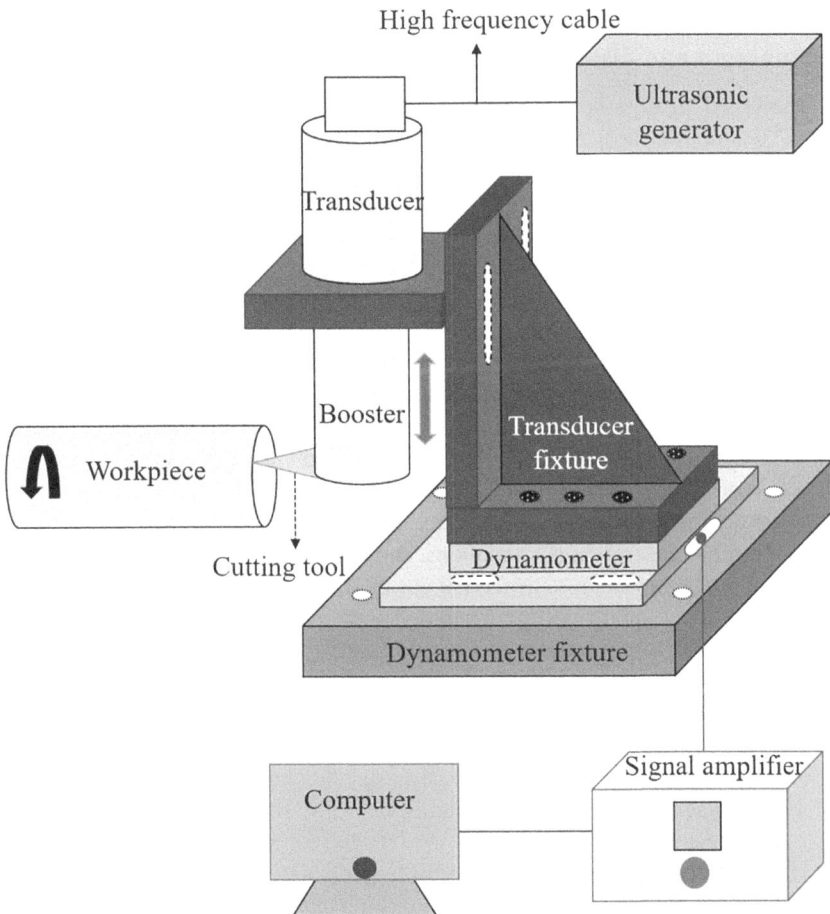

Figure 10.3 A schematic of the UVAT process.

is excited by the ultrasonic motion and starts vibrating in the direction of cutting, usually at a high frequency (15–40 kHz) with a low amplitude of 1–20 μm [88]. The one-dimensional (1D) UVAT was first introduced in the late 1950s to enhance the conventional machining process, aiming for a better surface finish, longer tool life, and lesser machining forces. This method was initially used for brass, cast iron, mild steel, and aluminum materials. But when this method was tested with other materials like difficult-to-cut materials and glass, it showed better results than did conventional machining. So, researchers explored this area more, and in the 1990s, two-dimensional (2D) UVAT was proposed. As expected, 2D UVAT exhibits improved performance in comparison to the 1D UVAT in terms of longer tool life, better surface finish, and lower machining forces. Nowadays, both 1D and 2D UVAT are being utilized to machine various materials and generate complex dimensional parts or surfaces in the manufacturing industries [89].

Advantages and disadvantages of the UVAT process

The UVAT process has various advantages over the conventional turning process. The important advantages are described as follows [90–95]:

1. The average machining forces than in the CT process due to the periodic separation of the cutting tool with the workpiece surface, which introduces aerodynamic lubrication.
2. The lubrication is enhanced due to the effect of aerodynamic lubrication, which reduces friction between the cutting tool and workpiece.
3. The burr formation and BUE are suppressed due to the reduction in pushing and bending stress in the deformation zone.
4. The chatter generation is minimized irrespective of the cutting tool geometry which ensures stable cutting compared to the CT process.
5. The tool life is enhanced due to the reduction in effective cutting time and better heat dissipation due to the periodic separation with the workpiece surface.
6. The chip generation is continuous, and fewer segmented chips are obtained due to the lower stress generated during the UVAT process.
7. The surface quality of the machined surface is enhanced due to the lower machining forces, minimization of burr and chatter, and lesser tool wear.
8. A better roundness profile is generated due to improved chip generation and quality of the machined surface.
9. The residual stresses are compressive in nature due to lower stress generation.
10. The critical depth of cut increases due to the reduction in machining forces, which prohibits the initiation of brittle fracture and ensures ductile regime cutting.
11. The process can be applied to any type of material; for example, ductile, brittle, soft, and other difficult-to-machine materials.
12. It is an eco-friendly and cost-effective process due to the elimination of cutting fluids.

Despite its numerous advantages, the UVAT process has some disadvantages, which are discussed as follows:

1. External fixtures are required to mount the system on the machine prior to machining operation, which increases the installation cost.
2. The selection of a suitable control unit is desirable to achieve efficient vibration cutting. A controlled resonant system can display better stability with strong nonlinearity.
3. The booster can be heated rapidly and the cutting tool can be broken at higher vibration amplitudes. Therefore, lower amplitudes are preferable.
4. Higher cutting speeds are not achieved due to the dependence on frequency and amplitude. At higher cutting speeds, the advantages of the UVAT process are not obtained and the process becomes similar to the CT process.
5. In certain cases, the flank face of the cutting tool collides with the machined surface, which leads to the generation of chipping and unusual wear. Hence, to avoid these changes the cutting and the vibratory direction of the cutting tool should remain the same.

10.7.4 Thermal and ultrasonic vibration assisted turning process

The thermal and ultrasonic vibration assisted turning process is the advanced modification of the UVAT process. In this process, the workpiece material is preheated before the turning operation and ultrasonic vibration is provided to the cutting tool, simultaneously in a single machine platform [96]. The researchers have termed this process as hot ultrasonic assisted turning (HUAT). The simultaneous application of heat and vibration to the workpiece and cutting tool respectively has shown lower machining forces, higher machining temperature, and lower surface roughness compared to conventional, heat, and vibration turning processes [97, 98]. Up to now, the heat sources used in the HUAT process include band-resistance heaters, plasma, and tunnel furnaces. The HUAT process is explored for Ti-15333 alloy, Inconel 718, Hastelloy-X alloy, aluminum alloy, and magnesium alloy [99–105].

10.8 SUMMARY

Industries are the backbone of any nation's progression towards urbanization, modernization, and economic growth. Within these industries, manufacturing operations play a pivotal role in the development of various products and components. Machining processes are a part of these manufacturing operations and are commonly utilized in manufacturing industries. However, these processes are the biggest consumers of natural resources and energy globally. Generally, cutting fluids are used during the machining process, which is responsible for harmful greenhouse gas emissions. They have significant adverse impacts on energy consumption, environmental pollution, human health, ozone layer depletion, higher machining cost, and more. Furthermore, production costs are also increased with the application of cutting fluids due to maintenance of the lubrication system, disposal, and cleaning of the cutting fluids. Therefore, the necessity for sustainable manufacturing has arisen on a global scale in order to address challenges like degradation of the ecological system, depletion of resources, higher energy consumption, higher greenhouse gases emission, waste minimization, and environmental pollution. Sustainable machining aims at developing a machining process which is energy efficient and eco-friendly. Hence, various techniques are explored to minimize the usage of cutting fluids for a sustainable process. Dry machining, minimum quantity lubrication (MQL), cryogenic machining, and hybrid machining are the various advancements in machining methods that aim to meet the goal of sustainable machining. Dry machining utilizes self-lubricating tools, coated tools, and modified machine tools to eliminate the usage of cutting fluids. The MQL technique uses various cooling lubricating fluids such as synthetic esters, fatty alcohols, emulsions, gaseous substances, and oil-water mixture by spraying minimum quantity at a low flow rate on the cutting zone. Cryogenic machining uses cryogenic cooling lubricating fluids like liquid nitrogen, air, water, steam, vapor, refrigerated carbon dioxide, and refrigerated compressed air to cool the machining area in an eco-friendly manner. Hybrid machining utilizes various energies during the machining process by eliminating the need for any fluids or other modification in machine tools or cutting tools. Thermal and ultrasonic vibration assisted turning processes are the most commonly utilized hybrid machining processes due to the significant benefits in the machining performance compared to the conventional turning process. The laser heat source is the most widely used thermal source due to its rapid, localized, and controlled heating at the workpiece surface. Additionally, thermal and ultrasonic vibration energies are combined in a single machine platform to obtain the benefits of both of these energy sources. Therefore, it can be concluded that hybrid machining processes play a crucial role in sustainable machining, offering substantial contributions to the manufacturing industries. These processes eliminate the need for cutting fluids, reduce the consumption of natural resources, minimize waste generation, lower environmental pollution, decrease machining costs, and promote eco-friendliness.

REFERENCES

[1] Li K, Lin B (2015) Impacts of urbanization and industrialization on energy consumption/CO_2 emissions: Does the level of development matter? Renew Sustain Energ Rev 52:1107–1122. https://doi.org/10.1016/j.rser.2015.07.185

[2] Zhou L, Li J, Li F, et al. (2016) Energy consumption model and energy efficiency of machine tools: A comprehensive literature review. J Clean Prod 112:3721–3734. https://doi.org/10.1016/j.jclepro.2015.05.093

[3] Azarpour A, Suhaimi S, Zahedi G, Bahadori A (2013) A review on the drawbacks of renewable energy as a promising energy source of the future. Arab J Sci Eng 38:317–328. https://doi.org/10.1007/s13369-012-0436-6

[4] Bruzzone AAG, Anghinolfi D, Paolucci M, Tonelli F (2012) Energy-aware scheduling for improving manufacturing process sustainability: A mathematical model for flexible flow shops. CIRP Ann Manuf Technol 61:459–462. https://doi.org/10.1016/j.cirp.2012.03.084

[5] Campitelli A, Cristóbal J, Fischer J, et al. (2019) Resource efficiency analysis of lubricating strategies for machining processes using life cycle assessment methodology. J Clean Prod 222:464–475. https://doi.org/10.1016/j.jclepro.2019.03.073

[6] Wickramasinghe KC, Sasahara H, Rahim EA, Perera GIP (2020) Green metalworking fluids for sustainable machining applications: A review. J Clean Prod 257:120552. https://doi.org/10.1016/j.jclepro.2020.120552

[7] Dambhare S, Deshmukh S, Borade A, et al. (2015) Sustainability issues in turning process: A study in Indian machining industry. Procedia CIRP 26:379–384. https://doi.org/10.1016/j.procir.2014.07.092

[8] Bhattacharya S, Protim Das P, Chatterjee P, Chakraborty S (2021) Prediction of reponses in a sustainable dry turning operation: A comparative analysis. Math Probl Eng 2021. https://doi.org/10.1155/2021/9967970

[9] Siniawski M, Bowman C (2009) Metal working fluids: Finding green in the manufacturing process. Ind Lubr Tribol 61:60–66. https://doi.org/10.1108/00368790910940374

[10] Fernando R, Gamage J, Karunathilake H (2021) Sustainable machining: Environmental performance analysis of turning. Int J Sustain Eng 1–20. https://doi.org/10.1080/19397038.2021.1995524

[11] Huang H, Ameta G (2014) Computational energy estimation tools for machining operations during preliminary design. Int J Sustain Eng 7:130–143.

[12] Dahmus JB, Gutowski TG (2016) An environmental analysis of machining. In: Proceedings of IMECE04, 2004 ASME International Mechanical Engineering Congress and Exposition, pp. 1–10.

[13] Schultheiss F, Zhou J, Gröntoft E, Ståhl JE (2013) Sustainable machining through increasing the cutting tool utilization. J Clean Prod 59:298–307. https://doi.org/10.1016/j.jclepro.2013.06.058

[14] Pervaiz S, Kannan S, Kishawy HA (2018) An extensive review of the water consumption and cutting fluid based sustainability concerns in the metal cutting sector. J Clean Prod 197:134–153. https://doi.org/10.1016/j.jclepro.2018.06.190

[15] Pusavec F, Krajnik P, Kopac J (2010) Transitioning to sustainable production—part I: Application on machining technologies. J Clean Prod 18:174–184. https://doi.org/10.1016/j.jclepro.2009.08.010

[16] Zhao F, Sharma A (2015) Environmentally friendly machining. In: HandBook of Manufacturing Engineering and Technology. Springer, London, pp. 1127–1154. https://doi.org/10.1007/978-1-4471-4670-4_14.

[17] Jamil M, Zhao W, He N, et al. (2021) Sustainable milling of Ti–6Al–4V: A trade-off between energy efficiency, carbon emissions and machining characteristics under MQL and cryogenic environment. J Clean Prod 281:125374. https://doi.org/10.1016/j.jclepro.2020.125374

[18] Lv L, Deng Z, Meng H, et al. (2020) A multi-objective decision-making method for machining process plan and an application. J Clean Prod 260:121072. https://doi.org/10.1016/j.jclepro.2020.121072

[19] Fratila D (2013) Sustainable manufacturing through environmentally-friendly machining. 1–21. https://doi.org/10.1007/978-3-642-33792-5_1

[20] Rajmohan T, Kalyan Chakravarthy VV, Nandakumar A, Satish Kumar SD (2020) Eco friendly machining processes for sustainability—review. IOP Conf Ser Mater Sci Eng 954. https://doi.org/10.1088/1757-899X/954/1/012044

[21] Sharma R, Jha BK, Pahuja V, Sharma S (2022) Role of environmental friendly machining on machinability. Mater Today Proc 50:640–648. https://doi.org/10.1016/j.matpr.2021.03.652

[22] Sharma R, Jha BK, Pahuja V (2021) Impact of environmental friendly machining on machinability: A review. Mater Today Proc 46:10362–10367. https://doi.org/10.1016/j.matpr.2020.12.498

[23] Weinert K, Inasaki I, Sutherland JW, Wakabayashi T (2004) Dry machining and minimum quantity lubrication. CIRP Ann Manuf Technol 53:511–537. https://doi.org/10.1016/S0007-8506(07)60027-4

[24] Grigoriev SN, Vereschaka AA, Vereschaka AS, Kutin AA (2012) Cutting tools made of layered composite ceramics with nano-scale multilayered coatings. Procedia CIRP 1:301–306. https://doi.org/10.1016/j.procir.2012.04.054

[25] Itoigawa F, Childs THC, Nakamura T, Belluco W (2006) Effects and mechanisms in minimal quantity lubrication machining of an aluminum alloy. Wear 260:339–344. https://doi.org/10.1016/j.wear.2005.03.035

[26] Supekar SD, Clarens AF, Stephenson DA, Skerlos SJ (2012) Performance of supercritical carbon dioxide sprays as coolants and lubricants in representative metalworking operations. J Mater Process Technol 212:2652–2658. https://doi.org/10.1016/j.jmatprotec.2012.07.020

[27] Ahmad-Yazid A, Taha Z, Almanar IP (2010) A review of cryogenic cooling in high speed machining (HSM) of mold and die steels. Sci Res Essays 5:412–427.

[28] Lauwers B (2011) Surface integrity in hybrid machining processes. Procedia Eng 19:241–251. https://doi.org/10.1016/j.proeng.2011.11.107

[29] Chavoshi SZ, Luo X (2015) Hybrid micro-machining processes: A review. Precis Eng 41:1–23. https://doi.org/10.1016/j.precisioneng.2015.03.001

[30] Lauwers B, Klocke F, Klink A, et al. (2014) Hybrid processes in manufacturing. CIRP Ann Manuf Technol 63:561–583. https://doi.org/10.1016/j.cirp.2014.05.003

[31] Rajurkar KP, Zhu D, McGeough JA, et al. (1999) New developments in electro-chemical machining. Ann CIRP 48:567–579. https://doi.org/10.1016/S0007-8506(07)63235-1

[32] Arka GN, Sahoo SK, Iqbal MM, Singh S (2021) Acoustic horn tool assembly design for ultrasonic assisted turning and its effects on performance potential. Mater Manuf Process 37:260–270. https://doi.org/10.1080/10426914.2021.2016819

[33] Bijanzad A, Munir T, Abdulhamid F (2022) Heat-assisted machining of superalloys: A review. Int J Adv Manuf Technol 118:3531–3557. https://doi.org/10.1007/s00170-021-08059-2

[34] Ulutan D, Pleta A, Mears L (2015) Electrically-assisted machining of titanium alloy Ti-6Al-4V and Nickel-based alloy IN-738: An investigation. In: Proceedings of the ASME 2015 International Manufacturing Science and Engineering Conference, Volume 1: Processing, Charlotte, NC, June 8–12, pp. 1–5.

[35] Hatefi S, Abou-El-Hossein K (2022) Review of magnetic-assisted single-point diamond turning for ultra-high-precision optical component manufacturing. Int J Adv Manuf Technol 120:1591–1607. https://doi.org/10.1007/s00170-022-08791-3

[36] Stevenson MG, Oxley PLB (1973) High temperature stress-strain properties of a low-carbon steel from hot machining tests. Proc Inst Mech Eng 187:263–272. https://doi.org/10.1243/PIME_PROC_1973_187_107_02

[37] Parida AK (2017) Heat assisted machining of nickel base alloys: Experimental and numerical analysis. PhD Dissertation, National Institute of Technology, Rourkela.

[38] Lei S, Pfefferkorn F (2007) A review on thermally assisted machining. In: Proceedings of the ASME 2007 International Manufacturing Science and Engineering Conference, Atlanta, GA, October 15–18, pp. 325–336.

[39] Shams OA, Pramanik A, Chandratilleke TT (2017) Thermal-assisted machining of titanium alloys. In: Advanced Manufacturing Technologies. Materials Forming, Machining and Tribology. 4th ed. Springer, Cham, pp. 49–76.

[40] Komanduri R, Flom DG, Lee M (1985) Highlights of the DARPA advanced machining research program. J Eng Ind 107:325–335. https://doi.org/10.1115/1.3186005

[41] Maity KP, Swain PK (2008) An experimental investigation of hot-machining to predict tool life. J Mater Process Technol 198:344–349. https://doi.org/10.1016/j.jmatprotec.2007.07.018

[42] Sun S, Brandt M, Dargusch MS (2010) Thermally enhanced machining of hard-to-machine materials-a review. Int J Mach Tools Manuf 50:663–680. https://doi.org/10.1016/j.ijmachtools.2010.04.008

[43] Ahmed N, Abdo BM, Darwish S, et al. (2017) Electron beam melting of titanium alloy and surface finish improvement through rotary ultrasonic machining. Int J Adv Manuf Technol 92:3349–3361. https://doi.org/10.1007/s00170-017-0365-3

[44] Kim EJ, Lee CM (2019) A study on the optimal machining parameters of the induction assisted milling with Inconel 718. Materials (Basel) 12:233. https://doi.org/10.3390/ma12020233

[45] Lee Y, Lee C (2019) A study on optimal machining conditions and energy efficiency in plasma assisted machining of Ti-6Al-4V. Materials (Basel) 12:2590. https://doi.org/10.3390/ma12162590

[46] Sun SJ, Brandt M, Mo JPT (2013) Current progresses of laser assisted machining of aerospace materials for enhancing tool life. Adv Mater Res 690–693:3359–3364. https://doi.org/10.4028/www.scientific.net/AMR.690-693.3359

[47] Nalbant M, Gökkaya H, Sur G (2007) Application of Taguchi method in the optimization of cutting parameters for surface roughness in turning. Mater Des 28:1379–1385. https://doi.org/10.1016/j.matdes.2006.01.008

[48] You SH, Lee JH, Oh SH (2019) A study on cutting characteristics in turning operations of titanium alloy used in automobile. Int J Precis Eng Manuf 20:209–216. https://doi.org/10.1007/s12541-019-00027-x

[49] Dutta S, Narala SKR (2021) Investigations on chip formation of turned novel AM alloy. Proc Inst Mech Eng Part E J Process Mech Eng 235:332–341. https://doi.org/10.1177/0954408920961196

[50] Venkatesan K, Ramanujam R, Kuppan P (2014) Laser assisted machining of difficult to cut materials: Research opportunities and future directions—a comprehensive review. Procedia Eng 97:1626–1636. https://doi.org/10.1016/j.proeng.2014.12.313

[51] Punugupati G, Kandi KK, Bose PSC, Rao CSP (2016) Laser assisted machining: A state of art review. IOP Conf Ser Mater Sci Eng 149:012014. https://doi.org/10.1088/1757-899X/149/1/012014

[52] Chryssolouris G, Anifantis N, Karagiannis S (1997) Laser assisted machining: An overview. J Manuf Sci Eng 119:766–769. https://doi.org/10.1115/1.2836822

[53] Kong XJ, Zhang HZ, Wu XF, Wang Y (2014) Laser-assisted machining of advanced materials. Mater Sci Forum 800–801:825–831. https://doi.org/10.4028/www.scientific.net/MSF.800-801.825

[54] Skvarenina S, Shin YC (2006) Laser-assisted machining of compacted graphite iron. Int J Mach Tools Manuf 46:7–17. https://doi.org/10.1016/j.ijmachtools.2005.04.013

[55] Kaselouris E, Baroutsos A, Papadoulis T, et al. (2020) A study on the influence of laser parameters on laser-assisted machining of Aisi H-13 steel. Key Eng Mater 827:92–97. https://doi.org/10.4028/www.scientific.net/KEM.827.92

[56] Rozzi JC, Pfefferkorn FE, Incropera FP, Shin YC (2000) Transient, three-dimensional heat transfer model for the laser assisted machining of silicon nitride: I. Comparison of predictions with measured surface temperature histories. Int J Heat Mass Transf 43:1409–1424. https://doi.org/10.1016/S0017-9310(99)00217-3

[57] Ding H, Shin YC (2013) Improvement of machinability of Waspaloy via laser-assisted machining. Int J Adv Manuf Technol 64:475–486. https://doi.org/10.1007/s00170-012-4012-8

[58] Yang N, Brandt M, Sun S (2009) Numerical and experimental investigation of the heat-affected zone in a laser-assisted machining of Ti-6Al-4V alloy process. Mater Sci Forum 618–619:143–146. https://doi.org/10.4028/www.scientific.net/MSF.618-619.143

[59] Ahn DG, Byun KW, Kang MC (2010) Thermal characteristics in the cutting of Inconel 718 superalloy using CW Nd:YAG laser. J Mater Sci Technol 26:362–366. https://doi.org/10.1016/S1005-0302(10)60059-X

[60] Venkatesan K, Ramanujam R, Kuppan P (2014) A review on conventional and laser assisted machining of aluminium based metal matrix composites. Eng Rev 34:75–84.

[61] Chen X, Liu C, Ke J, et al. (2020) Subsurface damage and phase transformation in laser-assisted nanometric cutting of single crystal silicon. Mater Des 190:108524. https://doi.org/10.1016/j.matdes.2020.108524

[62] Deswal N, Kant R (2022) Machinability analysis during laser assisted turning of aluminium 3003 alloy. Lasers Manuf Mater Process 9:56–71. https://doi.org/10.1007/s40516-022-00163-9

[63] Deswal N, Kant R (2022) Laser-assisted turning of aluminium 3003 alloy. Lasers Eng 53:197–214

[64] Kim KS, Kim JH, Choi JY, Lee CM (2011) A review on research and development of laser assisted turning. Int J Precis Eng Manuf 12:753–759. https://doi.org/10.1007/s12541-011-0100-1

[65] Jeon Y, Lee CM (2012) Current research trend on laser assisted machining. Int J Precis Eng Manuf 13:311–317. https://doi.org/10.1007/s12541-012-0040-4

[66] Li L (2000) Advances and characteristics of high-power diode laser materials processing. Opt Lasers Eng 34:231–253. https://doi.org/10.1016/S0143-8166(00)00066-X

[67] Kim IW, Lee CM (2016) A study on the machining characteristics of specimens with spherical shape using laser-assisted machining. Appl Therm Eng 100:636–645. https://doi.org/10.1016/j.applthermaleng.2016.02.005

[68] Pfefferkorn FE, Rozzi JC, Incropera FP, Shin YC (1997) Surface temperature measurement in laser-assisted machining processes. Exp Heat Transf 10:291–313. https://doi.org/10.1080/08916159708946549

[69] Rozzi JC, Pfefferkorn FE, Incropera FP, Shin YC (1998) Transient thermal response of a rotating cylindrical silicon nitride workpiece subjected to a translating laser heat source, part II: Parametric effects and assessment of a simplified model. J Heat Transfer 120:907–915. https://doi.org/10.1115/1.2825910

[70] Rozzi JC, Pfefferkorn FE, Shin YC, Incropera FP (2000) Experimental evaluation of the laser assisted machining of silicon nitride ceramics. J Manuf Sci Eng Trans ASME 122:666–670. https://doi.org/10.1115/1.1286556

[71] Pfefferkorn FE, Incropera FP, Shin YC (2005) Heat transfer model of semi-transparent ceramics undergoing laser-assisted machining. Int J Heat Mass Transf 48:1999–2012. https://doi.org/10.1016/j.ijheatmasstransfer.2004.10.035

[72] Tian Y, Shin YC (2006) Thermal modeling for laser-assisted machining of silicon nitride ceramics with complex features. J Manuf Sci Eng Trans ASME 128:425–434. https://doi.org/10.1115/1.2162906

[73] Zhao F, Bernstein WZ, Naik G, Cheng GJ (2010) Environmental assessment of laser assisted manufacturing: Case studies on laser shock peening and laser assisted turning. J Clean Prod 18:1311–1319. https://doi.org/10.1016/j.jclepro.2010.04.019

[74] Dixit US, Pandey PM, Verma GC (2019) Ultrasonic-assisted machining processes: A review. Int J Mechatronics Manuf Syst 12:227–254. https://doi.org/https://doi.org/10.1504/IJMMS.2019.103479

[75] Lotfi M, Akbari J (2021) Finite element simulation of ultrasonic-assisted machining: A review. Int J Adv Manuf Technol 116:2777–2796. https://doi.org/https://doi.org/10.1007/s00170-021-07205-0

[76] Sonia P, Jain JK, Saxena KK (2021) Influence of ultrasonic vibration assistance in manufacturing processes : A review. Mater Manuf Process 36:1451–1475. https://doi.org/10.1080/10426914.2021.1914843

[77] Luo H, Wang Y, Zhang P (2020) Effect of cutting and vibration parameters on the cutting performance of 7075-T651 aluminum alloy by ultrasonic vibration. Int J Adv Manuf Technol 107:371–384. https://doi.org/10.1007/s00170-020-05098-z

[78] Khajehzadeh M, Boostanipour O, Razfar MR (2020) Finite element simulation and experimental investigation of residual stresses in ultrasonic assisted turning. Ultrasonics 108:106208. https://doi.org/10.1016/j.ultras.2020.106208

[79] Llanos I, Campa Á, Iturbe A, et al. (2018) Experimental analysis of cutting force reduction during ultrasonic assisted turning of Ti6Al4V. Procedia CIRP 77:86–89. https://doi.org/10.1016/j.procir.2018.08.227

[80] Gan J, Wang X, Zhou M, et al. (2003) Ultraprecision diamond turning of glass with ultrasonic vibration. Int J Adv Manuf Technol 21:952–955. https://doi.org/10.1007/s00170-002-1416-x

[81] Suzuki N, Masuda S, Haritani M, Shamoto E (2004) Ultraprecision micromachining of brittle materials by applying ultrasonic elliptical vibration cutting. In: Micro-Nanomechatronics and Human Science, 2004 and The Fourth Symposium Micro-Nanomechatronics for Information-Based Society, pp. 133–138.

[82] Liu K, Li XP, Rahman M (2008) Characteristics of ultrasonic vibration-assisted ductile mode cutting of tungsten carbide. Int J Adv Manuf Technol 35:833–841. https://doi.org/10.1007/s00170-006-0761-6

[83] Weber H, Herberger J, Pilz R, Ddr THK (1984) Turning of machinable glass ceramics with an ultrasonically vibrated tool. CIRP Ann Manuf Technol 33:85–87. https://doi.org/10.1016/S0007-8506(07)61385-7

[84] Zhong ZW, Lin G (2006) Ultrasonic assisted turning of an aluminium-based metal matrix composite reinforced with SiC particles. Int J Adv Manuf Technol 27:1077–1081. https://doi.org/10.1007/s00170-004-2320-3

[85] Xu W, Zhang LC (2014) On the mechanics and material removal mechanisms of vibration-assisted cutting of unidirectional fibre-reinforced polymer composites. Int J Mach Tools Manuf 80–81:1–10. https://doi.org/10.1016/j.ijmachtools.2014.02.004

[86] Deswal N, Kant R (2022) Experimental investigation on magnesium AZ31B alloy during ultrasonic vibration assisted turning process. Mater Manuf Process 37:1708–1714. https://doi.org/10.1080/10426914.2022.2039701

[87] Deswal N, Kant R (2022) FE analysis of ultrasonic vibration assisted turning of magnesium AZ31B alloy. Mater Today Proc 62. https://doi.org/10.1016/j.matpr.2022.03.507

[88] Sharma V, Pandey PM (2016) Recent advances in ultrasonic assisted turning: A step towards sustainability. Cogent Eng 3:1–20. https://doi.org/10.1080/23311916.2016.1222776

[89] Xu S, Kuriyagawa T, Shimada K, Mizutani M (2017) Recent advances in ultrasonic-assisted machining for the fabrication of micro/nano-textured surfaces. Front Mech Eng 12:33–45. https://doi.org/10.1007/s11465-017-0422-5

[90] Xiao M, Karube S, Soutome T, Sato K (2002) Analysis of chatter suppression in vibration cutting. Int J Mach Tools Manuf 42:1677–1685. https://doi.org/10.1016/S0890-6955(02)00077-9

[91] Zhou M, Wang XJ, Ngoi BKA, Gan JGK (2002) Brittle—ductile transition in the diamond cutting of glasses with the aid of ultrasonic vibration. J Mater Process Technol 121:243–251. https://doi.org/10.1016/S0924-0136(01)01262-6

[92] Ma C, Shamoto E, Moriwaki T, et al. (2005) Suppression of burrs in turning with ultrasonic elliptical vibration cutting. Int J Mach Tools Manuf 45:1295–1300. https://doi.org/10.1016/j.ijmachtools.2005.01.011

[93] Mitrofanov AV, Ahmed N, Babitsky VI, Silberschmidt VV (2005) Effect of lubrication and cutting parameters on ultrasonically assisted turning of Inconel 718. J Mater Process Technol 162–163:649–654. https://doi.org/10.1016/j.jmatprotec.2005.02.170

[94] Zhong ZW, Lin G (2005) Diamond turning of a metal matrix composite with ultrasonic vibrations. Mater Manuf Process 20:727–735. https://doi.org/10.1081/AMP-200055124

[95] Khajehzadeh M, Razfar MR, Akhlaghi M (2014) Experimental investigation of tool temperature during ultrasonically assisted turning of aerospace aluminum. Mater Manuf Process 29:1453–1460. https://doi.org/10.1080/10426914.2014.930962

[96] Muhammad R, Maurotto A, Roy A, Silberschmidt VV (2011) Analysis of forces in vibro-impact and hot vibro-impact turning of advanced alloys. Appl Mech Mater 70:315–320. https://doi.org/10.4028/www.scientific.net/AMM.70.315

[97] Deswal N, Kant R (2021) A study on ultrasonic vibration and laser assisted turning of aluminium alloy. In: Advances in Forming, Machining and Automation. Springer, Singapore, pp. 165–172. https://doi.org/10.1007/978-981-19-3866-5_14.

[98] Deswal N, Kant R (2022) Hybrid turning process by interacting ultrasonic vibration and laser energies. Mater Manuf Process 38:570–576. https://doi.org/10.1080/10426914.2022.2065014

[99] Muhammad R, Maurotto A, Roy A, Silberschmidt VV (2012) Hot ultrasonically assisted turning of β-Ti alloy. Procedia CIRP 1:336–341. https://doi.org/10.1016/j.procir.2012.04.060

[100] Feyzi T, Safavi SM (2013) Improving machinability of Inconel 718 with a new hybrid machining technique. Int J Adv Manuf Technol 66:1025–1030. https://doi.org/10.1007/s00170-012-4386-7

[101] Sofuoğlu MA, Çakır FH, Gürgen S, et al. (2018) Numerical investigation of hot ultrasonic assisted turning of aviation alloys. J Brazilian Soc Mech Sci Eng 40:1–12. https://doi.org/10.1007/s40430-018-1037-4

[102] Deswal N, Kant R (2023) Radial direction ultrasonic-vibration and laser assisted turning of Al3003 alloy. Mater Res Express 10:044004. https://doi.org/10.1088/2053-1591/acce24

[103] Deswal N, Kant R (2022) Synergistic effect of ultrasonic vibration and laser energy during hybrid turning operation in magnesium alloy. Int J Adv Manuf Technol 121:857–876. https://doi.org/10.1007/s00170-022-09384-w

[104] Deswal N, Kant R (2023) Machinability and surface integrity analysis of magnesium AZ31B alloy during laser assisted turning. J Manuf Process 101:527–545. https://doi.org/10.1016/j.jmapro.2023.06.022

[105] Deswal N, Kant R (2022) Surface integrity analysis of aluminum 3003 alloy during ultrasonic-vibration-laser assisted turning. Proc Inst Mech Eng Part B J Eng Manuf 00:1–12. https://doi.org/10.1177/09544054231178951

Chapter 11

Machining performance and optimization of process parameters of monel alloy 400 using ECM process

P. Sudhakar Rao, Akhilesh Sinha, Shashwath G. Patil,
Ravi Kant and Sri Phani Sushma

11.1 INTRODUCTION

Electrochemical machining (ECM) is a novel machining technique based on electrochemistry. ECM is commonly employed in sectors where different components that are hard to cut and where complicated contours are required. A direct current (DC) (5–30 volt) voltage is provided throughout the inter-electrode gap across the pre-shaped electrode tool and the job material. Electrolyte moves quickly across the IEG, and typical density of current is about 20–200 A/cm². The electrochemical characteristics of metal, electrolytic properties, and electric voltage/current provided all influence the anode dissolution that is determined by Faraday's equations of electrolysis. ECM produces a nearly similar imprint of the tool electrode upon that work item [1–3].

ECM clearly outperforms other traditional machining methods in terms of application irrespective of work hardness, greater MRR, reduced tool wear, plane and brilliant surface, and manufacturing of parts with complicated structures having crack-free and stress-free surfaces [4, 5]. High residual stresses are created during conventional machining with Monel 400 alloys, resulting in a quick hardening which slows down the operation and causes tool electrode failure. As a result, machining these metals using ordinary machine tools is extremely challenging. On the other hand, we could compare machines almost with every alloy made with specific hardness or tensile strength. As a result, it may be a profitable option to machine Monel alloys, and its importance may grow in future. The goal here is to create comprehensive computational equations for relating interactive as well as higher-order effects of different process variables, including applied voltage (V), electrolyte concentration (EC), and inter-electrode gap (IEG), in relation to key machining parameters such as material removal rate (MRR), tool wear rate (TWR) as well as surface roughness (SR). This is aimed at maximizing the potential of the ECM process [6, 7]. To organize and analyse the studies, response surface methodology (RSM) is used. The goal of employing RSM is not only to study the response across the entire factor space, but also to identify the area of concern in which response achieves an optimal or near-optimal value [8–10].

DOI: 10.1201/9781032703046-11

11.2 CURRENT STATUS OF RESEARCH

Haisch et al. [11]: Anodic metal dissoluteness for steel-100Cr6 was considered in aqueous $NaNO_3$ as well as NaCl solutions. Flow of passage investigations were conducted with a large density of current up to 70 A/cm². Unsolvable carbide particles enhanced apparent current efficiency for NaCl by more than 100% and in $NaNO_3$ by more than 67%. Munda and Bhattacharyya [12]: This paper aimed to obtain an overall mathematical prototype using for relating the collective and high-order factors of various machining variables such as applied voltage pulse off/on ratio, electrolyte concentration, and frequency of voltage, as well as frequency of tool vibration upon most significant machining standards, namely the MRR as well as overcut. Santhi et al. [13]: Their objective was to develop a revolutionary method for optimizing titanium alloy process parameters (Ti6Al4V). Design, strategy, and method—A desire function study, the fuzzy set concept, and order preference method through similarity to ideal solution methods were used in order to construct ECM processing parameters for titanium alloys. Bahre et al. [14]: The feasibility of precision electrochemical machining (PECM) for lamellar cast iron machining was investigated in terms of machinability with $NaNO_3$ as the electrolyte and stainless steel as the tool material. The accuracy of created geometries, as well as the possibility of producing certain surface properties, was investigated in this study. Muthukumar et al. [15]: The core composite design technique was used in this experiment. After analysing 30 tests, a mathematical model was created to connect the machining factors with the rate of cut (ROC). The relevant coefficients were calculated using ANOVA with a 5% significance level. Kalaimathi et al. [16]: Different process parameters affect machining outcomes in terms of MRR and surface roughness (SR) for Monel 400 alloy. MRR and SR were improved by using an electrolyte concentration of 15 g/l and also an IEG of 0.4 mm. ANOVA was used to evaluate the results. Tiwari et al. [17] attempted to build a mathematical prototype for characteristics like MRR and SR for ECM over EN 19 steel using regression analysis. The analysis of variance was utilized to evaluate the sufficiency of developed prototypes. Krishnamurthy et al. [18]: MRR was discovered to be very important in ECM, and employing the best ECM process parameters typically save operating, tool, and maintenance costs while generating higher-accuracy products. This paper explored the effect and direct development of several process factors for titanium ECM. Li et al. [19]: A 10% sodium nitrate slurry was employed in the ECM process to make a number of holes in a titanium alloy sheet. In order to understand the electric characteristics of titanium alloy (Ti 6Al 4V) within a 10% $NaNO_3$ electrolyte, a current efficiency graph that included a polarization graph of the alloy was generated. Chen et al. [20]: In this study, orthogonal experiments were performed to test Ti60 ECM in order to regulate the impacts of different electrochemical method factors on SR. The most relevant characteristic was found to be the frequency of a voltage source. It has been observed that using correctly adjusted ECM settings may greatly reduce a work item's surface roughness. Rao et al. [21]: Based on a combination of the ECM process with a high MRR and a standard honing process that generates a controlled smooth surface finish, electrochemical honing was investigated as the maximum precise machining technology discovered for machining gears and cylinders. Jeykrishnan et al. [22]: This study focuses on the application of three critical factors in the vehicle industry: current, voltage, and electrolyte concentration. As a result, updated methods must be implemented to achieve the best results. Soundrapandian et al. [23]: The microscopic system was used in their research to measure overcut and conicity for drilled holes, and a profilometer was utilized to analyse the SR of the machined region to discover the optimal machine variable. ANOVA was used to evaluate the impact of each variable on MRR, overcut, and drilled hole circularity. Geethapriyan et al. [24]: In his work, the researchers used the Taguchi–Grey relational analysis technique to discover the relevance of various process factors and to examine the variables for the machining of titanium. Sharma et al. [25]: The outcomes of process variables like applied voltage, IEG, electrolyte concentration and electrolyte flow upon MRR and radial overcut were examined. A Taguchi (orthogonal array)-based approach was used to optimize several input parameters and final responses in ECM. Kumar [26]: ECM is a novel machining technique that employs Faraday's law to eliminate metal from a workpiece. Titanium alloys are employed in the manufacture of jet engines as well as other aerospace components. In this review study, the process properties of ECM, like MRR, surface finish, and overcut, were examined when titanium alloy was deposited in different electrolytes. Khan et al. [27] sought to improve ECM method variables using a combination comprising SS 316 (workpiece) and copper as a tool (tool material). A total of 27 tests were performed to assess the effect of ECM variables like electrolyte concentration, applied voltage, and current, as well as the tool's feed rate upon the work's MRR and SR. Khan et al. [28]: Other than the mechanical qualities of metal, it was discovered that thermal properties play a more important role; the forces created in this technique are modest, and thus burrs were not obtained. Khan et al. [29] explored the possibility of utilizing jatropha oil as a source of dielectric medium to replace kerosene oil, which is environmentally benign. Other dielectric fluids, such as hydrocarbons, have a negative effect on environment. Khan et al. [30]: Micro-machining was investigated, and combining micro-EDM and laser produced a decrease in machining time. They also discovered that a mixture of ultrasonic machining and the EDM process is commonly employed. Yadav et al. [31]: The researchers examined multi-objective optimization of process variables using principal component examination and grey relational examination to improve metal removal and to minimize roughness of Ti 6Al 4V. Daniel et al. [32]: In this study, the researchers used a variety of SiC combinations based on their weight percentage and particle size. In addition, a preset weight proportion of molybdenum disulphide was used. The aluminium composite was created by stir casting. Aluminium was fused using silicon carbide elements sized 10, 20, and 40 µm that are 5%, 10%, and 15% by weight, accordingly, and 2% molybdenum disulphide. A Taguchi L-27 array was utilized to achieve the desired target levels and investigate the influence of ECM parameters on MRR and SR. Debta et al. [33] investigated the most intriguing of magnesium alloys, AZ91D (90% magnesium), because of its remarkable toughness, high strength-to-weight ratio, and exceptional resistance to environmental rust. It can also be cast well, enabling it to be employed in the automobile and aerospace sectors. Kumar et al. [34] examined the behaviour of Incoloy during EDM machining, with response variables based on input power, pulse on/off duration, and voltage applied. They also demonstrated that variable optimization in EDM may be accomplished by combining dielectric fluid with powder. Sahu et al. [35] discovered variable optimization in the EDM process. Because ANOVA was applied, optimal findings were achieved. Suitable optimization was also used to optimize machining. Thakur et al. [36] considered the use of graphite tool electrodes in suitable dielectric media for the machining of titanium alloy. Machining performance was designed using appropriate methodologies for performance evaluation. Following adequate testing in the EDM process, optimal variables were determined. Sharma et al. [37] examined electrochemical machining of titanium alloy using various electrolytes and a Cu tool in their experimental study. The electrolytes KCl, $NaNO_3$, and NaCl were employed. Gobikrishnan et al. [38]: One of the most important aspects of removing MRR was determining electrolyte content. As input variables, voltage, current, and electrolyte concentration were used. The output parameters were obtained using MRR. In their study, the value of the trial was based on the orthogonal array of Taguchi's technique. Ramana et al. [39]: This study focused on the ability to machine nickel A286 by utilizing PVD-coated materials over a lengthy period. The machining performance of nickel A286 was evaluated utilizing experimental tests that adjusted the speed of cutting and depth of cut, as well as tool electrode feed rate with taking chip shape, flank wear rate, and surface roughness into consideration. Jerin et al. [40]: Special materials, like steel 12X18H10T, can withstand extreme temperatures while retaining their inherent properties. In order to widen its usefulness, this study evaluated its machining capabilities using the ECM technique for surface finishing. Thangamani et al. [41]: This research investigated how three different NaCl-based electrolytes affect the cutting of alloy Ti 6Al 4V, and the Taguchi model of experimentation was used to find features on the basis of electro chemical micro machining (ECMM) constraints such as voltage, electrolyte concentration, frequency of voltage, and duty cycle. Khan et al. [42] examined how vegetable oil contributes to sustain EDM machining and prevent environmental harm, although no such adjustment was necessary to support other types of biodegradable lubricants. Wasif et al. [43]: Al 5454 alloys, which are widely used in industrial works to create tool dies, moulds, blocks for engine, aviation parts, and a range of other components, were studied. Nagarajan et al. [44]: In this work, the researchers discovered that by varying the voltage (ranging from 11 to 15 volts), the electrolyte flow rate (from 1 to 3.0 L/min), and the electrolyte concentration (ranging from 120 to 190 g/l), they could utilize algorithms such as MFO, GWO, and PSO to determine key parameters like MRR and the presence of nickel in the sludge.

Vengatajalapathi et al. [45]: In this work, nickel hydroxide was cleaned from sludge in an environmentally friendly way during Monel 400 ECM. Filter materials included powdered coconut shell and wood dust. RSM's central composite design (CCD) was used for experiments. Singh et al. [46]: In this study, a higher MRR was obtained with a electrolyte concentration of 200 g/l and feed rate of about 18 micron/min. To obtain increased MRR with regulated input settings, a little expenditure of tool life was incurred.

11.3 EXPERIMENTAL METHODOLOGY

Pilot experiments conducted utilizing the one factor at a time (OFAT) technique aided in providing appropriate ranges of input variables for the ECM machine. Before starting the machining, a certain layer of input variables is chosen and an associated range is established, accompanied by a systematic layout of the experimentation performance. An accurate evaluation of variables for input, as shown in Table 11.1, and a proper setting of their ranges are essential for obtaining appropriate results. This allows for a reduction in the required number of performance tests during the machining process. This basic experiment served as a platform for future investigation [47, 48]. As a result, the OFAT approach determines the input parameters to be employed as well as the optimal range of primary experiments. The multiple pieces of Monel alloy 400 with dimensions of 25 mm × 25 mm × 5 mm depicted in Figure 11.1 are used for pilot and experimental operations. The notable ranges observed are as follows. To develop the design of experiment, the model is created utilizing the three stages of experiment design based on the Box–Behnken design (BBD) approach. As the electrode, copper is the most appropriate metal to use. The electrode has a diameter of 3 mm and a length of 60 mm, as illustrated in Figure 11.2. A surface roughness tester is utilized to assess the integrity of surface. MRR as well as TWR are computed by the weight change between them (initial weight – final weight).

11.4 RESULTS AND DISCUSSIONS

This study was carried out with the usage of copper tool and the RSM-based BBD method approach. Three levels of each input variable were chosen for testing

Table 11.1 Ranges fixed for main experiment based on pilot experiment

S. No.	Parameter	Level (−1)	Level (0)	Level (+1)
1.	Voltage	15	17	19
2.	Electrolyte conc.	40	50	60
3.	IEG	0.15	0.20	0.25

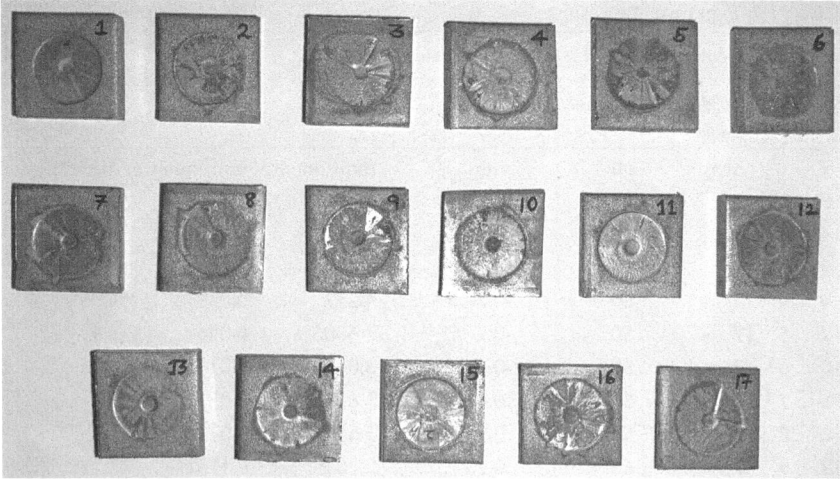

Figure 11.1 Machined workpiece.

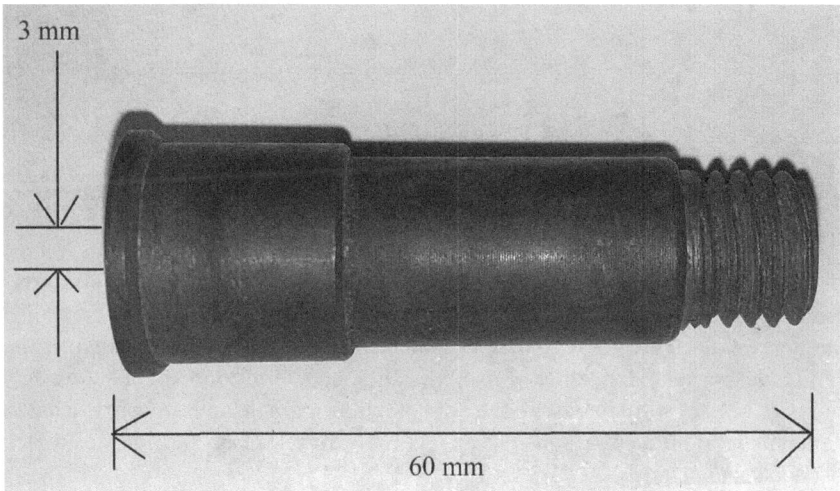

Figure 11.2 Copper tool.

performance by running every run for 10 minutes, and the outcomes are displayed in Table 11.2. The weight difference between the MRR and TWR was calculated after each experimental run. Each experimental run includes three levels of parameter values for analysis and optimization [49, 50].

Table 11.2 Experimental results

		Parameter 1	Parameter 2	Parameter 3	Response 1	Response 2	Response 3
Std	Run	Applied voltage	Electrolyte conc.	IEG	MRR	TWR	SR
		volts	g/l	mm	mm³/minute	mm³/minute	micro meter
8	1	19	50	0.25	6.8613	0.1858	1.51
14	2	17	50	0.2	3.56	0.0669	3.44
9	3	17	40	0.15	2.4545	0.0732	3.85
4	4	19	60	0.2	4.242	0.0389	4.06
17	5	17	50	0.2	2.6386	0.0468	3.42
10	6	17	60	0.15	3.01	0.0145	4.28
6	7	19	50	0.15	3.61	0.029	3.2
16	8	17	50	0.2	3.5534	0.01	3.4
12	9	17	60	0.25	5.3034	0.1101	3.229
7	10	15	50	0.25	5.79	0.0569	2.03
1	11	15	40	0.2	4.7545	0.0111	3.35
3	12	15	60	0.2	2.52	0.0256	4.28
15	13	17	50	0.2	4.1727	0.0044	3.05
2	14	19	40	0.2	3.54	0.0234	4.006
11	15	17	40	0.25	6.51	0.0322	2.529
13	16	17	50	0.2	3.2295	0.0145	3.41
5	17	15	50	0.15	2.625	0.114	2.896

11.4.1 Analysis and optimization for MRR, TWR, and SR

In Table 11.3, an F-value of 13.71 shows that the model looks significant. A large F-value is might occur due to noise just 0.12% of the time. Significant model terms have p-values less than 0.0500. In this situation, C, AB, and C^2 are critical model variables. Values greater than 0.1000 indicate that model terms are insignificant. If a model has a large number of insignificant model terms, model reduction may improve it. Lack of fit has an F-value of 0.42, indicating that it is not significant in comparison to pure error. Noise has a 74.66% chance of causing significant lack of fit F-value.

Table 11.4 showing an F-value of 8.09 suggests that the model seems significant. Only 0.58% of the time will an F-value this large be caused by noise. Significant model terms have p-values less than 0.0500. In this situation, crucial model terms include C, AC, BC, and C^2. Values greater than 0.1000 indicate that model terms were insignificant. If a model has a large number of insignificant model terms, model reduction may improve it. The F-value for lack of fit is 0.14, indicating that the lack of fit is small in contrast to the pure error. Noise has a 93.00% risk of causing a significant lack of fit F-value.

Table 11.5 showing an F-value of 40.69 suggests that the model is significant. A large F-value might occur due to noise just 0.01% of the time. Significant model

Table 11.3 ANOVA for toolwear rate

Source	Sum of squares	df	Mean square value	F-value	p-value	
Model type	29.05	9	3.23	13.71	0.0012	significant
A-Applied voltage	0.8216	1	0.8216	3.49	0.1039	
B-electrolyte conc.	0.5960	1	0.5960	2.53	0.1556	
C-IEG	20.37	1	20.37	86.53	< 0.0001	
AB	2.16	1	2.16	9.16	0.0192	
AC	0.0019	1	0.0019	0.0079	0.9316	
BC	0.7762	1	0.7762	3.30	0.1122	
A^2	0.5693	1	0.5693	2.42	0.1639	
B^2	0.0050	1	0.0050	0.0212	0.8884	
C^2	3.59	1	3.59	15.24	0.0059	
Residuals	1.65	7	0.2354			
Lack of fit	0.3975	3	0.1325	0.4239	0.7466	not significant
Pure errors	1.25	4	0.3126			
Total Cor	30.69	16				

Table 11.4 ANOVA for material removal rate

Source	Sum of squares	df	Mean square	F-value	p-value	
Model type	0.0337	9	0.0037	8.09	0.0058	significant
A-Applied voltage	0.0006	1	0.0006	1.30	0.2909	
B-electrolyte conc.	0.0003	1	0.0003	0.6539	0.4453	
C-IEG	0.0030	1	0.0030	6.43	0.0389	
AB	2.500E-07	1	2.500E-07	0.0005	0.9821	
AC	0.0114	1	0.0114	24.72	0.0016	
BC	0.0047	1	0.0047	10.08	0.0156	
A^2	0.0013	1	0.0013	2.81	0.1375	
B^2	0.0019	1	0.0019	4.15	0.0812	
C^2	0.0107	1	0.0107	23.05	0.0020	
Residuals	0.0032	7	0.0005			
Lack of fit	0.0003	3	0.0001	0.1415	0.9300	not significant
Pure errors	0.0029	4	0.0007			
Total Cor	0.0369	16				

terms have p-values less than 0.0500. In this situation, the model variables B, C, AB, AC, A^2, B^2, and C^2 are critical. Values greater than 0.1000 indicate that the model terms are insignificant. If a model has a large number of insignificant model terms, model reduction may improve it. The F-value for lack of fit is 0.73, indicating that the lack of fit is negligible in contrast to the pure error. Noise has a 58.34% risk of causing a significant lack of fit F-value.

Table 11.5 ANOVA for surface roughness

Source	Sum of squares	df	Mean square	F-value	p-value	
Model	8.84	9	0.9822	40.69	< 0.0001	significant
A-voltage	0.0061	1	0.0061	0.2507	0.6320	
B-electrolyte concentration	0.5586	1	0.5586	23.14	0.0019	
C-IEG	3.04	1	3.04	125.77	< 0.0001	
AB	0.1918	1	0.1918	7.95	0.0258	
AC	0.1697	1	0.1697	7.03	0.0328	
BC	0.0182	1	0.0182	0.7551	0.4137	
A^2	0.2456	1	0.2456	10.17	0.0153	
B^2	2.84	1	2.84	117.73	< 0.0001	
C^2	2.03	1	2.03	83.90	< 0.0001	
Residuals	0.1690	7	0.0241			
Lack of fit	0.0600	3	0.0200	0.7349	0.5834	not significant
Pure errors	0.1089	4	0.0272			
Total Cor	9.01	16				

11.4.2 Mathematical modelling and regression analysis

MRR = 59.2866 + −4.84361 * voltage + −0.440684 * electrolyte conc. + −75.3891 * IEG + 0.0367062 *voltage*electrolyte conc. + 0.21575 * voltage*IEG + −0.88105 * electrolyte conc.*IEG + 0.0919231 * voltage^2 + −0.000344075 * electrolyte conc^2 + 369.217 * IEG^2

The real equation is derived after doing the experiment; therefore, the metal removal rate is significantly dependent on the main voltage.

TWR = 3.90008 + −0.25264 * voltage + 0.00809 * electrolyte conc. + −20.1724 * IEG + 1.2505 * voltage*electrolyte conc. + 0.53475 * voltage*IEG+ 0.0683 * electrolyte conc.*IEG + 0.00439437 * voltage^2 + −0.000213475 * electrolyte conc.^2 + 20.131 * IEG^2

Because the tool wear rate is significantly dependent on the voltage, the following real equation was generated while completing the experiment.

SR = −18.7154 + 3.026 * voltage + −0.635925 * electrolyte conc. + 126.91 * IEG + −0.01095 * voltage*electrolyte conc. + −2.06 * AC + 0.135 * electrolyte conc.*IEG + −0.060375 * voltage^2 + 0.008215 * electrolyte conc.^2 + −277.4 * IEG^2

The real equation is derived after doing the experiment, so the surface roughness is significantly dependant on the voltage.

Figure 11.3 depicts the projected vs real value of material removal rate where the graph evidently illustrates that the real model of MRR established differs from theoretical values predicted during the experimental performance, which can be easily checked by examining the spread of the true values to the predict line.

Figure 11.4 depicts the expected vs the real value of the outcome parameter, which is the amount of tool wear rate. The graph clearly illustrates that the real

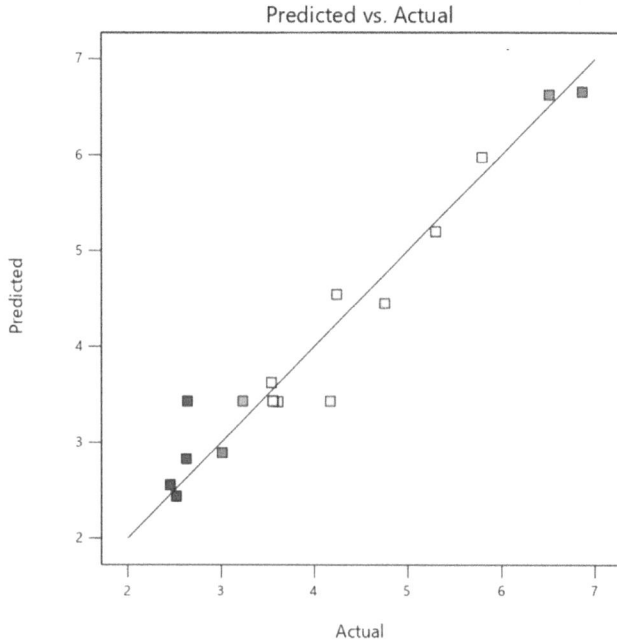

Figure 11.3 Predicted vs actual values of material removal rate.

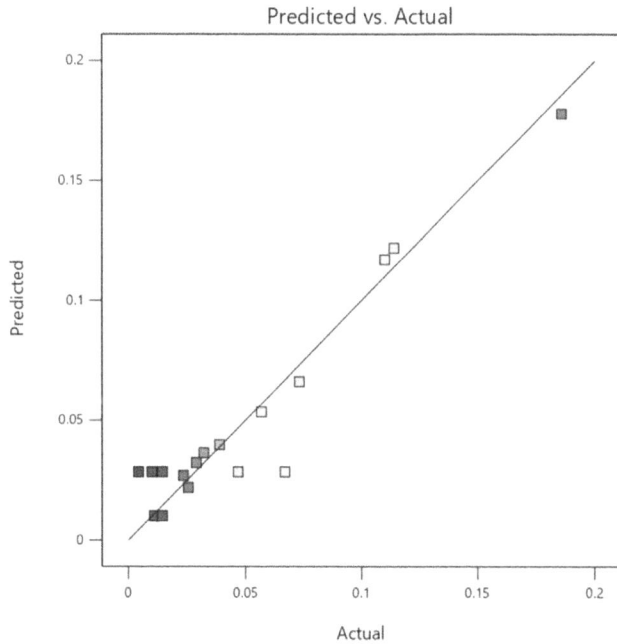

Figure 11.4 Predicted vs actual value of tool wear rate.

model of tool wear rate created is next to and near to the expected theoretical values obtained during the experimental performance, which can be easily checked by examining the spread of the true values to the expected actual line.

Figure 11.5 depicts the expected vs experimental result of surface integrity. This graph clearly illustrates that the real model of surface roughness established is very near to the expected theoretical values obtained during the experimental performance, which can be readily checked by looking at the dispersion of the real values to the estimated actual line.

Figure 11.6 displays a perturbation graph for MRR, which may be used to compare the effect on various variables in an appropriate point position in the design area. The output is distinguished by the variation of each element within its assigned holdings. A strong inclination in the characteristic curve indicates that the reaction is sensitive to the output parameter factor MRR. Closeness to the flat line indicates insensitivity to change upon that particular element.

Perturbation plot for rate of tool wear shown in Figure 11.7 will aid in comparing the effect on various factors in an appropriate point position in the design region. Variation of each element within its assigned holdings and other static factors characterizes the result. A strong inclination in the slope indicates that the reaction is sensitive to the specified output variable electrode wear rate. Closeness to the flat line indicates insensitivity to change upon that particular element.

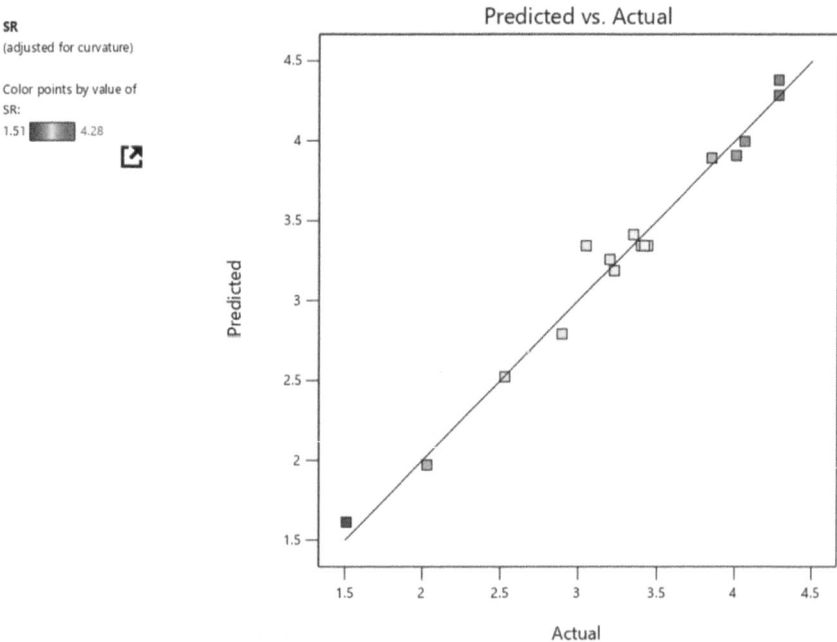

Figure 11.5 Predicted vs actual value of the surface roughness.

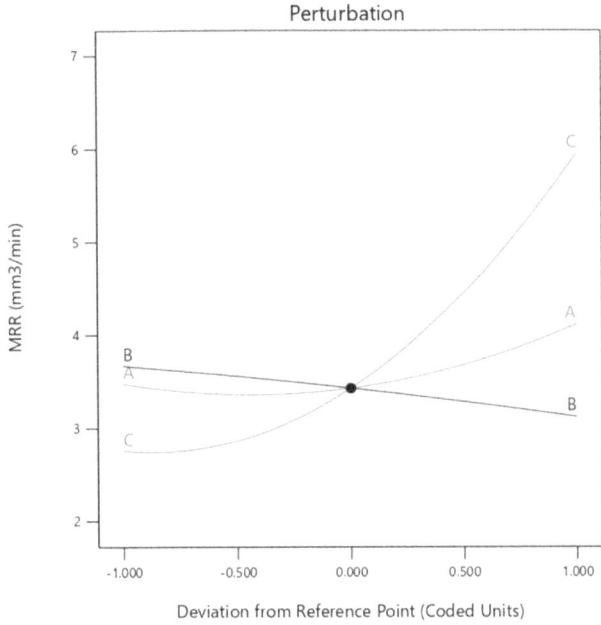

Figure 11.6 Perturbation graph for the material removal rate.

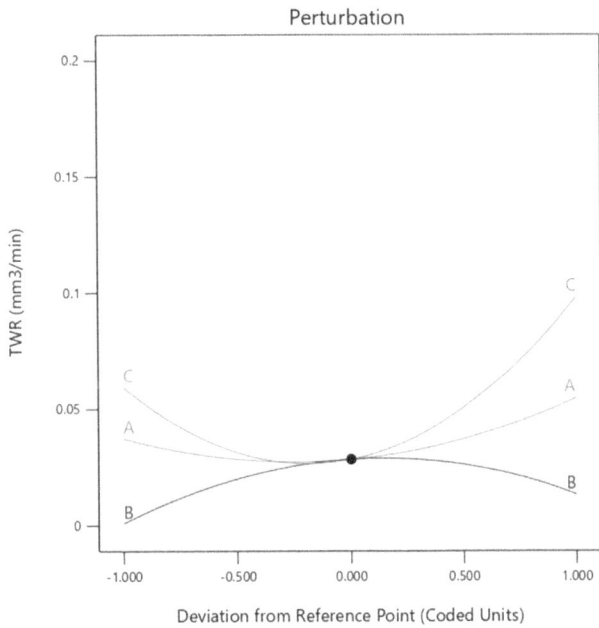

Figure 11.7 Perturbation graph of tool wear rate.

Figure 11.8 depicts a perturbation plot for surface roughness, which will aid in assessing the effect on various factors in a suitable point position in the design area. The outcome is defined by altering each element within its specified limitations. A strong inclination in the characteristic curve indicates that the reaction is sensitive to the output variable factor surface roughness. Closeness to the flat line indicates insensitivity to change upon that particular element.

11.4.3 Multiresponse optimization

Design Expert 13 software can perform multiple response optimization. The suggested model has three response variables and three input parameters, which are listed in Table 11.6. The instance illustrates how to optimize all output variables by evaluating each outcome and optimizing the variables. The optimization produced from the numeric report comprises Tables 11.6 and 11.7, the first of which presents a description of the constraints utilized to construct the next table of optimal process solutions.

The optimal value of the different parameters achieved during the electrochemical machining of Monel 400 alloy for output parameters were voltage = 15 volts, electrolyte concentration = 41.54 g/l, and IEG = 0.25 by machining it with a copper tool.

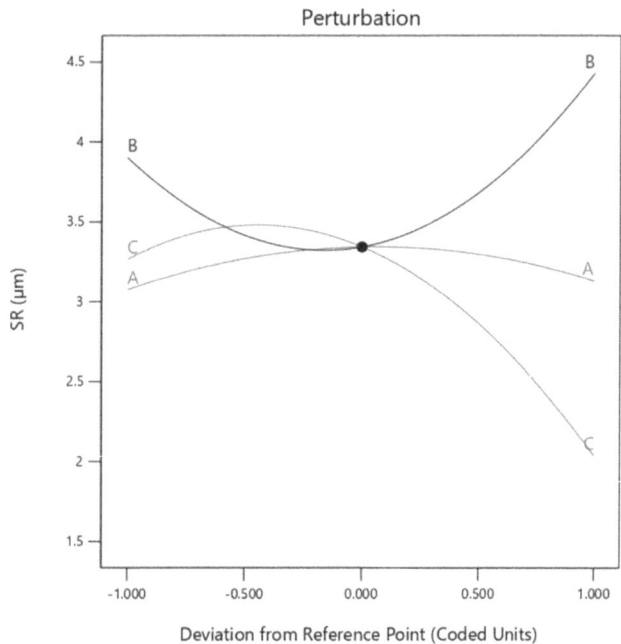

Figure 11.8 Perturbation graph for the surface roughness.

Table 11.6 Optimization of parameters

Name	Aim	Lower limit	Upper limit	Lower weight	Upper weight	Significance
A: applied voltage	In range	15	19	1	1	3
B: electrolyte conc.	In range	40	60	1	1	3
C: IEG	In range	0.15	0.25	1	1	3
Material removal rate	To maximize	2.4545	6.8613	1	1	3
Tool wear rate	To decrease	0.0044	0.1858	1	1	3
Surface roughness	To decrease	1.51	4.28	1	1	3

Table 11.7 Optimized results

S. No.	Voltage (V)	Electrolyte conc.(gm/l)	IEG	MRR (mm³/minute)	TWR (mm³/minute)	SR(μm)	Desirability
1.	15	41.54	0.25	7.175	0.004	2.093	0.924

11.5 CONCLUSIONS

The Box–Behnken design is utilized at three levels in order to build mathematical models for calculating the surface integrity factor of Monel alloy 400. The results were synthesized using a variety of multiple objective optimization techniques. The conclusions drawn are to be achieved:

- Influential factors for output results are as follows: applied voltage, electrolyte concentration, and IEG.
- Surface roughness was more noticeable at substantially higher voltages. The anticipated optimal value for surface roughness is 2.093 μm.
- It was discovered that a voltage of 15 volts, an electrolyte concentration of 41.54 g/l, and an IEG of 0.25 mm worked well for machining. Surface roughness of 2.093 m, material removal rate of 7.175 mm³/min, and tool wear rate of 0.004 mm³/min were measured as reaction metrics.
- On the Monel alloy 400, the copper tool produced the best reaction results.

Terminology

ANOVA   Analysis of variance
BBD   Box–Behnken design
CCD   Central composite design
ECM   Electrochemical machining
EDX   Energy-dispersive X-ray spectroscopy
GWO   Grey wolf optimizer
IEG   Inter-electrode gap
MFO   Moth-flame optimization

MRR   Material removal rate
RSM   Response surface methodology
SR   Surface roughness
PECM   Precision electrochemical machining
TWR   Tool wear rate

REFERENCES

[1] Fortana, M.G., 1986. *Corrosion Engineering*. McGraw-Hill.
[2] Moarrefzadeh, A., 2011. Finite element simulation of dimensional limitation of electro chemical machining (ECM) process. *International Journal of Multidisciplinary
Sciences and Engineering, 2*, pp. 39–43.
[3] McGeough, J.A., 1974. *Principles of Electrochemical Machining*. Chapman & Hall.
[4] Tam, S.C., Loh, N.L., Mah, C.P.A. and Loh, N.H., 1992. Electrochemical polishing of
biomedical titanium orifice rings. *Journal of Materials Processing Technology, 35*(1),
pp. 83–91.
[5] Kumar, J., Mandal, S. and Mahato, N., 1999. Study of electro chemical machining
etching effect on surface roughness and variation with chemical etching process,
07–09 Dec 2017, COPEN 2017, IIT Madras, pp. 646–649.
[6] Verma, N.K., 2013. *Experimental investigation of process parameters in electrochemical machining* (Doctoral dissertation), Indian institute of technology, IIT Mandi, HP.
[7] Muttamara, A. and Purktong, S., 2010. Improving the quality of groove in electro
chemical machining (ECM) processed by Taguchi method. *First TSME International
Conference on Mechanical Engineering, The International Journal of Advanced
Manufacturing Technology, 35*(7), pp. 821–832..
[8] Munda, J. and Bhattacharyya, B., 2008. Investigation into electrochemical micromachining (EMM) through response surface methodology based approach. *The International Journal of Advanced Manufacturing Technology, 35*(7), pp. 821–832.
[9] Selvakumar, G., Sarkar, S. and Mitra, S., 2012. Experimental investigation on die
corner accuracy for wire electrical discharge machining of Monel 400 alloy. *Proceedings of the Institution of Mechanical Engineers, Part B: Journal of Engineering
Manufacture, 226*(10), pp. 1694–1704.
[10] Parida, A.K. and Maity, K., 2018. Comparison the machinability of Inconel 718, Inconel 625 and Monel 400 in hot turning operation. *Engineering Science and Technology: An International Journal, 21*(3), pp. 364–370.
[11] Haisch, T., Mittemeijer, E. and Schultze, J.W., 2001. Electrochemical machining of
the steel 100Cr6 in aqueous NaCl and NaNO$_3$ solutions: Microstructure of surface
films formed by carbides. *Electrochimica Acta, 47*(1–2), pp. 235–241.
[12] Munda, J. and Bhattacharyya, B., 2008. Investigation into electrochemical micromachining (EMM) through response surface methodology based approach. *The International Journal of Advanced Manufacturing Technology, 35*(7), pp. 821–832.
[13] Santhi, M., Ravikumar, R. and Jeyapaul, R., 2013. Optimization of process parameters
in electro chemical machining (ECM) using DFA-fuzzy set theory-TOPSIS for titanium alloy. *Multidiscipline Modeling in Materials and Structures 9*(2), pp. 243–255.
[14] Bähre, D., Weber, O. and Rebschläger, A., 2013. Investigation on pulse electrochemical machining characteristics of lamellar cast iron using a response surface methodol-
ogy-based approach. *Procedia CIRP, 6*, pp. 362–367.

[15] Muthukumar, V., Rajesh, N., Venkatasamy, R., Sureshbabu, A. and Senthilkumar, N., 2014. Mathematical modeling for radial overcut on electrical discharge machining of Incoloy 800 by response surface methodology. *Procedia Materials Science*, 6, pp. 1674–1682.

[16] Kalaimathi, M., Venkatachalam, G. and Sivakumar, M., 2014. Experimental investigations on the electrochemical machining characteristics of Monel 400 alloys and optimization of process parameters. *Jordan Journal of Mechanical & Industrial Engineering*, 8(3), pp. 30–36.

[17] Tiwari, A., Mandal, A. and Kumar, K., 2015. Multi-objective optimization of electrochemical machining by non-dominated sorting genetic algorithm. *Materials Today: Proceedings*, 2(4–5), pp. 2569–2575.

[18] Krishnamurthy, V.P., Sivakumar, E.R. and Manikandan, M., 2015. Optimization of machining parameters of titanium alloy grade 2 in ECM by RSM method. *South Asian Journal of Engineering and Technology*, 22, pp. 256–267.

[19] Li, H., Gao, C., Wang, G., Qu, N. and Zhu, D., 2016. A study of electrochemical machining of Ti-6Al-4V in NaNO$_3$ solution. *Scientific Reports*, 6(1), pp. 1–11.

[20] Chen, X., Xu, Z., Zhu, D., Fang, Z. and Zhu, D., 2016. Experimental research on electrochemical machining of titanium alloy Ti60 for a blisk. *Chinese Journal of Aeronautics*, 29(1), pp. 274–282.

[21] Rao, P.S., Jain, P.K. and Dwivedi, D.K., 2015. Electro chemical honing (ECH) of external cylindrical surfaces of titanium alloys. *Procedia Engineering*, 100, pp. 936–945.

[22] Jeykrishnan, J., Ramnath, B.V., Elanchezhian, C. and Akilesh, S., 2017. Optimization of process parameters in electro-chemical machining (ECM) of D3 die steels using Taguchi technique. *Materials Today: Proceedings*, 4(8), pp. 7884–7891.

[23] Soundrapandian, E., Tajdeen, A. and Chari, K.K.B.A., 2017. Experimental study of electrochemical micro machining on titanium (Ti-6Al-4V) Alloy. In *Proceedings of 10th International Conference on Precision, Meso, Micro and Nano Engineering (copen 10), COPEN 2017, IIT Madras*, pp. 949–952.

[24] Geethapriyan, T., Kalaichelvan, K., Muthuramalingam, T. and Rajadurai, A., 2018. Performance analysis of process parameters on machining α–β titanium alloy in electrochemical micromachining process. *Proceedings of the Institution of Mechanical Engineers, Part B: Journal of Engineering Manufacture*, 232(9), pp. 1577–1589.

[25] Sharma, L.K., Sharma, S., Dubey, Y. and Parwani, L., 2019. Taguchi method approach for multi factor optimization of S1 tool steel in electrochemical machining. *International Journal of Research and Analytical Reviews (IJRAR)*, 6(2), pp. 403–409.

[26] Kumar, K.K. and Prasanna, J., 2019. A study on the electrochemical machining performance of titanium alloy in different electrolytes—a review. *International Journal of Emerging Technologies and Innovative Research*, 5, pp. 73–75.

[27] Khan, I.A., Rani, M., Deb, R.K. and Bundel, B.R., 2019. Effect on material removal rate and surface finish in ECM process when machining stainless steelV316 with Cu electrode. *International Journal of Recent Technology and Engineering*, 8(4).

[28] Khan, M.Y. and Rao, P.S., 2019. Electrical discharge machining: Vital to manufacturing industries. *International Journal of Innovative Technology and Exploring Engineering*, 8(11), pp. 1696–1701.

[29] Khan, M.Y., Rao, P.S. and Pabla, B.S., 2020. Investigations on the feasibility of Jatropha curcas oil based biodiesel for sustainable dielectric fluid in EDM process. *Materials Today: Proceedings*, 26, pp. 335–340.

[30] Khan, M.Y., Rao, P.S. and Pabla, B.S., 2020. Review of electrical discharge machining process with nanopowder and CNT mixed dielectric fluid. In *Proceedings of 4th*

international online multidisciplinary research conference (IOMRC-2020), Hyderabad, October, Vol. 23.

[31] Yadav, S.K. and Yadav, S.K.S., 2020. Multi-objective optimization of electrochemical cut-off grinding process of Ti-6Al-4V using PCA based grey relational analysis. *Materials Today: Proceedings*, *22*, pp. 3089–3099.

[32] Daniel, S.A.A., Ananth, S.V., Parthiban, A. and Sivaganesan, S., 2020. Optimization of machining parameters in electro chemical machining of Al5059/SiC/MoS$_2$ composites using Taguchi method. *Materials Today: Proceedings*, *21*, pp. 738–743.

[33] Debta, M.K., Mishra, R. and Masanta, M., 2020. Experimental investigation on the machining performance of AZ91D (90% Mg) alloy by wire-cut EDM. *Materials Today: Proceedings*, *33*, pp. 5557–5560.

[34] Kumar, N. and Rao, P.S., 2019. Analysis and optimization of process parameters during electric discharge machining of Incoloy 800: A review. International Journal of Scientific & Technology Research, IJSTR, Volume 9, Issue 02, February 2020, pp. 6002–6005.

[35] Sahu, P.K. and Rao, P.S., 2019. A review on parametric investigation and improvement of electrical discharge machining process. International Journal for Research in Applied Science & Engineering Technology (IJRASET), Volume 8 Issue XI Nov 2020, pp. 582–585.

[36] Thakur, A., Rao, P.S. and Khan, M.Y., 2021. Study and optimization of surface roughness parameter during electrical discharge machining of titanium alloy (Ti-6246). *Materials Today: Proceedings*, *44*, pp. 838–847.

[37] Sharma, V., Rao, P.S., Kumar, H. and Khan, M.Y., 2021. An experimental investigation ECM of titanium 6AL-4V-ELI alloy with different electrolytes, Journal of University of Shanghai for Science and Technology, JUSST, 23(2), pp. 292–303.

[38] Gobikrishnan, U., Suresh, P. and Kumaravel, P., 2021. Drilling investigations on Inconel alloy 625 material of material removal rate using micro electrochemical machining. *Materials Today: Proceedings*, *37*, pp. 1629–1633.

[39] Ramana, M.V., Rao, G.K.M., Sagar, B., Panthangi, R.K. and Kumar, B.R.R., 2021. Optimization of surface roughness and tool wear in sustainable dry turning of iron based Nickel A286 alloy using Taguchi's method. *Cleaner Engineering and Technology*, *2*, pp. 100–114.

[40] Jerin, A. and Karunakaran, K., 2021. Machinability investigation and optimizing process parameters in ECM of stainless steel − 12X18H10T for minimizing surface roughness. *Materials Today: Proceedings*. Volume 81, Part 2, 2023, pp. 443–448.

[41] Thangamani, G., Thangaraj, M., Moiduddin, K., Mian, S.H., Alkhalefah, H. and Umer, U., 2021. Performance analysis of electrochemical micro machining of titanium (Ti-6Al-4V) alloy under different electrolytes concentrations. *Metals*, *11*(2), p. 247.

[42] Khan, M.Y., Rao, P.S. and Pabla, B.S. 2022. Biodiesel as dielectric fluid for electrical discharge machining-A. *Turkish Journal of Physiotherapy and Rehabilitation*, *32*, p. 3.

[43] Wasif, M., Khan, Y.A., Zulqarnain, A. and Iqbal, S.A., 2022. Analysis and optimization of wire electro-discharge machining process parameters for the efficient cutting of Aluminum 5454 alloy. *Alexandria Engineering Journal*, *61*(8), pp. 6191–6203.

[44] Nagarajan, V., Solaiyappan, A., Mahalingam, S.K., Nagarajan, L., Salunkhe, S., Nasr, E.A., Shanmugam, R. and Hussein, H.M.A.M., 2022. Meta-heuristic technique-based

parametric optimization for electrochemical machining of Monel 400 alloys to investigate the material removal rate and the sludge. *Applied Sciences*, *12*(6), p. 2793.

[45] Vengatajalapathi, N., Ayyappan, S. and Rajasekar, V., 2022. Eco-friendly filtration of nickel from the sludge during electrochemical machining of Monel 400 alloys. Global NEST Journal, Vol 24, Issue 02, pp. 203–211.

[46] Singh, G., Singh, R. and Rao, P.S., 2022. Electrochemical Machining of Cu substrate with Cu tool,. *Advances in Manufacturing Technology: Computational Materials Processing and Characterization*, CRC Press, pp. 89–93.

[47] Rao, P.S., Jain, P.K., Dwivedi, D.K., 2015. "Precision Finishing of External Cylindrical Surfaces of Eon Finishing of External Cylindrical Surfaces of EN8 Steel By Electro Chemical Honing (ECH) Process Using OFAT Technique", Elsevier Publishers, J. of Materials Today Proceedings, 2 (2015), pp. 3220–3229.

[48] Rao, P.S., Jain, P.K., Dwivedi, D.K., 2015. "Electro Chemical Honing (ECH) of External Cylindrical Surfaces-An Innovative Step", DAAAM International Vienna Publishers, 09 Chapter, DAAAM International Scientific Book (2015), pp. 097–116.

[49] Rao, P.S., Jain, P.K., Dwivedi, D.K., 2016. "Electro Chemical Honing (ECH)-A New Paradigm in Hybrid Machining Process", DAAAM International Vienna Publishers, 26 Chapter, DAAAM International Scientific Book (2016), pp. 287–306.

[50] T. Singh, J.P. Misra, V. Upadhyay, P. Sudhakar Rao, 2018. "An adaptive neuro-fuzzy inference system (ANFIS) for Wire-EDM of ballistic grade aluminium alloy", International Journal of Automotive and Mechanical Engineering, Volume 15, Issue 2, (2018), pp. 5295–5307.

Index

For Product Safety Concerns and Information please contact our EU representative GPSR@taylorandfrancis.com Taylor & Francis Verlag GmbH, Kaufingerstraße 24, 80331 München, Germany

Printed and bound by CPI Group (UK) Ltd, Croydon, CR0 4YY
05/08/2025
01930894-0004